T0234835

LONDON MATHEMATICAL SOCIETY LECTURE NOTE SERIES

Managing Editor: Professor M. Reid, Mathematics Institute, University of Warwick, Coventry CV4 7AL, United Kingdom

The titles below are available from booksellers, or from Cambridge University Press at www.cambridge.org/mathematics

Partial Differential Equations and Fluid Mechanics

Edited by

JAMES C. ROBINSON & JOSÉ L. RODRIGO

University of Warwick

CAMBRIDGE
UNIVERSITY PRESS

CAMBRIDGE
UNIVERSITY PRESS

University Printing House, Cambridge CB2 8BS, United Kingdom

One Liberty Plaza, 20th Floor, New York, NY 10006, USA

477 Williamstown Road, Port Melbourne, VIC 3207, Australia

314-321, 3rd Floor, Plot 3, Splendor Forum, Jasola District Centre, New Delhi - 110025, India

103 Penang Road, #05-06/07, Visioncrest Commercial, Singapore 238467

Cambridge University Press is part of the University of Cambridge.

It furthers the University's mission by disseminating knowledge in the pursuit of education, learning and research at the highest international levels of excellence.

www.cambridge.org
Information on this title: www.cambridge.org/9780521125123

© Cambridge University Press 2009

First published 2009

A catalogue record for this publication is available from the British Library

ISBN 978-0-521-12512-3 Paperback

To Tania and Elizabeth

Contents

Preface

This volume is the result of a workshop, "Partial Differential Equations and Fluid Mechanics", which took place in the Mathematics Institute at the University of Warwick, May 21st–23rd, 2007.

Several of the speakers agreed to write review papers related to their contributions to the workshop, while others have written more traditional research papers. All the papers have been carefully edited in the interests of clarity and consistency, and the research papers have been externally refereed. We are very grateful to the referees for their work. We believe that this volume therefore provides an accessible summary of a wide range of active research topics, along with some exciting new results, and we hope that it will prove a useful resource for both graduate students new to the area and to more established researchers.

We would like to express their gratitude to the following sponsors of the workshop: the London Mathematical Society, the Royal Society, via a University Research Fellowship awarded to James Robinson, the North American Fund and Research Development Fund schemes of Warwick University, and the Warwick Mathematics Department (via MIR@W). JCR is currently supported by the EPSRC, grant EP/G007470/1.

Finally it is a pleasure to thank Yvonne Collins and Hazel Higgens from the Warwick Mathematics Research Centre for their assistance during the organization of the workshop.

Warwick,
December 2008

James C. Robinson
José L. Rodrigo

Contributors

Those contributors who presented their work at the Warwick meeting are indicated by a star in the following list.

Mahdi Boukrouche
Laboratory of Mathematics, University of Saint-Etienne, LaMUSE EA-3989, 23 rue du Dr Paul Michelon, Saint-Etienne, 42023. France.
Mahdi.Boukrouche@univ-st-etienne.fr

Miroslav Bulíček
Charles University, Faculty of Mathematics and Physics, Mathematical Institute, Sokolovská 83, 186 75 Prague 8. Czech Republic.
mbul8060@karlin.mff.cuni.cz

Masoumeh Dashti *
Mathematics Department, University of Warwick, Coventry, CV47AL. United Kingdom.
M.Dashti@warwick.ac.uk

Enrique Fernández-Cara *
Departamento de Ecuaciones Diferenciales y Análisis Numérico, Facultad de Matemáticas, Universidad de Sevilla, Apartado 1160, 41080 Sevilla. Spain.
cara@us.es

Francesco Gargano
Department of Mathematics, Via Archirafi 34, 90123 Palermo. Italy.
gargano@math.unipa.it

Igor Kukavica *
Department of Mathematics, University of Southern California, Los Angeles, CA 90089. USA.
kukavica@usc.edu

Maria Carmela Lombardo
Department of Mathematics, Via Archirafi 34, 90123 Palermo. Italy.
lombardo@math.unipa.it

Grzegorz Łukaszewicz *
University of Warsaw, Mathematics Department, ul. Banacha 2, 02-957, Warsaw. Poland.
glukasz@mimuw.edu.pl

Josef Málek *
Charles University, Faculty of Mathematics and Physics, Mathematical Institute, Sokolovská 83, 186 75 Prague 8. Czech Republic.
malek@karlin.mff.cuni.cz

Marius Paicu
Université Paris-Sud and CNRS, Laboratoire de Mathématiques, Orsay Cedex, F-91405. France.
Marius.Paicu@math.u-psud.fr

Kumbakonam R. Rajagopal
Department of Mechanical Engineering, Texas A&M University, College Station, TX 77843. USA.
krajagopal@mengr.tamu.edu

Geneviève Raugel *
CNRS and Université Paris-Sud, Laboratoire de Mathématiques, Orsay Cedex, F-91405. France.
Genevieve.Raugel@math.u-psud.fr

James C. Robinson
Mathematics Institute, University of Warwick, Coventry, CV4 7AL. United Kingdom.
J.C.Robinson@warwick.ac.uk

José L. Rodrigo *
Mathematics Department, University of Warwick, Coventry, CV4 7AL. United Kingdom.
J.Rodrigo@warwick.ac.uk

Ricardo M.S. Rosa *
Instituto de Matemática, Universidade Federal do Rio de Janeiro , Caixa
Postal 68530 Ilha do Fundão, Rio de Janeiro, RJ 21945-970. Brazil.
rrosa@ufrj.br

Witold Sadowski *
Faculty of Mathematics, Informatics and Mechanics, University of
Warsaw, Banacha 2, 02-097 Warszawa. Poland.
witeks@hydra.mimuw.edu.pl

Marco Sammartino *
Department of Mathematics, Via Archirafi 34, 90123 Palermo. Italy.
marco@math.unipa.it

Vincenzo Sciacca
Department of Mathematics, Via Archirafi 34, 90123 Palermo. Italy.
sciacca@math.unipa.it

1

Shear flows and their attractors

Mahdi Boukrouche

Laboratory of Mathematics, University of Saint-Etienne,
LaMUSE EA-3989, 23 rue du Dr Paul Michelon,
Saint-Etienne, 42023. France.
`Mahdi.Boukrouche@univ-st-etienne.fr`

Grzegorz Łukaszewicz

University of Warsaw, Mathematics Department,
ul. Banacha 2, 02-957 Warsaw. Poland.
`glukasz@mimuw.edu.pl`

Abstract

We consider the problem of the existence and finite dimensionality of attractors for some classes of two-dimensional turbulent boundary-driven flows that naturally appear in lubrication theory. The flows admit mixed, non-standard boundary conditions and time-dependent driving forces. We are interested in the dependence of the dimension of the attractors on the geometry of the flow domain and on the boundary conditions.

1.1 Introduction

This work gives a survey of the results obtained in a series of papers by Boukrouche & Łukaszewicz (2004, 2005a,b, 2007) and Boukrouche, Łukaszewicz, & Real (2006) in which we consider the problem of the existence and finite dimensionality of attractors for some classes of two-dimensional turbulent boundary-driven flows (Problems I–IV below). The flows admit mixed, non-standard boundary conditions and also time-dependent driving forces (Problems III and IV). We are interested in the dependence of the dimension of the attractors on the geometry of the flow domain and on the boundary conditions. This research is motivated by problems from lubrication theory. Our results generalize some earlier ones devoted to the existence of attractors and estimates of their dimensions for a variety of Navier–Stokes flows. We would like to mention a few results that are particularly relevant to the problems we consider.

Most earlier results on shear flows treated the autonomous Navier–Stokes equations. In Doering & Wang (1998), the domain of the flow is

Published in *Partial Differential Equations and Fluid Mechanics*, edited by James C. Robinson and José L. Rodrigo. © Cambridge University Press 2009.

an elongated rectangle $\Omega = (0, L) \times (0, h)$, $L \gg h$. Boundary conditions of Dirichlet type are assumed on the bottom and the top parts of the boundary and a periodic boundary condition is assumed on the lateral part of the boundary. In this case the attractor dimension can be estimated from above by $c\frac{L}{h}Re^{3/2}$, where c is a universal constant, and $Re = \frac{Uh}{\nu}$ is the Reynolds number. Ziane (1997) gave optimal bounds for the attractor dimension for a flow in a rectangle $(0, 2\pi L) \times (0, 2\pi L/\alpha)$, with periodic boundary conditions and given external forcing. The estimates are of the form $c_0/\alpha \leq dim\mathcal{A} \leq c_1/\alpha$, see also Miranville & Ziane (1997). Some free boundary conditions are considered by Ziane (1998), see also Temam & Ziane (1998), and an upper bound on the attractor dimension established with the use of a suitable anisotropic version of the Lieb-Thirring inequality, in a similar way to Doering & Wang (1998). Dirichlet-periodic and free-periodic boundary conditions and domains with more general geometry were considered by Boukrouche & Łukaszewicz (2004, 2005a,b) where still other forms of the Lieb-Thirring inequality were established to study the dependence of the attractor dimension on the shape of the domain of the flow. The Navier slip boundary condition and the case of an unbounded domain were considered recently by Mucha & Sadowski (2005).

Boundary-driven flows in smooth and bounded two-dimensional domains for a non-autonomous Navier–Stokes system are considered by Miranville & Wang (1997), using an approach developed by Chepyzhov & Vishik (see their 2002 monograph for details). An extension to some unbounded domains can be found in Moise, Rosa, & Wang (2004), cf. also Łukaszewicz & Sadowski (2004).

Other related problems can be found, for example, in the monographs by Chepyzhov & Vishik (2002), Doering & Gibbon (1995), Foias et al. (2001), Robinson (2001), and Temam (1997), and the literature quoted there.

Formulation of the problems considered.

We consider the two-dimensional Navier–Stokes equations,

$$u_t - \nu\Delta u + (u \cdot \nabla)u + \nabla p = 0 \tag{1.1}$$

and

$$\mathrm{div}\, u = 0 \tag{1.2}$$

in the channel

$$\Omega_\infty = \{x = (x_1, x_2) : -\infty < x_1 < \infty, \ 0 < x_2 < h(x_1)\},$$

where the function h is positive, smooth, and L-periodic in x_1.

Let

$$\Omega = \{x = (x_1, x_2) : 0 < x_1 < L, \ 0 < x_2 < h(x_1)\}$$

and $\partial\Omega = \bar{\Gamma}_0 \cup \bar{\Gamma}_L \cup \bar{\Gamma}_1$, where Γ_0 and Γ_1 are the bottom and the top, and Γ_L is the lateral part of the boundary of Ω.

We are interested in solutions of (1.1)–(1.2) in Ω that are L-periodic with respect to x_1 and satisfy the initial condition

$$u(x, 0) = u_0(x) \quad \text{for} \quad x \in \Omega, \tag{1.3}$$

together with the following boundary conditions on the bottom and on the top parts, Γ_0 and Γ_1, of the domain Ω.

Case I. We assume that

$$u = 0 \quad \text{on} \quad \Gamma_1 \tag{1.4}$$

(non-penetration) and

$$u = U_0 e_1 = (U_0, 0) \quad \text{on} \quad \Gamma_0. \tag{1.5}$$

Case II. We assume that

$$u.n = 0 \quad \text{and} \quad \tau \cdot \sigma(u, p) \cdot n = 0 \quad \text{on} \quad \Gamma_1, \tag{1.6}$$

i.e. the tangential component of the normal stress tensor $\sigma \cdot n$ vanishes on Γ_1. The components of the stress tensor σ are

$$\sigma_{ij}(u, p) = \nu\left(\frac{\partial u_i}{\partial x_j} + \frac{\partial u_j}{\partial x_i}\right) - p\,\delta_{ij}, \qquad 1 \le i, j \le 3, \tag{1.7}$$

where δ_{ij} is the Kronecker symbol. As for case I, we set

$$u = U_0 e_1 = (U_0, 0) \quad \text{on} \quad \Gamma_0. \tag{1.8}$$

Case III. We assume that

$$u = 0 \quad \text{on} \quad \Gamma_1 \quad \text{and} \tag{1.9}$$

$$u = U_0(t) e_1 = (U_0(t), 0) \quad \text{on} \quad \Gamma_0, \tag{1.10}$$

where $U_0(t)$ is a locally Lipschitz continuous function of time t.

Case IV. We assume that

$$u = 0 \quad \text{on} \quad \Gamma_1. \tag{1.11}$$

We also impose no flux across Γ_0 so that the normal component of the velocity on Γ_0 satisfies

$$u \cdot n = 0 \quad \text{on} \quad \Gamma_0, \tag{1.12}$$

and the tangential component of the velocity u_η on Γ_0 is unknown and satisfies the Tresca law with a constant and positive friction coefficient k. This means (Duvaut & Lions, 1972) that on Γ_0

$$|\sigma_\eta(u,p)| < k \Rightarrow u_\eta = U_0(t)e_1 \quad \text{and}$$
$$|\sigma_\eta(u,p)| = k \Rightarrow \exists\, \lambda \geq 0 \text{ such that } u_\eta = U_0(t)e_1 - \lambda\sigma_\eta(u,p), \tag{1.13}$$

where σ_η is the tangential component of the stress tensor on Γ_0 (see below) and

$$t \mapsto U_0(t)e_1 = (U_0(t), 0)$$

is the time-dependent velocity of the lower surface, producing the driving force of the flow. We suppose that U_0 is a locally Lipschitz continuous function of time t.

If $n = (n_1, n_2)$ is the unit outward normal to Γ_0, and $\eta = (\eta_1, \eta_2)$ is the unit tangent vector to Γ_0 then we have

$$\sigma_\eta(u,p) = \sigma(u,p) \cdot n - ((\sigma(u,p) \cdot n) \cdot n)n, \tag{1.14}$$

where $\sigma_{ij}(u,p)$ is the stress tensor whose components are defined in (1.7).

Each problem is motivated by a flow in an infinite (rectified) journal bearing $\Omega \times (-\infty, +\infty)$, where $\Gamma_1 \times (-\infty, +\infty)$ represents the outer cylinder, and $\Gamma_0 \times (-\infty, +\infty)$ represents the inner, rotating cylinder. In the lubrication problems the gap h between cylinders is never constant. We can assume that the rectification does not change the equations as the gap between cylinders is very small with respect to their radii.

This article is organized as follows. In Sections 1.2 and 1.3 we consider Problem I: (1.1)–(1.5), and Problem II: (1.1)–(1.3), (1.6), and (1.8). In Section 1.4 we consider Problem III: (1.1)–(1.3), (1.9), and (1.10). In Section 1.5 we consider Problem IV: (1.1)–(1.3), and (1.11)–(1.13).

1.2 Time-independent driving: existence of global solutions and attractors

In this section we consider Problem I: (1.1)–(1.5), and Problem II: (1.1)–(1.3), (1.6), and (1.8) and present results on the existence of unique global-in-time weak solutions and the existence of the associated global attractors.

Homogenization and weak solutions.

Let u be a solution of Problem I or Problem II, and set

$$u(x_1, x_2, t) = U(x_2)e_1 + v(x_1, x_2, t),$$

with

$$U(0) = U_0, \quad U(h(x_1)) = 0, \quad \text{and} \quad U'(h(x_1)) = 0, \quad x_1 \in (0, L).$$

Then v is L-periodic in x_1 and satisfies

$$v_t - \nu\Delta v + (v.\nabla)v + Uv_{,x_1} + (v)_2 U'e_1 + \nabla p = \nu U''e_1 \qquad (1.15)$$

and

$$\text{div } v = 0,$$

together with the initial condition

$$v(x, 0) = v_0(x) = u_0(x) - U(x_2)e_1.$$

By $(v)_2$ in (1.15) we have denoted the second component of v. The boundary conditions are

$$v = 0 \quad \text{on} \quad \Gamma_0 \cup \Gamma_1$$

for Problem I, and

$$v = 0 \quad \text{on} \quad \Gamma_0, \quad v \cdot n = 0 \quad \text{and} \quad \tau \cdot \sigma(v) \cdot n = 0 \quad \text{on} \quad \Gamma_1$$

for Problem II.

Now we define a weak form of the homogenized problem above. To this end we need some notation. Let $\mathcal{C}_L^\infty(\Omega_\infty)^2$ denote the class of functions in $\mathcal{C}^\infty(\Omega_\infty)^2$ that are L-periodic in x_1; define

$$\tilde{V} = \{v \in \mathcal{C}_L^\infty(\Omega_\infty)^2 : \text{ div } v = 0, \ v = 0 \text{ at } \Gamma_0 \cup \Gamma_1\}$$

for Problem I, and

$$\tilde{V} = \{v \in \mathcal{C}_L^\infty(\Omega_\infty)^2 : \text{ div } v = 0, \ v|_{\Gamma_0} = 0, \ v \cdot n|_{\Gamma_1} = 0\}$$

for Problem II; and let

$$V = \text{closure of } \tilde{V} \text{ in } H^1(\Omega) \times H^1(\Omega), \qquad \text{and}$$
$$H = \text{closure of } \tilde{V} \text{ in } L^2(\Omega) \times L^2(\Omega).$$

We define the scalar product and norm in H as

$$(u, v) = \int_\Omega u(x)v(x)\, dx \quad \text{and} \quad |v| = (v, v)^{1/2},$$

and in V the scalar product and norm are

$$(\nabla u, \nabla v) \quad \text{and} \quad |\nabla v|^2 = (\nabla v, \nabla v).$$

We use the notation $\langle \cdot, \cdot \rangle$ for the pairing between V and its dual V', i.e. $\langle f, v \rangle$ denotes the action of $f \in V'$ on $v \in V$.

Let

$$a(u, v) = \nu(\nabla u, \nabla v) \quad \text{and} \quad B(u, v, w) = ((u \cdot \nabla)v, w).$$

Then the natural weak formulation of the homogenized Problems I and II is as follows.

Problem 1.2.1 *Find*

$$v \in \mathcal{C}([0, T]; H) \cap L^2(0, T; V)$$

for each $T > 0$, such that

$$\frac{\mathrm{d}}{\mathrm{d}t}(v(t), \Theta) + a(v(t), \Theta) + B(v(t), v(t), \Theta) = F(v(t), \Theta),$$

for all $\Theta \in V$, and

$$v(x, 0) = v_0(x),$$

where

$$F(v, \Theta) = -a(\xi, \Theta) - B(\xi, v, \Theta) - B(v, \xi, \Theta),$$

and $\xi = Ue_1$ is a suitable background flow.

We have the following existence theorem (the proof is standard, see, for example, Temam, 1997).

Theorem 1.2.2 *There exists a unique weak solution of Problem 1.2.1 such that for all η, T, $0 < \eta < T$, $v \in L^2(\eta, T; H^2(\Omega))$, and for each $t > 0$ the map $v_0 \mapsto v(t)$ is continuous as a map from H into itself. Moreover, there exists a global attractor for the associated semigroup $\{S(t)\}_{t \geq 0}$ in the phase space H.*

1.3 Time-independent driving: dimensions of global attractors

The standard procedure for estimating the global attractor dimension, which we use here, is based on the theory of dynamical systems (Doering & Gibbon, 1995; Foias et al., 2001; Temam, 1997) and involves two important ingredients: an estimate of the time-averaged energy dissipation rate ϵ and a Lieb–Thirring-like inequality. The precision and physical soundness of an estimate of the number of degrees of freedom of a given flow (expressed by an estimate of its global attractor dimension) depends directly on the quality of the estimate of ϵ and a good choice of the Lieb–Thirring-like inequality which depends, in particular, on the geometry of the domain and on the boundary conditions of the flow.

In this section we continue to consider the time-independent Problems I and II. First, we present an estimate of the time-averaged energy dissipation rate of these two flows and then present two versions of the Lieb–Thirring inequality for functions defined on a non-rectangular domain. Finally we use these inequalities to give an upper bound on the global attractor dimension in terms of the data and the geometry of the domain. We use the fractal (or upper box-counting) dimension: for a subset X of a Banach space B, this is given by

$$d_f(X) = \limsup_{\epsilon \to 0} \frac{\log N(X, \epsilon)}{-\log \epsilon},$$

where $N(X, \epsilon)$ is the minimum number of B-balls of radius ϵ required to cover X, see Falconer (1990) for more details.

We define the time-averaged energy dissipation rate per unit mass ϵ of weak solutions u of Problems I and II as follows,

$$\epsilon = \frac{\nu}{|\Omega|} \langle |\nabla u|^2 \rangle := \limsup_{T \to +\infty} \frac{\nu}{|\Omega|} \frac{1}{T} \int_0^T |\nabla u(t)|^2 \, dt. \tag{1.16}$$

Let $h_0 = \min_{0 \leq x_1 \leq L} h(x_1)$. We define the Reynolds number of the flow u by $Re = (h_0 U_0)/\nu$. Then we have (Boukrouche & Lukaszewicz, 2004, 2005a):

Theorem 1.3.1 *For the Navier–Stokes flows u of Problems I and II with $Re \gg 1$ the time-averaged energy dissipation rate per unit mass ϵ defined in (1.16) satisfies*

$$\epsilon \leq C \frac{U_0^3}{h_0}, \tag{1.17}$$

where C is a numerical constant.

Observe that the above estimate coincides with a Kolmogorov-type bound on the time-averaged energy-dissipation rate which is independent of viscosity at large Reynolds numbers (Doering & Gibbon, 1995;

Foias et al., 2001). Estimate (1.17) is the same as that obtained earlier for a rectangular domain by Doering & Constantin (1991) who used a background flow suitable for the channel case (see also Doering & Gibbon, 1995).

To find upper bounds on the dimension of global attractors in terms of the geometry of the flow domain Ω we use the following versions of the anisotropic Lieb–Thirring inequality (Boukrouche & Łukaszewicz, 2004, 2005a).

Let

$$\widetilde{H}^1 = \{v \in \mathcal{C}_L^\infty(\Omega_\infty)^2 : v = 0 \text{ on } \partial\Omega_\infty\}$$

and

$$H^1 = \text{closure of } \widetilde{H}^1 \text{ in } H^1(\Omega) \times H^1(\Omega).$$

Lemma 1.3.2 *Let $\varphi_j \in H^1$, $j = 1, \ldots, m$ be an orthonormal family in $L^2(\Omega)$ and let $h_M = \max\limits_{0 \le x_1 \le L} h(x_1)$. Then*

$$\int_\Omega \left(\sum_{j=1}^m \varphi_j^2 \right)^2 dx \le \sigma \left[1 + \left(\frac{h_M}{L} \right)^2 \right] \sum_{j=1}^m \int_\Omega |\nabla\varphi_j|^2 dx,$$

where σ is an absolute constant.

Rather than proving this lemma here, we give the full argument for the following result whose proof is more involved. Let

$$\widetilde{H}_f^1 = \{v \in \mathcal{C}_L^\infty(\Omega_\infty)^2 : v|_{\Gamma_0} = 0, \quad v \cdot n|_{\Gamma_1} = 0\}$$

and

$$H_f^1 = \text{closure of } \widetilde{H}_f^1 \text{ in } H^1(\Omega) \times H^1(\Omega).$$

Lemma 1.3.3 *Let $\varphi_j \in H_f^1$, $j = 1, \ldots, m$ be an sub-orthonormal family in $L^2(\Omega)$, i.e.*

$$\sum_{i,j=1}^m \xi_i \xi_j \int_{\Omega_1} \varphi_i \varphi_j \, dy \le \sum_{k=1}^m \xi_k^2 \quad \forall \, \xi \in \mathbb{R}^m.$$

Then

$$\int_\Omega \left(\sum_{j=1}^m \varphi_j^2 \right)^2 dx \le \sigma_1 \sum_{j=1}^m \int_\Omega |\nabla\varphi_j|^2 \, dx + \sigma_2 m + \sigma_3, \qquad (1.18)$$

where $\sigma_1 = \kappa_1(1 + \max\limits_{0 \le x_1 \le L} |h'(x_1)|^2)$, $\sigma_2 = \kappa_2(\frac{1}{L^2} + \frac{1}{h_0^2})$,

$$\sigma_3 = \kappa_3 \int_\Omega \left(\frac{h'(x_1)}{h(x_1)}\right)^4 (1 + h'(x_1)^4)\, dx,$$

and κ_1, κ_2, *and* κ_3 *are some absolute constants.*

Proof (Boukrouche & Łukaszewicz, 2005b) Let $\Omega_1 = (0, L) \times (0, h_0)$, and let $\psi_j \in H^1(\Omega_1)$, $j = 1, \dots, m$, be a family of functions that are sub-orthonormal in $L^2(\Omega_1)$. Ziane (1998) showed that

$$\int_{\Omega_1} \left(\sum_{j=1}^m \psi_j^2\right)^2 dy \le C_0 \left(\sum_{j=1}^m \int_{\Omega_1} \left(\frac{\partial \psi_j}{\partial y_1}\right)^2 dy + \frac{|\psi_j|^2_{L^2(\Omega_1)}}{L^2}\right)^{\frac{1}{2}}$$

$$\times \left(\sum_{j=1}^m \int_{\Omega_1} \left(\frac{\partial \psi_j}{\partial y_2}\right)^2 dy + \frac{|\psi_j|^2_{L^2(\Omega_1)}}{h_0^2}\right)^{\frac{1}{2}}$$

for some absolute constant C_0. Now, for our family φ_j defined in Ω, we set

$$\psi_j(y_1, y_2) = \varphi_j(x_1, x_2)\sqrt{\frac{h(x_1)}{h_0}},$$

where $h_0 = \min\limits_{0 \le x_1 \le L_1} h(x_1)$, $y_1 = x_1$, and $y_2 = x_2 h_0/h(x_1)$. For $x = (x_1, x_2)$ in Ω, $y = (y_1, y_2)$ is in Ω_1, and the family ψ_j, $j = 1, \dots, m$, in Ω_1 has the required properties. Changing variables in the above inequality and observing that

$$dy_1 dy_2 = \frac{h_0}{h(x_1)} dx_1 dx_2,$$

$$\frac{\partial \psi_j}{\partial y_1} = \left(\frac{\partial \varphi_j}{\partial x_1}\sqrt{\frac{h(x_1)}{h_0}} + \varphi_j \frac{h'(x_1)}{2\sqrt{h_0 h(x_1)}}\right) + \sqrt{\frac{h(x_1)}{h_0}}\frac{\partial \varphi_j}{\partial x_2}\frac{h'(x_1)}{h(x_1)}x_2, \quad \text{and}$$

$$\frac{\partial \psi_j}{\partial y_2} = \frac{\partial \varphi_j}{\partial x_2}\sqrt{\frac{h(x_1)}{h_0}},$$

with $h(x_1)/h_0 \geq 1$, we obtain

$$\int_\Omega \left(\sum_{j=1}^m \varphi_j^2 \right)^2 dx$$

$$\leq C_0 \left(\sum_{j=1}^m \int_\Omega \left(\frac{\partial \varphi_j}{\partial x_1} a + \varphi_j b + a\mu \frac{\partial \varphi_j}{\partial x_2} x_2 \right)^2 \frac{dx}{a^2} + \frac{|\varphi_j|_{L^2(\Omega)}^2}{L^2} \right)^{\frac{1}{2}}$$

$$\times \left(\sum_{j=1}^m \int_\Omega \left(\frac{\partial \varphi_j}{\partial x_2} \right)^2 dx + \frac{|\varphi_j|_{L^2(\Omega)}^2}{h_0^2} \right)^{\frac{1}{2}},$$

where

$$a(x) = \sqrt{\frac{h(x_1)}{h_0}}, \quad b(x) = \frac{h'(x_1)}{2\sqrt{h_0 h(x_1)}}, \quad \text{and} \quad \mu(x) = \frac{h'(x_1)}{h(x_1)}.$$

After simple calculations we get

$$\int_\Omega \left(\sum_{j=1}^m \varphi_j^2 \right)^2 dx \leq \frac{C_0}{2} \sum_{j=1}^m \int_\Omega \left(\left(\frac{\partial \varphi_j}{\partial x_1} \right)^2 + \left(\frac{\partial \varphi_j}{\partial x_2} \right)^2 \right) dx$$

$$+ \quad C_0 |\varphi_j|_{L^2(\Omega)}^2 \left(\frac{1}{L^2} + \frac{1}{h_0^2} \right) + \frac{C_0}{2} \int_\Omega \sum_{j=1}^m \frac{\partial \varphi_j}{\partial x_1} \varphi_j \mu \, dx$$

$$+ \quad \frac{C_0}{8} \int_\Omega \left(\sum_{j=1}^m \varphi_j^2 \right) \mu^2 \, dx + C_0 \int_\Omega \sum_{j=1}^m \frac{\partial \varphi_j}{\partial x_1} \frac{\partial \varphi_j}{\partial x_2} \mu x_2 \, dx$$

$$+ \quad \frac{C_0}{2} \int_\Omega \sum_{j=1}^m \varphi_j \frac{\partial \varphi_j}{\partial x_2} \mu^2 x_2 \, dx + \frac{C_0}{2} \int_\Omega \sum_{j=1}^m \left(\frac{\partial \varphi_j}{\partial x_2} \right)^2 \mu^2 x_2^2 \, dx. \quad (1.19)$$

When $h' = 0$, only the first two terms on the right hand side are not zero. We estimate the additional terms as follows.

$$\frac{C_0}{2} \int_\Omega \sum_{j=1}^m \frac{\partial \varphi_j}{\partial x_1} \varphi_j \mu \, \mathrm{d}x \le \frac{C_0}{2} \int_\Omega \left(\sum_{j=1}^m \left(\frac{\partial \varphi_j}{\partial x_1} \right)^2 \right)^{\frac{1}{2}} \left(\sum_{j=1}^m \varphi_j^2 \right)^{\frac{1}{2}} \mu \, \mathrm{d}x$$

$$\le \frac{C_0}{2} \int_\Omega \sum_{j=1}^m \left(\frac{\partial \varphi_j}{\partial x_1} \right)^2 \mathrm{d}x + \frac{C_0}{8} \int_\Omega \left(\sum_{j=1}^m \varphi_j^2 \right) \mu^2 \mathrm{d}x$$

$$\le \frac{C_0}{2} \int_\Omega \sum_{j=1}^m \left(\frac{\partial \varphi_j}{\partial x_1} \right)^2 \mathrm{d}x + \frac{1}{16} \int_\Omega \left(\sum_{j=1}^m \varphi_j^2 \right)^2 \mathrm{d}x + \frac{(C_0)^2}{16} \int_\Omega \mu^4 \, \mathrm{d}x,$$

and

$$\frac{C_0}{8} \int_\Omega \left(\sum_{j=1}^m \varphi_j^2 \right) \mu^2 \mathrm{d}x \le \frac{1}{16} \int_\Omega \left(\sum_{j=1}^m \varphi_j^2 \right)^2 \mathrm{d}x + \frac{(C_0')^2}{16} \int_\Omega \mu^4 \, \mathrm{d}x.$$

Now,

$$C_0 \int_\Omega \sum_{j=1}^m \frac{\partial \varphi_j}{\partial x_1} \frac{\partial \varphi_j}{\partial x_2} \mu x_2 \, \mathrm{d}x$$

$$\le C_0 \int_\Omega \left(\sum_{j=1}^m \left(\frac{\partial \varphi_j}{\partial x_1} \right)^2 \right)^{\frac{1}{2}} \left(\sum_{j=1}^m \left(\frac{\partial \varphi_j}{\partial x_2} \right)^2 \right)^{\frac{1}{2}} \mu x_2 \, \mathrm{d}x$$

$$\le \frac{C_0}{2} \int_\Omega \mu^2 x_2^2 \sum_{j=1}^m \left(\frac{\partial \varphi_j}{\partial x_1} \right)^2 \mathrm{d}x + \frac{C_0}{2} \int_\Omega \sum_{j=1}^m \left(\frac{\partial \varphi_j}{\partial x_2} \right)^2 \mathrm{d}x,$$

and

$$\frac{C_0}{2} \int_\Omega \sum_{j=1}^m \varphi_j \frac{\partial \varphi_j}{\partial x_2} \mu^2 x_2 \, \mathrm{d}x$$

$$\le \frac{C_0}{2} \int_\Omega \left(\sum_{j=1}^m \varphi_j^2 \right)^{\frac{1}{2}} \left(\sum_{j=1}^m \left(\frac{\partial \varphi_j}{\partial x_2} \right)^2 \right)^{\frac{1}{2}} \mu^2 x_2 \, \mathrm{d}x$$

$$\le \frac{C_0}{8} \int_\Omega \left(\sum_{j=1}^m \varphi_j^2 \right) \mu^4 x_2^2 \, \mathrm{d}x + \frac{C_0}{2} \int_\Omega \sum_{j=1}^m \left(\frac{\partial \varphi_j}{\partial x_2} \right)^2 \mathrm{d}x$$

$$\le \frac{1}{16} \int_\Omega \left(\sum_{j=1}^m \varphi_j^2 \right)^2 \mathrm{d}x + \frac{(C_0)^2}{16} \int_\Omega \mu^8 x_2^4 \, \mathrm{d}x + \frac{C_0}{2} \int_\Omega \sum_{j=1}^m \left(\frac{\partial \varphi_j}{\partial x_2} \right)^2 \mathrm{d}x.$$

Applying the above inequalities in (1.19) and replacing x_2 by $h(x_1)$ in some integrals we obtain the elegant estimate

$$\int_\Omega \left(\sum_{j=1}^m \varphi_j^2 \right)^2 dx \ \leq \ C_0'' \sum_{j=1}^m \int_\Omega (1 + h'(x_1)^2) |\nabla \varphi_j|^2 \, dx$$

$$+ \ C_0 \left(\frac{1}{L^2} + \frac{1}{h_0^2} \right) \sum_{j=1}^m |\varphi_j|_{L^2(\Omega)}$$

$$+ \ C_3'' \int_\Omega \left(\frac{h'(x_1)}{h(x_1)} \right)^4 (1 + h'(x_1)^4) \, dx;$$

since

$$\sum_{j=1}^m |\varphi_j|_{L^2(\Omega)} = m,$$

(1.18) follows.   □

Now, to estimate from above the dimension of the global attractor we follow the standard procedure (Robinson, 2001; Temam, 1997). Using Lemmas 1.3.2 and 1.3.3 to estimate the usual trace operator we obtain the following results (Boukrouche & Łukaszewicz, 2004, 2005a).

Theorem 1.3.4 *Problem I. Assume that the domain Ω is thin and that the flow is strongly turbulent, namely*

$$\frac{h_M}{L} << 1 \quad and \quad Re >> 1.$$

Then the fractal dimension of the global attractor $\mathcal{A}_{\mathrm{NSE}}$ can be estimated as follows,

$$d_f(\mathcal{A}_{\mathrm{NSE}}) \leq \kappa \frac{|\Omega|}{h_0^2} (Re)^{3/2}, \tag{1.20}$$

where κ is an absolute constant. For a rectangular domain $\Omega = (0, L) \times (0, h_0)$ we obtain, in particular,

$$d_f(\mathcal{A}_{\mathrm{NSE}}) \leq \kappa \frac{L}{h_0} (Re)^{3/2}. \tag{1.21}$$

Theorem 1.3.5 *Problem II. Assume that the domain Ω is thin and that the flow is strongly turbulent, namely*

$$\frac{h_M}{L} << 1 \quad and \quad Re >> 1.$$

Then the fractal dimension of the global attractor $\mathcal{A}_{\mathrm{NSE}}$ can be estimated as follows,

$$d_f(\mathcal{A}_{\mathrm{NSE}}) \leq \kappa \max \left\{ \sigma_2 |\Omega|, \sqrt{2\sigma_3 |\Omega| + \sigma_1^2 \left(\frac{L h_M}{h_0^2} \right)^2 (Re)^3} \right\} \quad (1.22)$$

where $\sigma_1 = \kappa_1(1 + \max\limits_{0 \leq x_1 \leq L} |h'(x_1)|^2)$, $\sigma_2 = \kappa_2(\frac{1}{L^2} + \frac{1}{h_0^2})$, and

$$\sigma_3 = \kappa_3 \int_\Omega \left(\frac{h'(x_1)}{h(x_1)} \right)^4 (1 + h'(x_1)^4) \, dx,$$

with κ, κ_1, κ_2, and κ_3 some numerical constants. For a rectangular domain $\Omega = (0, L) \times (0, h_0)$ we obtain, in particular,

$$d_f(\mathcal{A}_{\mathrm{NSE}}) \leq \kappa \frac{L}{h_0} (Re)^{3/2}.$$

Estimate (1.21) was obtained by Doering & Wang (1998). Estimate (1.20) is its direct generalization for more general geometry of the flow domain. Estimate (1.22) reduces to that obtained earlier for a rectangle and agrees with our expectations about the behaviour of strongly turbulent shear flows in thin domains met in lubrication theory. It helps us to understand the influence of geometry of the flow and roughness of the boundary (as measured by h') on the behaviour of the fluid.

1.4 Time-dependent driving: dimension of the pullback attractor

In this section we consider Problem III written in a weak form, and present a result about the existence of a unique global in time solution. Then we show the existence of a pullback attractor for the corresponding evolutionary process by using the energy equation method developed recently by Caraballo, Łukaszewicz, & Real (2006a,b) to cover the pullback attractor case. We also obtain an upper bound on the dimension of the pullback attractor in terms of the data, by using the method proposed by Caraballo, Langa, & Valero (2003).

The weak formulation of Problem III is similar to that of Problem I, the only difference being that now the problem is non-autonomous. This comes from the time-dependent boundary condition on the bottom part of the boundary. Accordingly, the background flow now depends on time,

$$u(x_1, x_2, t) = U(x_2, t)e_1 + v(x_1, x_2, t), \quad (1.23)$$

with

$$U(0,t) = U_0(t) \quad \text{and} \quad U(h(x_1),t) = 0, \quad x_1 \in (0,L), \, t \in (-\infty,\infty).$$
$$(1.24)$$

Let H and V be the same function spaces as for Problem I. Then the natural weak formulation of the homogenized Problem III is as follows.

Problem 1.4.1 *Find*

$$v \in \mathcal{C}([\tau,T];H) \cap L^2(\tau,T;V)$$

for each $T > \tau$, such that

$$\frac{\mathrm{d}}{\mathrm{d}t}(v(t),\Theta) + \nu a(v(t),\Theta) + b(v(t),v(t),\Theta) = F(v(t),\Theta), \quad t > \tau, \quad (1.25)$$

for all $\Theta \in V$, and

$$v(x,\tau) = v_0(x),$$

where

$$F(v,\Theta) = -\nu a(\xi,\Theta) - b(\xi,v,\Theta) - b(v,\xi,\Theta) - (\xi_{,t},\Theta), \quad (1.26)$$

and $\xi = U e_1$ is a suitable background flow.

We have the following existence and uniqueness theorem (Boukrouche et al., 2006).

Theorem 1.4.2 *Let U_0 be a locally Lipschitz continuous function on the real line. Then there exists a unique weak solution of Problem 1.4.1 such that for all η, T, $\tau < \eta < T$, $v \in L^2(\eta,T;H^2(\Omega))$, and for each $t > \tau$ the map $v_0 \mapsto v(t)$ is continuous as a map from H into itself.*

We shall now study the existence of the pullback attractor for the evolutionary process associated with this problem. First, we recall some basic notions about pullback attractors.

Let us consider an evolutionary process U on a metric space X, i.e. a family $\{U(t,\tau); \, -\infty < \tau \le t < +\infty\}$ of continuous mappings $U(t,\tau) : X \to X$, such that $U(\tau,\tau)x = x$, and

$$U(t,\tau) = U(t,r)U(r,\tau) \quad \text{for all } \tau \le r \le t.$$

Suppose that $\mathcal{D}$ is a nonempty class ('universe') of parameterized sets $\widehat{D} = \{D(t); \, t \in \mathbb{R}\} \subset \mathcal{P}(X)$, where $\mathcal{P}(X)$ denotes the family of all nonempty subsets of X, with the property that if $D \in \mathcal{D}$ and $\widetilde{D}(t) \subseteq D(t)$ for every $t \in \mathbb{R}$ then $\widetilde{D} \in \mathcal{D}$.

Definition 1.4.3 *A process $U(t, \tau)$ is said to be pullback $\mathcal{D}$-asymptotically compact if for each $t \in \mathbb{R}$ and $\widehat{D} \in \mathcal{D}$, any sequence $\tau_n \to -\infty$, and any sequence $x_n \in D(\tau_n)$, the sequence $\{U(t, \tau_n)x_n\}$ $(\tau_n \le t)$ is relatively compact in X.*

Definition 1.4.4 *A family $\widehat{B} \in \mathcal{D}$ is said to be pullback $\mathcal{D}$-absorbing for the process $U(t, \tau)$ if for any $t \in \mathbb{R}$ and any $\widehat{D} \in \mathcal{D}$, there exists a $\tau_0(t, \widehat{D}) \le t$ such that*

$$U(t, \tau)D(\tau) \subset B(t) \quad \text{for all } \tau \le \tau_0(t, \widehat{D}).$$

Definition 1.4.5 *A family $\widehat{A} = \{A(t); \, t \in \mathbb{R}\} \subset \mathcal{P}(X)$ is said to be a pullback $\mathcal{D}$-attractor for $U(\cdot, \cdot)$ if*
(a) $A(t)$ is compact for all $t \in \mathbb{R}$,
(b) $\widehat{A}$ is pullback $\mathcal{D}$-attracting, i.e.

$$\lim_{\tau \to -\infty} \text{dist}(U(t, \tau)D(\tau), A(t)) = 0 \quad \text{for all } \widehat{D} \in \mathcal{D} \text{ and all } t \in \mathbb{R},$$

(c) $\widehat{A}$ is invariant, i.e.

$$U(t, \tau)A(\tau) = A(t) \quad \text{for all } \tau \le t.$$

We have the following result (Caraballo et al., 2006b):

Theorem 1.4.6 *Suppose that the process $U(t, \tau)$ is pullback $\mathcal{D}$-asymptotically compact, and that $\widehat{B} \in \mathcal{D}$ is a family of pullback $\mathcal{D}$-absorbing sets for $U(\cdot, \cdot)$. Then the family $\widehat{A} = \{A(t); \, t \in \mathbb{R}\} \subset \mathcal{P}(X)$ defined by*

$$A(t) = \Lambda(\widehat{B}, t), \quad t \in \mathbb{R},$$

where for each $\widehat{D} \in \mathcal{D}$

$$\Lambda(\widehat{D}, t) = \bigcap_{s \le t} \left(\overline{\bigcup_{\tau \le s} U(t, \tau)D(\tau)} \right),$$

is a pullback $\mathcal{D}$-attractor for $U(\cdot, \cdot)$. In addition

$$A(t) = \overline{\bigcup_{\widehat{D} \in \mathcal{D}} \Lambda(\widehat{D}, t)}.$$

Furthermore, $\widehat{A}$ is minimal in the sense that if $\{C(t); \, t \in \mathbb{R}\} \subset \mathcal{P}(X)$ is a family of closed sets such that for every $B \in \mathcal{D}$

$$\lim_{\tau \to -\infty} \text{dist}(U(t, \tau)B(\tau), C(t)) = 0,$$

then $A(t) \subseteq C(t)$.

Now, we come back to the context of Problem 1.4.1. For $t \geq \tau$ let us define the map $U(t, \tau)$ in H by

$$U(t, \tau)v_0 = v(t; \tau, v_0), \quad t \geq \tau, \quad v_0 \in H, \tag{1.27}$$

where $v(t; \tau, v_0)$ is the solution of Problem 1.4.1. From the uniqueness of solutions to this problem, one immediately obtains

$$U(t, \tau)v_0 = U(t, r)(U(r, \tau)v_0), \quad \text{for all } \tau \leq r \leq t, \ v_0 \in H.$$

From Theorem 1.4.2 it follows that for all $t \geq \tau$, the process mapping $U(t, \tau) : H \to H$, defined by (1.27), is continuous. Consequently, the family $\{U(t, \tau), \ \tau \leq t\}$ defined by (1.27) is a process in H.

We define the universe of the parameterized families of sets as follows: for $\sigma = \nu\lambda_1$ and $|D(t)|^+ = \sup\{|y| : y \in D(t)\}$, let

$$\mathcal{D}_\sigma = \{D : \mathbb{R} \to \mathcal{P}(H); \ \lim_{t \to -\infty} e^{\sigma t}(|D(t)|^+)^2 = 0\}.$$

Then we have the following (Boukrouche et al., 2006):

Theorem 1.4.7 *Let U_0 be a locally Lipschitz continuous function on the real line such that*

$$\int_{-\infty}^t e^{\sigma s}(|U_0(s)|^3 + |U_0'(s)|^2) \, ds < +\infty \quad \text{for all } t \in \mathbb{R}.$$

Then, there exists a unique pullback $\mathcal{D}_\sigma$-attractor $\widehat{A} \in \mathcal{D}_\sigma$ for the process $U(t, \tau)$ defined by (1.27).

We can also show that the dimension of the attractor is finite:

Theorem 1.4.8 *Let U_0 be a locally Lipschitz continuous function on the real line such that for some real $t^\star$, $r > 0$, $M_b > 0$, $M > 0$, all $t \leq t^\star$ and all $s \leq t^\star - r$,*

$$|U_0(t)| \leq M_b \quad \text{and} \quad \int_s^{s+r} |U_0'(\eta)|^2 d\eta \leq M.$$

Then the attractor $\{A(t) : t \in \mathbb{R}\}$ from Theorem 1.4.7 has finite fractal dimension, namely,

$$d_f(A(t)) \leq d$$

for all $t \in \mathbb{R}$ and some constant d.

For the proof of Theorem 1.4.8 (Boukrouche et al., 2006) we have used a result due to Caraballo et al. (2003) which in our notation can be expressed as follows:

Theorem 1.4.9 *Suppose that there exist constants K_0, K_1, $\theta > 0$ such that*

$$|A(t)|^+ = \sup\{|y| : y \in A(t)\} \leq K_0|t|^\theta + K_1$$

for all $t \in \mathbb{R}$. Also assume that for any $t \in \mathbb{R}$ there exists $T = T(t)$, $l = l(t, T) \in [1, +\infty)$, $\delta = \delta(t, T) \in (0, 1/\sqrt{2})$, and $N = N(t)$ such that for any $u, v \in A(\tau)$, $\tau \leq t - T$,

$$|U(\tau + T, \tau)u - U(\tau + T, \tau)v| \leq l|u - v|,$$

$$|Q_N(U(\tau + T, \tau)u - U(\tau + T, \tau)v)| \leq \delta|u - v|,$$

where Q_N is the projector mapping H onto some subspace $H_N^\perp$ of co-dimension $N \in \mathbb{N}$. Then, for any $\eta = \eta(t) > 0$ such that $\sigma = \sigma(t) = (6\sqrt{2}l)^N(\sqrt{2}\delta)^\eta < 1$, the fractal dimension of $A(t)$ is bounded, with $d_f(A(t)) \leq N + \eta$.

The new element in Problem III in relation to Problem I is the allowance of the speed of rotation of the cylinder to depend on time. We emphasize that neither quasi-periodicity nor even boundedness of the non-autonomous term are required to prove the existence of the corresponding pullback attractor and to estimate its fractal dimension. The theory of pullback attractors allowed us to impose quite general assumptions on the velocity of the boundary.

To prove the existence of the pullback attractor we used the energy equation method, as applied recently by Caraballo et al. (2006a,b) to pullback attractors, which also works in the case of some unbounded domains of the flow as it bypasses the usual compactness argument. In turn, to estimate the pullback attractor dimension we used the method proposed by Caraballo et al. (2003), an alternative to the usual one based on Lyapunov exponents (Temam, 1997). Notice that to estimate the pullback attractor dimension no restriction was imposed on the non-autonomous term in the future, but the term had to be bounded in the past. While the latter property could seem a strong condition, at the moment there is no result in the literature on the finite dimensionality of pullback attractors that avoids this assumption.

1.5 Time-dependent driving with Tresca's boundary condition

In this section we consider Problem IV: (1.1)–(1.3), (1.11)–(1.13). First, we homogenize the boundary condition (1.13). Then we present a variational formulation of the homogenized problem. In the end we present results about the existence and uniqueness of a solution that is global in time, and about the existence of a pullback attractor.

To homogenize the boundary condition (1.13) let

$$u(x_1, x_2, t) = U(x_2, t)e_1 + v(x_1, x_2, t),$$

with

$$U(0, t) = U_0(t), \quad U(h(x_1), t) = 0, \quad \frac{\partial U(x_2, t)}{\partial x_2}\Big|_{x_2=0} = 0,$$

for $x \in (0, L)$ and $t \in (-\infty, \infty)$. We obtain

$$v = 0 \quad \text{on} \quad \Gamma_1,$$

and

$$v \cdot n = 0 \quad \text{on} \quad \Gamma_0.$$

The Tresca condition transforms to the following conditions on Γ_0:

$$|\sigma_\eta(v, p)| < k \quad \Rightarrow \quad v_\eta = 0,$$

while

$$|\sigma_\eta(v, p)| = k \quad \Rightarrow \quad \exists\, \lambda \geq 0 \text{ such that } v_\eta = -\lambda\sigma_\eta(v, p).$$

In the end the initial condition becomes

$$v(x, \tau) = v_0(x) = u_0(x) - U(x_2, \tau)e_1.$$

Let H and V be function spaces as for Problem II, and let us define the functional j on V by

$$j(u) = \int_{\Gamma_0} k|u(x_1, 0)|\mathrm{d}x_1.$$

The variational formulation of the homogenized problem is as follows.

Problem 1.5.1 *Given $\tau \in \mathbb{R}$ and $v_0 \in H$, find $v : (\tau, \infty) \to H$ such that:*
(i) for all $T > \tau$,

$$v \in \mathcal{C}([\tau, T]; H) \cap L^2(\tau, T; V), \quad \text{with} \quad v_t \in L^2(\tau, T; V'),$$

(ii) for all Θ *in* V, *all* $T > \tau$, *and for almost all* t *in the interval* $[\tau, T]$, *the following variational inequality holds*

$$\langle v_t(t), \Theta - v(t) \rangle + \nu\, a(v(t), \Theta - v(t)) + b(v(t), v(t), \Theta - v(t))$$
$$+ j(\Theta) - j(v(t)) \geq (\mathcal{L}(v(t)), \Theta - v(t)),$$

and

(iii) the initial condition

$$v(x, \tau) = v_0(x)$$

holds.

In (1.28) *the functional* $\mathcal{L}(v(t))$ *is defined for almost all* $t \geq \tau$ *by,*

$$(\mathcal{L}(v(t)), \Theta) = -\nu a(\xi, \Theta) - (\xi_{,t}(t), \Theta)$$
$$-b(\xi(t), v(t), \Theta) - b(v(t), \xi(t), \Theta),$$

where $\xi = U e_1$ *is a suitable smooth background flow.*

We have the following relation between classical and weak formulations (Boukrouche & Łukaszewicz, 2007).

Proposition 1.5.2 *Every classical solution of Problem IV is also a solution of Problem 1.5.1. On the other hand, every solution of Problem 1.5.1 that is smooth enough is also a classical solution of Problem IV.*

Theorem 1.5.3 (Boukrouche & Łukaszewicz, 2007) *Let* $v_0 \in H$ *and the function* $s \mapsto |U_0(s)|^3 + |U_0'(s)|^2$ *be locally integrable on the real line. Then there exists a solution of Problem 1.5.1.*

Proof We sketch here only the main steps of the proof. Observe that the functional j is convex but not differentiable. To overcome this difficulty we use the following approach (Haslinger, Hlaváček, & Necas, 1996). For $\delta > 0$ let $j_\delta : V \to \mathbb{R}$ be a functional defined by

$$\varphi \mapsto j_\delta(\varphi) = \frac{1}{1+\delta} \int_{\Gamma_0} k|\varphi|^{1+\delta} \mathrm{d}x,$$

which is convex, lower continuous and finite on V, and has the following properties:

(i) $\exists\, \chi \in V'$ and $\mu \in \mathbb{R}$ such that $j_\delta(\varphi) \geq \langle \chi, \varphi \rangle + \mu \quad \forall \varphi \in V$,

(ii) $\lim_{\delta \to 0^+} j_\delta(\varphi) = j(\varphi) \quad \forall \varphi \in V$, and

(iii) $v_\delta \rightharpoonup v$ (weakly) in $V \Rightarrow \lim_{\delta \to 0^+} j_\delta(v_\delta) \geq j(v)$.

The functional j_δ is Gâteaux differentiable in V, with

$$(j_\delta'(v)\,,\,\Theta) = \int_{\Gamma_0} k|v|^{\delta-1} v\,\Theta\,\mathrm{d}x \quad \forall\,\Theta \in V.$$

We consider the following equation

$$\left(\frac{\mathrm{d}v_\delta}{\mathrm{d}t}\,,\,\Theta\right) + \nu a(v_\delta(t)\,,\,\Theta) + b(v_\delta(t)\,,\,v_\delta(t),\Theta) + (j_\delta'(v_\delta)\,,\,\Theta)$$

$$= -\nu a(\xi(t)\,,\,\Theta) - (\xi_{,t}\,,\,\Theta) - b(\xi(t)\,,\,v_\delta(t),\Theta) - b(v_\delta(t)\,,\,\xi(t)\,,\Theta),$$

with initial condition

$$v_\delta(\tau) = v_0,$$

establish an *a priori* estimate for v_δ for $\delta > 0$, and then show that the limit function v is a solution to Problem 1.5.1. $\qquad\square$

Moreover, the solution is unique and depends continuously on the initial data, namely, we have the following:

Theorem 1.5.4 *Under the hypotheses of Theorem 1.5.3, the solution v of Problem 1.5.1 is unique and the map $v(\tau) \rightarrow v(t)$, for $t > \tau$, is continuous from H into itself.*

Now we shall study existence of the pullback attractor using a method based on the concept of the Kuratowski measure of non-compactness of a bounded set, developed by Song & Wu (2007). This method is very useful when one deals with variational inequalities as it overcomes obstacles coming from the usual methods. One needs neither compactness of the dynamics which results from the second energy inequality nor asymptotic compactness, see Boukrouche et al. (2006), Caraballo et al. (2006a,b), or Temam (1997), which results from the energy equation. In the case of variational inequalities it is sometimes very difficult to obtain the second energy inequality due to the presence of boundary functionals, on the other hand, we do not have any energy equation.

We now recast the theory of Song & Wu (2007) in the language of evolutionary processes, and then apply it to our problem. Recall that the Kuratowski measure of non-compactness (Kuratowski, 1930) of a bounded subset B of H, $\alpha(B)$, is defined as

$$\alpha(B) = \inf\{\delta : B \text{ admits a finite cover by sets of diameter } \leq \delta\}.$$

Definition 1.5.5 *The process $U(t, \tau)$ is said to be pullback ω-limit compact if for any $B \in B(H)$, for $t \in \mathbb{R}$,*

$$\lim_{\tau \to \infty} \alpha \left(\bigcup_{s \le t - \tau} U(t, s) B \right) = 0.$$

In fact U is pullback ω-limit compact if and only if it is pullback $\mathcal{D}$-asymptotically compact in the sense of Definition 1.4.3, where $\mathcal{D}$ is taken to be the collection of all time-independent bounded sets (see, for example, Kloeden & Langa, 2007).

Definition 1.5.6 *Let H be a Banach space. The process U is said to be norm-to-weak continuous on H if for all $(t, s, x) \in \mathbb{R} \times \mathbb{R} \times H$ with $t \ge s$ and for every sequence $(x_n) \in H$,*

$$x_n \to x \quad \text{strongly in } H \quad \Longrightarrow \quad U(t, s)x_n \rightharpoonup U(t, s)x \quad \text{weakly in } H.$$

Theorem 1.5.7 *Let H be a Banach space, and U a process on H. If U is norm-to-weak continuous and possesses a uniformly absorbing set B_0, then U possesses a pullback attractor $\mathcal{A} = \{\mathcal{A}(t)\}_{t \in \mathbb{R}}$, with*

$$\mathcal{A}(t) = \Lambda(B_0, t) \quad \forall\, t \in \mathbb{R},$$

if and only if it is pullback ω-limit compact.

We now state the main theorem from Song & Wu (2007). The terminology "flattening property" was coined by Kloeden & Langa (2007).

Theorem 1.5.8 (cf. Song & Wu, 2007) *Let H be a Banach space. If the process U has the pullback flattening property, i.e. if for any $t \in \mathbb{R}$, a bounded subset B of H, and $\varepsilon > 0$, there exists an $s_0(t, B, \varepsilon)$ and a finite-dimensional subspace E of H such that for some bounded projector $P : H \to E$,*

$$P \left(\bigcup_{s \le s_0} U(t, s) B \right) \quad \text{is bounded}$$

and

$$\left| (I - P) \left(\bigcup_{s \le s_0} U(t, s) B \right) \right| \le \varepsilon,$$

then U is pullback ω-limit compact.

Now let U be the evolutionary process associated with Problem 1.5.1.

Lemma 1.5.9 *Let*

$$\sup_{h\in\mathbb{R}}\int_h^{h+1} F(s)\,\mathrm{d}s < R(F) < \infty, \tag{1.28}$$

$\sigma > 0$ *and* $t \geq \tau$. *Then for every* $\varepsilon > 0$ *there exists* $\delta = \delta(\varepsilon) > 0$ *such that*

$$\int_\tau^t \mathrm{e}^{-\sigma(t-s)} F(s)\,\mathrm{d}s \leq \frac{\varepsilon}{2} + \frac{\mathrm{e}^{-\sigma\delta}}{1-\mathrm{e}^{-\sigma}}R(F).$$

Proof Let δ be such that $\int_{t-\delta}^t F(s)\mathrm{d}s \leq \frac{\varepsilon}{2}$, $\tau < \delta < t$. Then, by (1.28),

$$\int_\tau^t \mathrm{e}^{-\sigma(t-s)} F(s)\mathrm{d}s \;\leq\; \frac{\varepsilon}{2} + \sum_{k=1}^\infty \int_{t-(\delta+k+1)}^{t-(\delta+k)} \mathrm{e}^{-\sigma(t-s)} F(s)\mathrm{d}s$$

$$\leq\; \frac{\varepsilon}{2} + \frac{\mathrm{e}^{-\sigma\delta}}{1-\mathrm{e}^{-\sigma}}R(F). \qquad \square$$

Let $F(s) = |U_0(s)|^3 + |U_0'(s)|^2$, cf. Theorem 1.5.3. Then we have

Lemma 1.5.10 *Let the initial condition* v_0 *in Problem 1.5.1 belong to a ball* $B(0,\rho)$ *in* H. *Suppose that (1.28) holds. Then the solution* v *of Problem 1.5.1 satisfies*

$$\sup_{h\geq\tau}\int_h^{h+1} |\nabla v(s)|^2\mathrm{d}s \leq \frac{2}{\nu}\{\rho^2 + (1+\frac{\mathrm{e}^{-\sigma\delta}}{1-\mathrm{e}^{-\sigma}})R(F)\}. \tag{1.29}$$

Proof Taking $\Theta = 0$ in (1.28) we obtain,

$$\frac{1}{2}\frac{\mathrm{d}}{\mathrm{d}t}|v(t)|^2 + \frac{\nu}{2}|\nabla v(t)|^2 \leq F(t) \tag{1.30}$$

and, in consequence,

$$\frac{1}{2}\frac{\mathrm{d}}{\mathrm{d}t}|v(t)|^2 + \frac{\sigma}{2}|v(t)|^2 \leq F(t), \tag{1.31}$$

with $\sigma = \nu\lambda_1$. By Gronwall's inequality and Lemma 1.5.9 with $\varepsilon = \rho^2$ we conclude from the last inequality that for $t \geq \tau$,

$$|v(t)|^2 \leq 2\rho^2 + \frac{2\mathrm{e}^{-\sigma\delta}}{1-\mathrm{e}^{-\sigma}}R(F). \tag{1.32}$$

Integrating (1.30) we obtain the first energy estimate: for $\tau \leq \eta \leq t$,

$$|v(t)|^2 + \nu \int_\eta^t |\nabla v(s)|^2 \mathrm{d}s \leq 2 \int_\eta^t F(s)\mathrm{d}s + |v(\eta)|^2.$$

Using this estimate and (1.32) we obtain (1.29). $\qquad\qquad\qquad\qquad\square$

Theorem 1.5.11 *Let $v_0 \in H$ and U_0 be such that (1.28) holds, with $F(s) = |U_0(s)|^3 + |U_0'(s)|^2$. Then there exists a pullback attractor $\mathcal{A}$ in the sense of Theorem 1.5.7 for the evolutionary process U.*

Proof From (1.31), (1.32), the Gronwall inequality, and Lemma 1.5.9 we obtain

$$|U(s+t,t)v_0|^2 \leq \mathrm{e}^{-\sigma s}|v_0|^2 + \varepsilon + \frac{2\mathrm{e}^{-\sigma\delta}}{1-\mathrm{e}^{-\sigma}}R(F).$$

For v_0 in $B(0,\rho)$ and s large enough, $U(s+t,t)v_0 \in B(0,\rho_0)$, where ρ_0 depends only on ε, ρ, and $R(F)$, which means that there exists a uniformly absorbing ball in H.

From Theorem 1.5.4 it follows that the evolutionary process U is strongly continuous in H, whence, in particular, it is norm-to-weak continuous on H.

Thus, according to Theorem 1.5.7 and Theorem 1.5.8, to finish the proof we have to prove that U has the pullback flattening property.

Let A be the Stokes operator in H. Since A^{-1} is continuous and compact in H, there exists a sequence $\{\lambda_j\}_{j=1}^\infty$ such that $0 < \lambda_1 \leq \lambda_2 \leq \ldots \leq \lambda_j \leq \ldots$ with $\lim_{j\to+\infty} \lambda_j = \infty$, and a family of elements $\{\varphi_j\}_{j=1}^\infty$ of $D(A)$, which are orthonormal in H such that $A\varphi_j = \lambda_j \varphi_j$.

We define the m-dimensional subspace V_m, of V, and the orthogonal projection operator $P_m : V \to V_m$ by $V_m = \mathrm{span}\{\varphi_1, \ldots, \varphi_m\}$ and $P_m v = \sum_{j=1}^m (v, \varphi_j)\varphi_j$. For $v \in D(A) \subset V$ we can decompose v as $v = P_m v + (I - P_m)v = P_m v + v_2$.

Set $\Theta = v_1(t)$ in (1.28) to get

$$\frac{1}{2}\frac{\mathrm{d}}{\mathrm{d}t}|v_2(t)|^2 + \nu|\nabla v_2(t)|^2 \leq j(v_1(t)) - j(v(t)) - b(v(t), v(t), v_2(t)) + (\mathcal{L}(v(t)), v_2(t)).$$

From the continuity of the trace operator we have

$$j(v_1(t)) - j(v(t)) \leq j(v_2(t)) \leq C + \frac{\nu}{4}\|v_2(t)\|^2.$$

Using the anisotropic Ladyzhenskaya inequality $\|v\|_{L^4(\Omega)} \leq C|v|^{\frac{1}{2}}|\nabla v|^{\frac{1}{2}}$ (where $C = C(\Omega)$) we easily arrive at

$$\frac{\mathrm{d}}{\mathrm{d}t}|v_2(t)|^2 + \nu|\nabla v_2(t)|^2 \leq C_2(1 + F(t) + |\nabla v(t)|^2).$$

and

$$\frac{\mathrm{d}}{\mathrm{d}t}|v_2(t)|^2 + \nu\lambda_{m+1}|v_2(t)|^2 \leq C_2(1 + F(t) + |\nabla v(t)|^2).$$

Now, let $\varepsilon > 0$ be given. Using Lemmas 1.5.9 and 1.5.10, and taking m large enough, we obtain

$$|(I - P_m)U(s + t, t)v_0|^2 \leq \varepsilon$$

uniformly in t, for $v_0 \in B(0, \rho)$ and all $s \geq s_0(\rho, \varepsilon)$ large enough. This ends the proof of the theorem. $\qquad\square$

Acknowledgements

This research was supported by the Polish Government Grant MEiN 1 P303A 017 30 and Project FP6 EU SPADE2.

References

Boukrouche, M. & Łukaszewicz, G. (2004) An upper bound on the attractor dimension of a 2D turbulent shear flow in lubrication theory. *Nonlinear Analysis* **59**, 1077–1089.

Boukrouche, M. & Łukaszewicz, G. (2005a) An upper bound on the attractor dimension of a 2D turbulent shear flow with a free boundary condition, in *Regularity and other Aspects of the Navier–Stokes Equations*, Banach Center Publications **70**, 61–72, Institute of Mathematics, Polish Academy of Science, Warszawa.

Boukrouche, M. & Łukaszewicz, G. (2005b) Attractor dimension estimate for plane shear flow of micropolar fluid with free boundary. *Mathematical Methods in the Applied Sciences* **28**, 1673–1694.

Boukrouche, M. & Łukaszewicz, G. (2007) On the existence of pullback attractor for a two-dimensional shear flow with Tresca's boundary condition, in *Parabolic and Navier–Stokes Equations*. Banach Center Publications **81**, 81–93, Institute of Mathematics, Polish Academy of Science, Warszawa.

Boukrouche, M., Łukaszewicz, G., & Real, J. (2006) On pullback attractors for a class of two-dimensional turbulent shear flows. *International Journal of Engineering Science* **44**, 830–844.

Caraballo, T., Langa, J.A., & Valero, J. (2003) The dimension of attractors of non-autonomous partial differential equations. *ANZIAM J.* **45**, 207–222.

Caraballo, T., Łukaszewicz, G., & Real, J. (2006a) Pullback attractors for asymptotically compact non-autonomous dynamical systems. *Nonlinear Analysis, TMA* **64**, 484–498.

Caraballo, T., Łukaszewicz, G., & Real, J. (2006b) Pullback attractors for non-autonomous 2D-Navier–Stokes equations in some unbounded domains. *C. R. Acad. Sci. Paris, Ser. I* **342**, 263–268.

Chepyzhov, V.V. & Vishik, M.I. (2002) *Attractors for equations of mathematical physics*. Providence, RI.

Doering, C.R. & Constantin, P. (1991) Energy dissipation in shear driven turbulence. *Phys. Rev. Lett.* **69**, 1648–1651.

Doering, C.R. & Gibbon, J.D. (1995) *Applied Analysis of the Navier–Stokes Equations*. Cambridge University Press, Cambridge.

Doering, C.R. & Wang, X. (1998) Attractor dimension estimates for two-dimensional shear flows. *Physica D* **123**, 206–222.

Duvaut, G. & Lions, J.L. (1972) *Les inéquations en mécanique et en physique*. Dunod, Paris.

Falconer, K. (1990) *Fractal Geometry*. Wiley, Chichester.

Foias, C., Manley, O., Rosa, R., & Temam, R. (2001) *Navier–Stokes Equations and Turbulence*. Cambridge University Press, Cambridge.

Haslinger, J., Hlavácek, I., & Nečas, J. (1996) Numerical Methods for unilateral problems in solid mechanics, in Ciarlet, P.G. & Lions, J.L. (eds.), *Handbook of Numerical Analysis, Vol IV*, 313–485, North Holland, Amsterdam.

Kloeden, P.K. & Langa, J.A. (2007) Flattening, squeezing, and the existence of random attractors. *Proc. Roy. Soc. London A* **463**, 163–181.

Kuratowski, K. (1930) Sur les espaces complets. *Fund. Math.* **15**, 301–309.

Łukaszewicz, G. & Sadowski, W. (2004) Uniform attractor for 2D magneto-micropolar fluid flow in some unbounded domains. *Z. Angew. Math. Phys.* **55**, 1–11.

Miranville, A. & Wang, X. (1997) Attractor for non-autonomous nonhomogeneous Navier–Stokes equations. *Nonlinearity* **10**, 1047–1061.

Miranville, A. & Ziane, M. (1997) On the dimension of the attractor for the Bénard problem with free surfaces. *Russian J. Math. Phys.* **5**, 489–502.

Moise, I., Rosa, R., & Wang, X. (2004) Attractors for non-compact non-autonomous systems via energy equations. *Discrete and Continuous Dynamical Systems* **10**, 473–496.

Mucha, P. & Sadowski, W. (2005) Long time behaviour of a flow in infinite pipe conforming to slip boundary conditions. *Mathematical Methods in the Applied Sciences* **28**, 1867–1880.

Robinson, J.C. (2001) *Infinite-Dimensional Dynamical Systems*. Cambridge University Press, Cambridge.

Song, H. & Wu, H. (2007) Pullback attractors of non-autonomous reaction-diffusion equations. *J.Math. Anal. Appl.* **325**, 1200–1215.

Temam, R. (1997) *Infinite Dimensional Dynamical Systems in Mechanics and Physics*. Second Edition, Springer-Verlag, New York.

Temam, R. & Ziane, M. (1998) Navier–Stokes equations in three-dimensional thin domains with various boundary conditions. *Adv. in Differential Equations* **1**, 1–21.

Ziane, M. (1997) Optimal bounds on the dimension of the attractor of the Navier–Stokes equations. *Physica D* **105**, 1–19.

Ziane, M. (1998) On the 2D-Navier–Stokes equations with the free boundary condition. *Appl. Math. and Optimization* **38**, 1–19.

2

Mathematical results concerning unsteady flows of chemically reacting incompressible fluids

Miroslav Bulíček

Charles University, Faculty of Mathematics and Physics,
Mathematical Institute, Sokolovská 83, 186 75 Prague 8. Czech Republic.
mbul8060@karlin.mff.cuni.cz

Josef Málek

Charles University, Faculty of Mathematics and Physics,
Mathematical Institute, Sokolovská 83, 186 75 Prague 8. Czech Republic.
malek@karlin.mff.cuni.cz

Kumbakonam R. Rajagopal

Department of Mechanical Engineering, Texas A&M University,
College Station, TX 77843. USA.
krajagopal@mengr.tamu.edu

Abstract

We investigate the mathematical properties of unsteady three-dimensional internal flows of chemically reacting incompressible shear-thinning (or shear-thickening) fluids. Assuming that we have Navier's slip at the impermeable boundary we establish the long-time existence of a weak solution when the data are large.

2.1 Introduction

Even though 150 years have elapsed since Darcy (1856) published his celebrated study, the equation he introduced (or minor modifications of it) remains as the main model to describe the flow of fluids through porous media due to a pressure gradient. While the equation that Darcy provided in his study is referred to as a "law" it is merely an approximation, and a very simple one at that, for the flow of a fluid through porous media. The original equation due to Darcy can be shown to be an approximation of the equations governing the flow of a fluid through a porous solid within the context of the theory of mixtures by appealing to numerous assumptions (Atkin & Craine, 1976a,b; Bowen, 1975;

Published in *Partial Differential Equations and Fluid Mechanics*, edited by
James C. Robinson and José L. Rodrigo. © Cambridge University Press 2009.

Green & Naghdi, 1969; Adkins, 1963a,b). Hassanizadeh (1986) and Gray (1983) have also shown that Darcy's equation can be obtained using an averaging technique, but not within the context of mixtures. To obtain Darcy's equation within the context of the theory of mixtures, one ignores the balance of linear momentum for the solid (which is assumed to be rigid and thus the stress is whatever it needs to be in order for the flow to take place), assumes that the only interaction between the fluid and the solid is the frictional resistance at the pores of the solid, and that this resistance is proportional to the difference in the velocity between the fluid and solid. The frictional effects within the fluid and thus the dissipation within the fluid are ignored. If the frictional resistance at the pore, between the solid and the fluid, is not assumed to be proportional to the difference in velocity between the fluid and the solid, but depends, in another way, in a nonlinear manner on the difference, one obtains the Darcy–Forchheimer equation (see Forchheimer, 1901). If the viscosity of the fluid is not neglected, i.e. the viscous dissipation in the fluid is not ignored, and if it is assumed that it is like that in the classical Navier–Stokes fluid, then one obtains the equation developed by Brinkman (1947a,b) (see also Rajagopal, 2007).

An interesting counterpart presents itself when we consider a complex fluid such as blood which is maintained in a state of delicate balance by virtue of a myriad of chemical reactions that take place, some that cause the blood to coagulate, others that cause lysis, etc. In fact, even the development of a simple model for blood requires dozens of equations and these govern the biochemical reactions that have to be coupled to the balance equations (Anand, Rajagopal, & Rajagopal, 2003, 2005). As blood involves constituents that can be modelled by the Navier–Stokes model (for the plasma for instance) or a purely elastic model (say for platelets) and others that are viscoelastic solids (cells) or viscoelastic fluids, one would have a system of equations that would be totally intractable. Thus, it is absolutely necessary to simplify the model while capturing the quintessential feature of the mechanical response characteristics. Instead of keeping track of all the constituents of blood, even though it might be an oversimplification, one could consider blood as a homogeneous fluid whose properties change due to a chemical state variable, which we shall refer to as the concentration c, which is a consequence of all chemical reactions that take place. Thus, in essence, we are replacing a plethora of chemical reactions by a single equation that has the same effect, on average, as the system of reactions that actually occur.

It is then possible to model the flow of blood through a coupled system of equations: the balance of mass, linear and angular momentum for the homogeneous single fluid, and an advection-diffusion equation for the chemical state variable c, the concentration. It is also possible to think of blood as a single homogeneous fluid that is co-occupying the flow domain with another fluid that is capable of reacting with the homogeneous fluid and thereby changing its properties. This reacting fluid is carried along by the flowing fluid, and in the spirit of the development of Darcy's equation, we can choose to ignore the balance equation for the reacting fluid (similar to ignoring the balance equations for the porous solid). As the second fluid moves with the same velocity as the carrier, we do not have an interactive force like the "drag force" that is a consequence of the relative velocity between the two fluids, acting on the fluids. Bridges & Rajagopal (2006) associate the concentration with the ratio of the density of the reactant to the sum of the density of the reactant and the homogeneous fluid. While a concentration defined in such a manner tends to zero when the density of the second fluid tends to zero, it cannot tend to unity as the density of the carrier fluid (which they assume to be incompressible) is not zero and the density of the reacting fluid is finite. Though Bridges & Rajagopal (2006) motivate the concentration through such a ratio, as far as their study is concerned the concentration c is merely treated as a variable that can change (one could view it as an internal variable with a clear physical underpinning, namely the concentration of a second fluid that is undergoing a reaction).

As blood is a multi-constituent material, with the constituents distributed in the vessel in an inhomogeneous manner, it would be more appropriate to approximate it as a single constituent fluid that is inhomogeneous; that is, the homogenization of the multi-constituent body leads to an inhomogeneous body. It is important to recognize that while referring to the "homogenization" of the body we are referring to an averaging procedure of the multi-constituents, while when describing the body as inhomogeneous we are referring to the fact that the averaged body has properties that change from one material point to another. It is important to keep this distinction in mind. In the study that is being carried out, we are able to capture this inhomogeneity by allowing the material properties to change due to the presence of chemical reactions which arise from the concentration of reactants that are carried along by the homogeneous fluid (see the more detailed explanation that follows). It is also important to recognize that if the properties of a fluid vary in a particular configuration of the fluid this does not imply that the fluid is inhomogeneous as one merely needs a configuration (some configuration in which the body can be placed) in which the properties of a body are the same for the body to be homogeneous. A body is said to be homogeneous if there exists a configuration in which the properties of the body are the same at every point in the configuration. Consider a fluid with shear-rate-dependent viscosity. At rest, corresponding to zero shear rate, the viscosity of the fluid is the zero-shear-rate viscosity. However, during a flow in which the shear rate varies in the body, the viscosity will not be the same everywhere. We cannot conclude from the viscosity not being the same everywhere that the body is inhomogeneous. There is a configuration, namely that corresponding to the state of rest where the properties are the same, and hence the body is homogeneous. The monograph by Truesdell (1991) contains a general discussion of homogeneity and Anand & Rajagopal (2004) give examples of flows of inhomogeneous fluids where material properties other than just the density being non-constant are considered. There are numerous studies concerning fluids with non-constant density, going back to the seminal work of Lord Rayleigh (1883), and the books by Yih (1965, 1980) are devoted to the study of the flows of such fluids.

Bridges & Rajagopal (2006) studied the pulsating flow of a chemically reacting incompressible fluid in terms of the balance equations for a homogenized fluid (as explained above, the equation considered there is an averaged equation for a single "average" constituent, the averaged single constituent not being necessarily a homogeneous body in that its properties are the same at every material point) and a diffusion-convection equation for the concentration c.

If the fluid (reactant) that is being carried along and the carrier fluid (which is the fluid obtained by "homogenizing" the multi-constituent fluid such as blood) are of comparable density, then assigning the notion of the ratio of the density of the reactant to the total density would not be appropriate as the balance of linear momentum for the fluid that is carried, whose properties are changing due to the reaction, cannot be expressed merely in terms of its density as the inertial term would have a contribution due to the density of the fluid that is also carried along. We are primarily interested in the fluid that is carried along and reacting with our fluid of interest having associated with it a much smaller and in fact ignorable density. Thus, as mentioned earlier, in the work of Bridges & Rajagopal (2006) the concentration has to be interpreted in the sense of a variable that is a measure of the reaction rather than a ratio of densities. This point is not made clear in their work though they do mention that c could be an internal variable, which we refer to as a chemical state variable. Here, we shall choose to think of c as a chemical state variable rather than the ratio of densities, and we shall suppose that the reactant that is carried is not of comparable density to the fluid of interest.

While it would be preferable to study the problem of multi-constituent materials such as blood within the context of mixture theory such an approach is not without serious difficulties. Not only is it necessary to keep track of all the individual constituents and provide constitutive relations for them, as well as model all the interactions between the constituents, we have a far deeper problem, that of providing boundary conditions for each of the constituents. Usually, one is able to ascertain only the boundary conditions for the mixture as a whole and this is a basic problem that is inherent to mixture theory (see Rajagopal & Tao (1995) for a detailed discussion of the same topic).

It is well established that blood in large blood vessels like the aorta behaves essentially as a Navier–Stokes fluid while in narrower blood vessels it can be approximated as a single-constituent fluid that shear thins. In fact, the generalized viscosity associated with such a shear-thinning fluid can change by a factor of forty (Yeleswarapu, 1996; Yeleswarapu et al., 1998). This is a consequence of the diameter of the cells becoming significant with respect to the diameter of the blood vessel. In even narrower capillarities and arterioles where the diameter of the blood vessel is comparable or even smaller than the diameter of a cell, we would not be justified in modelling the flowing blood as a continuum. Experiments by Thurston (1972, 1973) also indicate that blood is capable of stress relaxation. This is not surprising as blood contains a considerable number of cells, platelets, etc. The cells have membranes that are elastic or viscoelastic. Thus, were we to model blood in an averaged sense as a single-constituent fluid we would have to take into account its ability for shear thinning and stress relaxation. In this study we shall not concern ourselves with fluids that are capable of stress relaxation.

We shall assume that the homogenized single-constituent fluid is incompressible. This means that the fluid can only undergo isochoric motions and thus

$$\text{div } v = 0 \, .$$

We shall further suppose that the viscosity of the fluid depends on the concentration c and the symmetric part of the velocity gradient to allow for the possibility that the properties of the fluid can change due to chemical reactions as well as shear-thinning or shear-thickenning[1], and thus the Cauchy stress **T** in our fluid of interest is given by

$$\mathbf{T} = -p\mathbf{I} + 2\mu(c, |\mathbf{D}(\boldsymbol{v})|^2)\mathbf{D} =: -p\mathbf{I} + \mathbf{S}(c, \mathbf{D}), \qquad (2.1)$$

where

$$\mathbf{D} = \frac{1}{2}\left[(\nabla\boldsymbol{v}) + (\nabla\boldsymbol{v})^T\right].$$

Finally, we assume that the flux vector $\boldsymbol{q}_c$ related to the chemical reactions is given by

$$\boldsymbol{q}_c = \boldsymbol{q}_c(c, \nabla c, |\mathbf{D}|^2) := -\mathbf{K}(c, |\mathbf{D}|^2)\nabla c. \qquad (2.2)$$

The specific form of the coefficients K_{ij} of the matrix **K** depends on the specific application (chemical reaction or system of reactions) under consideration. If we are interested in a fluid such as blood and use the equations developed here, we have to replace a host of chemical reactions by one "averaged" reaction and one cannot say much about the form of **K** unless we decide on which specific problem we are interested in. For instance, coagulation and lysis have totally different effects on the fluid, one leading to an increase in the viscosity and the other leading to a decrease in viscosity. Similarly, if we are interested in ATIII deficiency or Sickle cell anaemia we would have to consider other forms for the coefficient. Also, when dealing with a complex system like blood wherein one has numerous chemical reactions one has to be cognizant of the fact that each of these reactions take place at different rates and blood is maintained in a delicate state of balance while a myriad of biochemical reactions take place. On the other hand if we were dealing with a polymer melt undergoing some reaction we would have a totally different form for the diffusion coefficient. Thus, in this study we shall merely assume a specific form for the coefficient to illustrate our ideas.

Recently, Bulíček, Feireisl, & Málek (2009) considered an incompressible Navier–Stokes fluid whose density and thermal conductivity depend on the temperature. Under the assumption that the fluid meets Navier's slip and zero-heat-flux boundary conditions, they established the existence of weak (as well as suitable weak) solutions for long times when the data can be large. Around the same time, Bulíček, Málek, & Rajagopal (2008) considered unsteady flows of incompressible fluids whose viscosity depends on the temperature, shear-rate and pressure. Such a fluid model is markedly different from the classical incompressible Navier–Stokes

---

[1]When blood coagulates its viscosity increases while lysis leads the viscosity of the coagulated blood to decrease.

or incompressible Navier–Stokes–Fourier fluid in that the relationship between the stress and the symmetric part of the velocity gradient is implicit. Assuming Navier's slip at the solid boundary they established the existence of suitable weak solutions for long time, when the data is large.

If we were to ignore the dependence of the viscosity on the pressure, and replace the dependence of the viscosity on the temperature by its dependence on the chemical state variable c (the concentration), we have a problem that bears close relationship to the models studied by Bulíček et al. (2008). While in the problem wherein the viscosity depends on the temperature we have to satisfy the balance of energy, which leads to an equation for the temperature (or the internal energy), in the problem being considered here, we have a diffusion-convection equation for the chemical state variable c.

As the structure of the diffusion-convection equation is simpler in comparison to the equation representing the balance of energy, the problem studied here might, at the first glance, seem easier than that considered by Bulíček et al. (2008). On the other hand, the dependence of the material moduli on the pressure considered by Bulíček et al. (2008) allows one to consider only fluids that can shear thin, while in the present study we investigate unsteady flows of both shear-thinning and shear-thickening fluids. As a consequence, we can establish several new solutions and we also are able to make statements concerning the effect of the material parameters on the nature of these solutions.

The structure of the paper is the following. In the next section, we formulate the governing equations, and state the appropriate initial and boundary conditions. For the sake of simplicity, and in order to make some comparison with the earlier study by Bulíček et al. (2008), Navier's slip boundary conditions and $C^{1,1}$ domains are considered first. We also state the assumptions concerning the constitutive quantities S and q_c, define the notion of weak solution to our problem and formulate the result regarding its existence. Section 2.3 is focused on the proof (we merely provide the main steps and refer the reader to former studies for details). Section 2.4 contains several extensions of the main result in various directions (no-slip boundary conditions, qualitative properties of the solution, validity of the result for a large range of model parameters, etc). Here we do not provide any details, but rather refer to studies where results are established for similar problems, so the interested reader can (with some effort) deduce the validity of the results that are presented. An appendix at the end of this article includes several auxiliary assertions used in the proof of the main result.

2.2 Formulation of the problem and the results

2.2.1 Balance equations, boundary and initial conditions. Structure of S and q_c.

We are interested in understanding the mathematical properties relevant to unsteady flows of chemically reacting fluids whose transport coefficients depend on the concentration and on the shear rate, flowing in a bounded open set Ω in $\mathbb{R}^3$ with boundary $\partial\Omega$. We would like to establish the existence of a solution in the domain $Q := (0, T) \times \Omega$, where $(0, T)$ denotes the time interval of interest. It would be worthwhile to find how these solutions behave with time and in space, i.e. to carry out an analysis that discusses the nature of the solution. We shall however restrict ourselves to the question of existence of solutions in this study.

Motions of incompressible fluids that are reacting chemically are described in terms of the velocity field v, the pressure (mean normal stress) p, and the concentration c by means of a system of partial differential equations that are a consequence of the balance of mass, balance of linear and angular momentum, and the diffusion equation for c. The balance of angular momentum which leads to the Cauchy stress being symmetric is automatically met by virtue of the form chosen for the Cauchy stress T. The system governing the flows of interest takes the form

$$\operatorname{div} v = 0, \quad v_{,t} + \operatorname{div}(v \otimes v) - \operatorname{div} S = f - \nabla p,$$
$$c_{,t} + \operatorname{div}(cv) = -\operatorname{div} q_c, \tag{2.3}$$

where f represents the specific body forces. In this setting, for given functions v_0 and c_0 defined in Ω, we prescribe the initial conditions

$$v(0, x) = v_0(x) \quad \text{and} \quad c(0, x) = c_0(x) \qquad (x \in \Omega). \tag{2.4}$$

We assume that the boundary $\partial\Omega$ is completely described from outside by a finite number of overlapping $C^{0,1}$-mappings: for this we use the notation $\Omega \in C^{0,1}$. We also set $\Gamma := (0, T) \times \partial\Omega$. We prescribe the following boundary conditions

$$v \cdot n = 0, \quad v_\tau = -\frac{1}{\alpha}[Tn]_\tau = -\frac{1}{\alpha}[Sn]_\tau \quad \text{on } \Gamma,$$
$$q_c \cdot n = 0 \quad \text{on } \Gamma_N \quad \text{and} \quad c = c_b \quad \text{on } \Gamma_D, \tag{2.5}$$

where n is the unit outward normal and z_τ stands for the projection of the quantity z along the tangent plane, i.e. $z_\tau = z - (z \cdot n)n$ and Γ_D, Γ_N are regular surfaces (i.e. they are images of open two-dimensional intervals by a $C^{1,1}$ mapping) with $\Gamma_N \cap \Gamma_D = \emptyset$ and $\overline{\Gamma_N \cup \Gamma_D} = \Gamma$. The first line in (2.5) expresses the fact that the solid boundary is impervious and tangential components of the velocity fulfil Navier's slip boundary condition; the second line in (2.5) says that on Γ_N there is no flux of concentration through the boundary and a nonhomogeneous Dirichlet boundary condition for c holds on Γ_D.

We will also require that for some open $\Omega_0 \subseteq \Omega$

$$\int_{\Omega_0} p(t,x)\,dx = h(t) \quad \text{for all } t \in [0,T], \tag{2.6}$$

where h is a given function. The above requirement concerning the mean value of the pressure over a domain of non-zero area measure is completely consistent with how pressure measurements are made, namely a pressure gauge sensing the total force due to the normal stress on a small portion of the surface area on which the pressure gauge is mounted.

Regarding the admissible structure for the constitutive quantities S and q_c characterizing the specific fluid, we assume that $\mathsf{S} : [0,1] \times \mathbb{R}^{3\times 3}_{sym} \to \mathbb{R}^{3\times 3}_{sym}$ that appears in (2.1) is a continuous mapping such that for some $r > 1$

there are $C_1, C_2, C_3 \in (0,\infty)$ and a function $\gamma_1 \in L^\infty([0,1])$

such that for all $c \in [0,1]$ and $\mathsf{D} \in \mathbb{R}^{3\times 3}_{sym}$ we have: $\tag{2.7}$

$$C_1|\mathsf{D}|^r - C_3 \le \mathsf{S}(c,\mathsf{D}) \cdot \mathsf{D} \quad \text{and} \quad |\mathsf{S}(c,\mathsf{D})| \le C_2\gamma_1(c)|\mathsf{D}|^{r-1} + C_3$$

and

for all $\mathsf{D}_1, \mathsf{D}_2 \in \mathbb{R}^{3\times 3}_{sym}$, $\mathsf{D}_1 \ne \mathsf{D}_2$ and all $c \in [0,1]$ we have
$$(\mathsf{S}(c,\mathsf{D}_1) - \mathsf{S}(c,\mathsf{D}_2)) \cdot (\mathsf{D}_1 - \mathsf{D}_2) > 0. \tag{2.8}$$

We also assume that K that appears in (2.2) is a continuous mapping of $[0,1] \times \mathbb{R}^+_0$ into $\mathbb{R}^{3\times 3}$ such that for some $\beta \in \mathbb{R}$, and $c_4, c_5 \in (0,\infty)$ the flux vector $q_c = q_c(c,z,s) = -\mathsf{K}(c,s)z$ is such that for all $c \in [0,1]$, $z \in \mathbb{R}^3$, $s \in \mathbb{R}^+_0$

$$C_4(1+s)^\beta|z|^2 \le -q_c(c,z,s) \cdot z \quad \text{and}$$
$$|q_c(c,z,s)| \le C_5\gamma_1(c)(1+s)^\beta|z|. \tag{2.9}$$

We are unaware of any experimental evidence concerning the structure of the diffusion matrix K. In fact, one has to recognize that in order to have such information one needs to correlate raw experimental data with the model that we are using and since we are proposing a new model it is not surprising that such data is unavailable. In light of this we shall attempt to develop the theory for as large a range of β as possible.

2.2.2 Function spaces, definition of solution, main theorem

We first characterize the regularity of the domain by requiring that the boundary of Ω is smooth enough so that L^q-regularity for certain values of $q \in [1, \infty]$ holds. We shall make this statement more precise in the next paragraph.

Given $z \in L^q(\Omega)$ with $\int_\Omega z \, \mathrm{d}x = 0$, let the symbol $\mathcal{N}_{\Omega_0}^{-1}(z)$ denote the unique solution of the Neumann problem

$$\Delta u = z \text{ in } \Omega, \qquad \nabla u \cdot \boldsymbol{n} = 0 \text{ on } \partial\Omega, \qquad \int_{\Omega_0} u \, \mathrm{d}x = 0. \qquad (2.10)$$

For the special case $\Omega_0 = \Omega$ we use the abbreviation $\mathcal{N}^{-1}(z) := \mathcal{N}_\Omega^{-1}(z)$. Thus, in particular, denoting $g^{\boldsymbol{v}} := \mathcal{N}^{-1}(\operatorname{div} \boldsymbol{v})$ we can define the vector $\boldsymbol{v}_{\mathrm{div}}$ as

$$\boldsymbol{v}_{\mathrm{div}} := \boldsymbol{v} - \nabla g^{\boldsymbol{v}}, \qquad (2.11)$$

which implies the Helmholtz decomposition $\boldsymbol{v} = \boldsymbol{v}_{\mathrm{div}} + \nabla g^{\boldsymbol{v}}$.

It is known, see Novotný & Straškraba (2004, Lemma 3.17), that if $\Omega \in \mathcal{C}^{0,1}$, then for any $s \in (1, \infty)$

$$\|g^{\boldsymbol{v}}\|_{1,s} \le C(\Omega, s)\|\boldsymbol{v}\|_s \qquad \text{and}$$
$$\|\boldsymbol{v}_{\mathrm{div}}\|_s \le (C(\Omega, s) + 1)\|\boldsymbol{v}\|_s,$$

where $\|\cdot\|_s$ denotes the norm in $L^s(\Omega)$ and $\|\cdot\|_{1,s}$ the norm in $W^{1,s}(\Omega)$.

We say that a bounded domain $\Omega \subset \mathbb{R}^3$ with Lipschitz boundary is of class $\mathcal{R}$, and we then write $\Omega \in \mathcal{R}$, if the L^q-regularity theory for the Neumann problem (2.10) holds for $q = r$ and $q = \frac{5r}{5r-6}$ (where r is introduced in (2.7)), and consequently the following estimates are valid for the same qs:

$$\|g^{\boldsymbol{v}}\|_{2,q} \le C_{\mathrm{reg}}(\Omega, q)\|\operatorname{div} \boldsymbol{v}\|_q \qquad \text{and}$$
$$\|\boldsymbol{v}_{\mathrm{div}}\|_{1,q} \le (C_{\mathrm{reg}}(\Omega, q) + 1)\|\boldsymbol{v}\|_{1,q}. \qquad (2.12)$$

It is known, see (Grisvard, 1985, Proposition 2.5.2.3, p. 131), that if $\Omega \in \mathcal{C}^{1,1}$, then (2.12) holds for any $q \in (1, \infty)$, and thus $\Omega \in \mathcal{R}$. For another set of conditions that are sufficient for $\Omega \in \mathcal{R}$ we refer the reader to Bulíček et al. (2008) where this issue is discussed in detail.

Before giving the definition of what we mean by a solution to (2.3)–(2.6), we need to introduce subspaces (and their duals) of vector-valued

Sobolev functions from $W^{1,q}(\Omega)^3$ that have zero normal component on the boundary (note that $q' = q/(q-1)$). We define

$$W_{\boldsymbol{n}}^{1,q} := \{\boldsymbol{v} \in W^{1,q}(\Omega)^3; \operatorname{tr} \boldsymbol{v}\cdot\boldsymbol{n} = 0 \text{ on } \partial\Omega\}, \quad W_{\boldsymbol{n}}^{-1,q'} := \left(W_{\boldsymbol{n}}^{1,q}\right)^*,$$

$$W_{\boldsymbol{n},\mathrm{div}}^{1,q} := \left\{\boldsymbol{v} \in W_{\boldsymbol{n}}^{1,q}; \operatorname{div} \boldsymbol{v} = 0\right\}, \quad W_{\boldsymbol{n},\mathrm{div}}^{-1,q'} := \left(W_{\boldsymbol{n},\mathrm{div}}^{1,q}\right)^*,$$

$$W_{\Gamma_D}^{1,q}(\Omega) := \{h \in W^{1,q}(\Omega); \operatorname{tr} h = 0 \text{ on } \Gamma_D\}, \quad W_{\Gamma_D}^{-1,q'}(\Omega) := \left(W_{\Gamma_D}^{1,q}(\Omega)\right)^*,$$

$$L_{\boldsymbol{n},\mathrm{div}}^{q} := \overline{\left\{\boldsymbol{v} \in W_{\boldsymbol{n},\mathrm{div}}^{1,q}\right\}}^{\|\cdot\|_q}.$$

For $r, q \in [1, +\infty]$, we also introduce relevant spaces of a Bochner-type, namely,

$$X^{r,q} := \{\boldsymbol{u} \in L^r(0,T; W_{\boldsymbol{n}}^{1,r}) \cap L^q(0,T; L^q(\Omega)^3),$$
$$\operatorname{tr} \boldsymbol{u} \in L^2(0,T; (L^2(\partial\Omega))^3)\},$$

$$X_{\mathrm{div}}^{r,q} := \{\boldsymbol{u} \in X^{r,q}, \operatorname{div} \boldsymbol{u} = 0\},$$

$$\mathcal{C}(0,T; L_{\mathrm{w}}^q(\Omega)^3) := \{\boldsymbol{u} \in L^\infty(0,T; L^q(\Omega)^3);$$
$$(\boldsymbol{u}(t), \boldsymbol{\varphi}) \in \mathcal{C}([0,T]) \; \forall \, \boldsymbol{\varphi} \in \mathcal{C}(\overline{\Omega})^3\}.$$

In the last definition, we used the notation (f, g) for $\int_\Omega f(x)g(x) \, \mathrm{d}x$ if $fg \in L^1(\Omega)$. In an analogous manner, we shall use the symbols $(f, g)_Q$, $(f, g)_{\partial\Omega}$, $(f, g)_\Gamma$, etc. If $f \in X$ and $g \in X^*$ we often use the symbol $\langle g, f \rangle$ instead of $\langle g, f \rangle_{X^*, X}$. The same bracket notation is used for vector functions $\boldsymbol{f}$, $\boldsymbol{h}$ and tensor functions $\mathbf{F}$, $\mathbf{H}$ as well.

We assume that $\boldsymbol{f}$ that appears on the right hand side of $(2.3)_1$, the prescribed function h for the pressure (see (2.6)), and the initial values $\boldsymbol{v}_0$ and c_0 that appear in (2.4) satisfy

$$\boldsymbol{f} \in L^{r'}(0,T; W_{\boldsymbol{n}}^{-1,r'}), \qquad h \in L^{r'}(0,T),$$
$$\boldsymbol{v}_0 \in L_{\boldsymbol{n},\mathrm{div}}^2, \qquad 0 \le c_0(x) \le 1 \quad \text{for a.e. } x \in \Omega, \tag{2.13}$$

and we also require that the given concentration c_b on Γ_D appearing in $(2.5)_4$ is such that

for $n := \max\{2, r/(r-2\beta)\}$ there is a $\widetilde{c}_b \in L^\infty(Q) \cap L^n(0,T; W^{1,n}(\Omega))$, with $\widetilde{c}_{b,t} \in L^1(Q)$ such that $0 \le \operatorname{tr} \widetilde{c}_b = c_b \le 1$ on Γ_D.

$$\tag{2.14}$$

Definition 2.2.1 *Take* $\Omega \in \mathcal{R}$, $\alpha \in [0, \infty)$ *and assume that* (2.13) *and* (2.14) *hold. Assume that* $\mathbf{S}$ *satisfies* (2.7)–(2.8) *with* $r > \frac{6}{5}$, *and* $\boldsymbol{q}_c$ *satisfies* (2.9) *with* $-r < 2\beta < r$. *Setting*

$$m = \min\left\{\tfrac{5r}{6}, r'\right\}, \quad q = \min\left\{2, \tfrac{2r}{r-2\beta}\right\}, \; and \; s = \min\left\{2, \tfrac{2r}{r+2\beta}\right\}, \quad (2.15)$$

we say that (v, p, c) is a weak solution to (2.3)–(2.6) if

$$v \in \mathcal{C}([0,T]; L^2_w(\Omega)^3) \cap X_{\mathrm{div}}^{r,\frac{5r}{3}}, \quad v_{,t} \in L^m(0,T; W_n^{-1,m}), \quad (2.16)$$

$$p \in L^m(0,T; L^m(\Omega)) \; and$$

$$\int_{\Omega_0} p(x,t) \, \mathrm{d}x = h(t) \; for \; a.e. \; t \in (0,T), \tag{2.17}$$

$$c - \tilde{c}_b \in L^q(0,T; W^{1,q}_{\Gamma_D}(\Omega)), \quad c_{,t} \in L^{s'}(0,T; W^{-1,s'}_{\Gamma_D}(\Omega)), \quad (2.18)$$

$$0 \le c \le 1 \; a.e. \; in \; Q, \tag{2.19}$$

$$(1 + |\mathbf{D}(v)|^2)^{\frac{\beta}{2}} \nabla c \in L^2(0,T; L^2(\Omega)^3), \tag{2.20}$$

(v, p, c) *satisfy the following weak formulation of the equations*

$$\langle v_{,t}, \varphi \rangle - (v \otimes v, \nabla \varphi)_Q + \alpha(v, \varphi)_\Gamma + (\mathbf{S}(c, \mathbf{D}(v)), \mathbf{D}(\varphi))_Q$$
$$= \langle f, \varphi \rangle + (p, \mathrm{div}\,\varphi)_Q \qquad for \; all \; \varphi \in L^{m'}(0,T; W_n^{1,m'}), \tag{2.21}$$

$$(c_{,t}, \varphi)_Q - (cv, \nabla \varphi)_Q + \left(\mathbf{K}(c, |\mathbf{D}(v)|^2)\nabla c, \nabla \varphi\right)_Q = 0$$
$$for \; all \; \varphi \in L^s(0,T; W^{1,s}_{\Gamma_D}(\Omega)), \tag{2.22}$$

and (v, c) satisfy the initial conditions in the following sense

$$\lim_{t \to 0+} \|v(t) - v_0\|_2^2 + \|c(t) - c_0\|_2^2 = 0. \tag{2.23}$$

Theorem 2.2.2 *Let $r > \frac{8}{5}$. Then for any data fulfilling (2.13) and (2.14) and for any $T \in (0,\infty)$ there exists a weak solution to (2.3)–(2.6) in the sense of Definition 2.2.1.*

To the best of our knowledge, this is the first result concerning long-time existence of solutions to a model such as (2.3) where the material coefficients depend on c (concentration) and $|\mathbf{D}|^2$ (shear rate). In addition, the result holds for large data fulfilling (2.13)–(2.14) and the result concerns flows in general domains (with $C^{1,1}$ boundary, for example) under reasonable Navier's slip boundary conditions. Following Wolf (2007), the result can be extended to no-slip boundary conditions and to more general domains (we then, however, lose the integrability of the pressure), and it is also possible to include lower values of the power-law index r (this follows by applying the approach due to Diening,

Růžička, & Wolf, 2008). We discuss these possible extensions in more detail in Section 2.4.

The related studies that we are aware of wherein the existence of weak solutions are established for different models of chemically reacting fluids are due to Roubíček (2005, 2007) who in addition considers a system of chemical reactions together with electrical and thermal stimuli. However, he only treats the case $r \geq \frac{11}{5}$ and the fluxes related to the chemical reactions are independent of the shear rate.

The problem considered here shares certain similarities with the initial boundary value problems for inhomogeneous incompressible fluids driven by the system of equations

$$\operatorname{div} v = 0, \qquad \varrho_{,t} + \operatorname{div}(\varrho v) = 0,$$
$$(\varrho v)_{,t} + \operatorname{div}(\varrho v \otimes v) - \operatorname{div} S = \varrho f - \nabla p, \qquad (2.24)$$
$$S = 2\mu(\varrho, |D(v)|^2)D(v).$$

We refer the reader to Antontsev, Kazhikhov, & Monakhov (1990) and Lions (1996) for the analysis of models when the stress tensor is given by $S = 2\mu(\varrho)D(v)$. The models where the viscosity μ also depends in a polynomial way on $|D(v)|^2$ are analysed in Fernández-Cara, Guillén, & Ortega (1997) (for $r \geq \frac{12}{5}$), in Guillén-González (2004) (for $r \geq 2$ and the spatially periodic problem), and in Frehse & Růžička (2008) (for $r \geq \frac{11}{5}$ and the problem with no-slip boundary conditions). In contrast, the result presented here holds even for $r < 2$ and one needs to handle the additional nonlinear diffusion term in (2.3) that provides information on the gradient of the concentration (not currently available for ϱ in (2.24)).

2.3 A proof of Theorem 2.2.2

Before establishing the main existence result we would like to make a few comments about the construction of the proofs in this paper. Many of the lemmas and other results such as interpolation inequalities that are necessary to establish the main theorem have been proved elsewhere with regard to problems with similar mathematical structures. Thus, in order to avoid repetition we do not document detailed proofs for these but merely refer the reader to where such details can be found. In this paper we shall only focus on the main or new parts of the proofs.

2.3.1 An (ε, η)-approximate problem and uniform estimates

For positive and fixed ε and η, we consider the following approximation

$$- \varepsilon \Delta p + \operatorname{div} \boldsymbol{v} = 0 \text{ in } Q, \qquad \frac{\partial p}{\partial n} = 0 \text{ on } \Gamma, \qquad \int_{\Omega_0} p \, dx = h(t), \quad (2.25)$$

$$\boldsymbol{v}_{,t} + \operatorname{div}(\boldsymbol{v}_\eta \otimes \boldsymbol{v}) - \operatorname{div} \mathbf{S}(c, \mathbf{D}(\boldsymbol{v})) = -\nabla p + \boldsymbol{f}, \qquad (2.26)$$

$$c_{,t} + \operatorname{div}(\boldsymbol{v}_\eta c) = -\operatorname{div} \boldsymbol{q}_c(c, \nabla c, \mathbf{D}(\boldsymbol{v})), \qquad (2.27)$$

where for a given $\boldsymbol{v} \in L^r(0, T; W_{\boldsymbol{n}}^{1,r})$ we use $\boldsymbol{v}_\eta$ to denote a function defined through

$$\boldsymbol{v}_\eta := ((\boldsymbol{v}\omega_\eta) * r_\eta)_{\operatorname{div}},$$

where $r_\eta(x) := \frac{1}{\eta^3} r\left(\frac{x}{\eta}\right)$ with $r \in \mathcal{D}(\mathbb{R}^3)$ non-negative, radially symmetric, with $\int_{\mathbb{R}^3} r \, dx = 1$ (i.e. r is a regularization kernel); ω_η is a smooth function such that $\operatorname{dist}(\operatorname{supp} \omega_\eta, \partial\Omega) \geq \eta$ and $\omega_\eta = 1$ for all x satisfying $\operatorname{dist}(x, \partial\Omega) \geq 2\eta$; and $\boldsymbol{z}_{\operatorname{div}} := \boldsymbol{z} - \nabla \mathcal{N}^{-1}(\operatorname{div} \boldsymbol{z})$ (see (2.11)).

Rewriting (2.25) as $p - h/|\Omega_0| = \frac{1}{\varepsilon}\mathcal{N}_{\Omega_0}^{-1}(\operatorname{div} \boldsymbol{v})$, and inserting it into (2.26) we can also equivalently rewrite (2.25)–(2.27) as a system for $\boldsymbol{v}$ and c only.

We will assume (referring to Bulíček et al. (2008) for details concerning the solvability of a similar, yet more complicated, system) that for fixed $\varepsilon, \eta > 0$ there is a solution $(\boldsymbol{v}, p, c) = (\boldsymbol{v}^{\varepsilon,\eta}, p^{\varepsilon,\eta}, c^{\varepsilon,\eta})$ satisfying

$$\varepsilon(\nabla p, \nabla \psi)_\Omega = (\boldsymbol{v}, \nabla \psi)_\Omega \text{ for a.e. } t \in (0, T) \text{ and all } \psi \in W^{1,2}(\Omega), \quad (2.28)$$

$$\begin{aligned} \langle \boldsymbol{v}_{,t}, \boldsymbol{\varphi}\rangle - (\boldsymbol{v}_\eta \otimes \boldsymbol{v}, \nabla\boldsymbol{\varphi})_Q + \alpha(\boldsymbol{v}, \boldsymbol{\varphi})_\Gamma + (\mathbf{S}(c, \mathbf{D}(\boldsymbol{v})), \mathbf{D}(\boldsymbol{\varphi}))_Q \\ = (p, \operatorname{div}\boldsymbol{\varphi})_Q + \langle \boldsymbol{f}, \boldsymbol{\varphi}\rangle \quad \text{for all } \boldsymbol{\varphi} \in L^r(0, T; W_{\boldsymbol{n}}^{1,r}), \end{aligned} \quad (2.29)$$

and

$$\langle c_{,t}, \varphi\rangle - (\boldsymbol{v}_\eta c, \nabla\varphi) - (\boldsymbol{q}_c(c, \mathbf{D}(\boldsymbol{v}), \nabla c), \nabla\varphi)_Q = 0 \\ \text{for all } \varphi \in L^s(0, T; W_{\Gamma_D}^{1,s}(\Omega)). \quad (2.30)$$

Even more, we will require that we can take $\boldsymbol{\varphi} = \boldsymbol{v}$ in (2.29), $\varphi = c - \widetilde{c_b}$ in (2.30) and $\psi = p$ in (2.28). These additional assumptions can be easily verified on the level of Galerkin approximations; see Bulíček et al. (2008) for details. Here, these requirements enable us to derive uniform estimates that are independent of ε, η in the function spaces that appear in Definition 2.2.1.

Taking $\varphi = v$ in (2.29) and $\psi = p$ in (2.28) and adding the equations, we obtain (after using integration by parts)

$$\sup_{t \in (0,T)} \|v(t)\|_2^2 + \int_0^T (S(c, D(v)), D(v)), \, dt + \alpha \|v\|_{2,\partial\Omega}^2$$
$$+ \varepsilon \|\nabla p\|_2^2 \, d\tau \le \|v_0\|_2^2 + \int_0^T \langle f, v \rangle \, dt. \tag{2.31}$$

Applying (2.7) and Korn's inequality (Lemma 2.5.3 in the Appendix) to the second term, standard duality estimates and Young's inequality to the last term in (2.31), and also the interpolation inequality (2.75) with $q = \frac{5r}{3}$, we conclude that

$$\sup_{t \in (0,T)} \|v(t)\|_2^2 + \int_0^T \|v(t)\|_{1,r}^r + \|v(t)\|_{\frac{5r}{3}}^{\frac{5r}{3}} + \varepsilon \|\nabla p(t)\|_2^2 \, dt \le C. \tag{2.32}$$

Next, we define $\varphi_1 := \min\{0, c\}$ and $\varphi_2 := \max\{c, 1\} - 1$. Since $c = c_b$ on Γ_D, using (2.14) we get that $\varphi_1, \varphi_2 = 0$ on Γ_D. Thus, setting $\varphi = \varphi_i \chi_{[0,t]}$ in (2.30) we get

$$\|\varphi_i(t)\|_2^2 - 2 \int_0^t (v_\eta c, \nabla \varphi_i)_\Omega - 2 \int_0^t (q_c(c, D(\nabla v), \nabla c), \nabla \varphi_i)_\Omega = 2\|\varphi_i(0)\|_2^2.$$

Hence, using $\operatorname{div} v_\eta = 0$ and the assumption (2.9) we deduce that

$$\|\varphi_1(t)\|_2^2 + \|\varphi_2(t)\|_2^2 \le \|\varphi_1(0)\|_2^2 + \|\varphi_2(0)\|_2^2.$$

It follows from (2.14) that $\varphi_i(0, x) = 0$ for a.e. $x \in \Omega$. Combining all these results we finally obtain that

$$0 \le c(x, t) \le 1 \qquad \text{for a.e. } (x, t) \in Q. \tag{2.33}$$

Next, taking $\varphi = c - \tilde{c}_b$ in (2.30), we arrive at

$$\langle c_{,t}, c - \tilde{c}_b \rangle - (v_\eta c, \nabla(c - \tilde{c}_b))_Q - (q_c(c, D(v), \nabla c), \nabla(c - \tilde{c}_b))_Q = 0.$$

Adding and subtracting $\tilde{c}_b$ into the first term, integrating the result with respect to time, using (2.14), (2.33) and the facts that $\operatorname{div} v_\eta = 0$ and $v_\eta \cdot n = 0$ on $\partial\Omega$, we find that

$$-(q_c(c, D(v), \nabla c), \nabla c)_Q \le 1 + (\tilde{c}_{b,t}, \tilde{c}_b - c)_Q - (v_\eta c, \nabla \tilde{c}_b)_Q$$
$$- (q_c(c, D(v), \nabla c), \nabla \tilde{c}_b)_Q. \tag{2.34}$$

Using the assumptions (2.9) and (2.14), the uniform estimates (2.32) and (2.33), and the fact that $2\beta \leq r$ we deduce from (2.34) that

$$\int_Q (1 + |\mathbf{D}(v)|^2)^\beta |\nabla c|^2 \, dx \, dt \leq C + C \int_Q (1 + |\mathbf{D}(v)|^2)^\beta |\nabla c| |\nabla \widetilde{c_b}| \, dx \, dt$$

$$\leq C + \frac{1}{2} \int_Q (1 + |\mathbf{D}(v)|^2)^\beta |\nabla c|^2 \, dx \, dt + C \int_Q (1 + |\mathbf{D}(v)|^2)^\beta |\nabla \widetilde{c_b}|^2 \, dx \, dt$$

$$\overset{(2.14),(2.32)}{\leq} C + \frac{1}{2} \int_Q (1 + |\mathbf{D}(v)|^2)^\beta |\nabla c|^2 \, dx \, dt.$$

Thus

$$\int_Q (1 + |\mathbf{D}(v)|^2)^\beta |\nabla c|^2 \, dx \, dt \leq C. \tag{2.35}$$

Therefore, for $\beta \geq 0$ we conclude that

$$\int_0^T \|c\|_{1,2}^2 \, dt \leq C. \tag{2.36}$$

For $\beta < 0$, recalling that then $q = \frac{2r}{r-2\beta} > 1$ (see definition in (2.15)), we can compute

$$\int_Q |\nabla c|^q = \int_Q |\nabla c|^q (1 + |\mathbf{D}(v)|^2)^{\frac{\beta q}{2}} (1 + |\mathbf{D}(v)|^2)^{-\frac{\beta q}{2}}$$

$$\leq \int_Q |\nabla c|^2 (1 + |\mathbf{D}(v)|^2)^\beta + \int_Q (1 + |\mathbf{D}(v)|^2)^{-\frac{\beta q}{2-q}} \tag{2.37}$$

$$= \int_Q |\nabla c|^2 (1 + |\mathbf{D}(v)|^2)^\beta + \int_Q (1 + |\mathbf{D}(v)|^2)^{\frac{r}{2}} \overset{(2.32),(2.35)}{\leq} C.$$

It is a direct consequence of (2.32), (2.33), (2.36) and (2.37), and the assumptions (2.7) and (2.9), that

$$\int_Q |\mathbf{S}(c, \mathbf{D}(v))|^{r'} + |q_c(c, \mathbf{D}(v), \nabla c)|^s \, dx \, dt \leq C \tag{2.38}$$

with $s = \min\{2, \frac{2r}{r+2\beta}\} > 1$ defined in (2.15). Finally, it follows from (2.29) and (2.30) and the estimates established above that (for details see Málek et al. (1996) or Bulíček et al. (2008))

$$\|v_{,t}^\varepsilon\|_{(X_{\mathrm{div}}^{r,\frac{5r}{5r-8}})_*} + \|c_{,t}^\varepsilon\|_{L^{s'}(0,T;W^{-1,s'}(\Omega))} \leq C. \tag{2.39}$$

To obtain uniform estimates on p, we assume, without loss of generality, that $h(t) \equiv 0$. We introduce $p_0 := p - \frac{1}{|\Omega|} \int_\Omega p \, dx$ and observe (by contradiction) that since $\int_{\Omega_0} p(t) \, dx = h(t) = 0$ there is a constant C independent of ε, η such that

$$\|p\|_q \leq C \|p_0\|_q. \tag{2.40}$$

Consequently, it suffices to find uniform estimates for p_0. For this purpose, we consider φ in (2.29) first of the form

$$\varphi = \nabla \mathcal{N}^{-1}(|p_0|^{r'-2} p_0 - \frac{1}{|\Omega|} \int_\Omega |p_0|^{r'-2} p_0 \, dx)$$

and then

$$\varphi = \nabla \mathcal{N}^{-1}(|p_0|^{m-2} p_0 - \frac{1}{|\Omega|} \int_\Omega |p_0|^{m-2} p_0 \, dx)$$

(with m defined in (2.15) as $m = \min\{r', \frac{5r}{6}\}$). Such a choice of φ clearly leads to

$$(p, \operatorname{div} \varphi)_Q = \int_0^T \|p_0\|_\alpha^\alpha \, dt \qquad \text{(first with } \alpha = r' \text{ and then with } \alpha = m\text{)}.$$

Replacing the left-hand side by means of (2.29), proceeding step by step as in Bulíček et al. (2008), and in particular using the fact that

$$\int_0^T \langle v_{,t}, \varphi \rangle \, dt = -\frac{2\varepsilon}{\alpha} \|p_0(T)\|_\alpha^\alpha \leq 0 \qquad (\alpha = r' \text{ or } m),$$

we conclude that

$$\int_0^T \|p_0\|_{r'}^{r'} dt \leq C(\eta) \overset{(2.40)}{\Longrightarrow} \int_0^T \|p\|_{r'}^{r'} dt \leq C(\eta),$$

$$\int_0^T \|p_0\|_m^m dt \leq C \overset{(2.40)}{\Longrightarrow} \int_0^T \|p\|_m^m dt \leq C. \tag{2.41}$$

Using (2.29) and (2.32), these estimates imply that

$$\int_0^T \|v_{,t}\|_{W_n^{-1,r'}}^{r'} \, dt \leq C(\eta) \quad \text{and} \quad \int_0^T \|v_{,t}\|_{W_n^{-1,m}}^m \, dt \leq C; \tag{2.42}$$

see Bulíček et al. (2008) for details.

2.3.2 Limit as $\varepsilon \to 0$

Let $(v^\varepsilon, c^\varepsilon, p^\varepsilon)$ be used in this subsection to denote $(v^{\varepsilon,\eta}, c^{\varepsilon,\eta}, p^{\varepsilon,\eta})$, the solution of the (ε, η)-approximation (2.28)–(2.30). We introduce the notation

$$q_c^\varepsilon := q_c(c^\varepsilon, \mathbf{D}(v^\varepsilon), \nabla c^\varepsilon) \quad \text{and} \quad \mathbf{S}^\varepsilon := \mathbf{S}(c^\varepsilon, \mathbf{D}(v^\varepsilon)).$$

It is then a consequence of the estimates (2.32)–(2.42) (which are all independent of ε), Corollary 2.5.2, and Aubin–Lions Lemma (Lemma 2.5.5) that we can find (again labelled by ε) subsequences of $\{v^\varepsilon, c^\varepsilon, p^\varepsilon, \mathbf{S}^\varepsilon, q_c^\varepsilon\}$ such that

$$v_{,t}^\varepsilon \rightharpoonup v_{,t} \qquad \text{weakly in } L^{r'}(0, T; W_{\boldsymbol{n}}^{-1,r'}), \tag{2.43}$$

$$v^\varepsilon \rightharpoonup v \qquad \text{weakly in } L^r(0, T; W_{\boldsymbol{n}}^{1,r}), \tag{2.44}$$

$$v^\varepsilon \to v \qquad \text{strongly in } L^h(0, T; L^h(\Omega)^3) \text{ for all } h < \tfrac{5r}{3}, \tag{2.45}$$

$$\operatorname{tr} v^\varepsilon \to \operatorname{tr} v \qquad \text{strongly in } L^2(0, T; L^2(\partial\Omega)^3), \tag{2.46}$$

$$c_{,t}^\varepsilon \rightharpoonup c_{,t} \qquad \text{weakly in } L^{s'}(0, T; W^{-1,s'}), \tag{2.47}$$

$$c^\varepsilon \rightharpoonup c \qquad \text{weakly in } L^q(0, T; W^{1,q}(\Omega)), \tag{2.48}$$

$$c^\varepsilon \to c \qquad \text{strongly in } L^s(0, T; L^s(\Omega)) \text{ for all } s < \infty, \tag{2.49}$$

$$p^\varepsilon \rightharpoonup p \qquad \text{weakly in } L^{r'}(0, T; L^{r'}(\Omega)), \tag{2.50}$$

$$\mathbf{S}^\varepsilon \rightharpoonup \overline{\mathbf{S}} \qquad \text{weakly in } L^{r'}(0, T; L^{r'}(\Omega)^{3\times3}), \quad \text{and} \tag{2.51}$$

$$q_c^\varepsilon \rightharpoonup \overline{q_c} \qquad \text{weakly in } L^s(0, T; L^s(\Omega)^3). \tag{2.52}$$

Since $\sqrt{\varepsilon}\nabla p^\varepsilon$ is uniformly bounded in $L^2(0, T; L^2(\Omega)^3)$, applying (2.44) to (2.28) we immediately conclude that $\operatorname{div} v = 0$ in Q.

In order to take the limit in (2.29) and (2.30) we first identify the limits of $\mathbf{S}^\varepsilon$ and q_c^ε. To prove that $\overline{\mathbf{S}} = \mathbf{S}(c, \mathbf{D}(v))$, it is enough to establish almost everywhere convergence for ∇v^ε. To show this, we observe that (2.51), (2.49), (2.7), and Lebesgue's Dominated Convergence Theorem (used to show that $\|\mathbf{S}(c^\varepsilon, \mathbf{D}(v)) - \mathbf{S}(c, \mathbf{D}(v))\|_{r'} \to 0$ as $\varepsilon \to 0$) imply that $(\mathbf{S}(c^\varepsilon, \mathbf{D}(v)), \mathbf{D}(v^\varepsilon - v))_Q \to 0$ as $\varepsilon \to 0$. Next, taking $\varphi := v^\varepsilon - v$ as a test function in (2.29) it is easy to observe, using the above convergence results, that $\limsup_{\varepsilon \to 0}(\mathbf{S}^\varepsilon, \mathbf{D}(v^\varepsilon - v))_Q \le 0$. Thus we obtain

$$0 \overset{(2.8)}{\le} (\mathbf{S}(c^\varepsilon, \mathbf{D}(v^\varepsilon)) - \mathbf{S}(c^\varepsilon, \mathbf{D}(v)), \mathbf{D}(v^\varepsilon - v))_Q \le o(1),$$

where $o(1)$ denotes a quantity that vanishes as $\varepsilon \to 0$. Then the assumption (2.8) on the strict monotonicity of $\mathbf{S}$ implies that at least for a (labelled again by ε) subsequence $\nabla v^\varepsilon \to \nabla v$ a.e. in Q. Therefore,

$$v^\varepsilon \to v \text{ strongly in } L^h(0, T; W_{\boldsymbol{n}}^{1,h}) \quad \text{for all } h < r, \tag{2.53}$$

which is sufficient to prove (modulo a subsequence) that $\overline{\mathbf{S}} = \mathbf{S}(c, \mathbf{D}(v))$.

To show that $\overline{q_c} = q_c(c, \mathbf{D}(v), \nabla c)$, we first observe (using (2.49), (2.53), and (2.9)) that

$$\mathbf{K}(c^\varepsilon, |\mathbf{D}(v^\varepsilon)|^2) \to \mathbf{K}(c, |\mathbf{D}(v)|^2) \text{ strongly in } L^h(0, T; L^h(\Omega)^{3\times3}) \tag{2.54}$$

for all $h \in [1, \frac{r}{2\beta})$ if $\beta > 0$ and all $h \in (1, \infty)$ if $\beta \leq 0$. Thus, for $\beta \leq 0$ it is a direct consequence of the convergence properties shown above that $\overline{q_c} = q_c(c, \mathbf{D}(v), \nabla c)$. For $\beta \geq 0$ we proceed more carefully. Using the notation $\kappa^\varepsilon := |\mathbf{K}(c^\varepsilon, \mathbf{D}(v^\varepsilon))|$, (2.54) implies that for all $h \in (1, \infty)$

$$\frac{\mathbf{K}(c^\varepsilon, |\mathbf{D}(v^\varepsilon)|^2)}{\kappa^\varepsilon} \to \frac{\mathbf{K}(c, |\mathbf{D}(v)|^2)}{\kappa} \quad \text{strongly in } L^h(0, T; L^h(\Omega)^{3\times 3}).$$

Thus, to identify $\overline{q_c}$, it is enough to identify $\overline{\kappa \nabla c}$ (here the symbol $\overline{\kappa \nabla c}$ denotes the weak limit of $\kappa^\varepsilon \nabla c^\varepsilon$). Then (2.9), (2.33), and (2.35) imply (modulo a subsequence) that $\sqrt{\kappa^\varepsilon} \nabla c^\varepsilon \rightharpoonup \sqrt{\kappa} \nabla c$ weakly in $L^2(Q)$. Thus, having $2\beta < r$ and the strong convergence (2.54), we know that $\sqrt{\kappa^\varepsilon} \to \sqrt{\kappa}$ strongly in $L^2(Q)$. Hence, to identify the limit $\overline{q_c}$ it is enough to identify the limit $\sqrt{\kappa} \nabla c$. Using this procedure inductively, we see that it is enough that for some $k \in \mathbb{N}$ we have

$$\left(\kappa^\varepsilon\right)^{2^{-k}} \nabla c^\varepsilon \rightharpoonup \kappa^{2^{-k}} \nabla c \quad \text{weakly in } L^1(Q). \tag{2.55}$$

Hence, we find k such that $\left(\frac{r2^{k-1}}{\beta}\right)' > q'$. Then it is a consequence of (2.54) and (2.48) that the convergence (2.55) is valid and this finishes the proof of the convergence of q_c^ε.

Finally, having all these convergence results (2.43)–(2.52) in hand together with the identification of the limit of the nonlinear terms, we can take the limit $\varepsilon \to 0$ in (2.29) and in (2.30) in a standard way and conclude that for any $\eta > 0$ the triplet $(v, c, p) := (v^\eta, c^\eta, p^\eta)$ satisfies

$$\langle v_{,t}, \varphi \rangle - (v_\eta \otimes v, \nabla \varphi)_Q + (\mathbf{S}(c, \mathbf{D}(v)), \nabla \varphi)_Q$$
$$+ \alpha(v, \varphi)_\Gamma - (p, \operatorname{div} \varphi)_Q = \langle f, \varphi \rangle \tag{2.56}$$

$$\text{for all } \varphi \in L^r(0, T; W_n^{1,r}) \text{ such that } \operatorname{tr} \varphi \in L^2(\Gamma),$$

and

$$\langle c_{,t}, \varphi \rangle - (v_\eta c, \nabla \varphi)_Q = (q_c(c, \mathbf{D}(v), \nabla c), \nabla \varphi)_Q$$
$$\text{for all } \varphi \in L^{m'}(0, T; W^{1,m'}(\Omega)) \text{ such that } \varphi|\Gamma_D = 0. \tag{2.57}$$

The attainment of initial conditions is again standard and can be proved by using the same methods as those described in Málek et al. (1996).

2.3.3 *Limit as $\eta \to 0$*

Let $(v^\eta, c^\eta, p^\eta, \mathbf{S}^\eta, q_c^\eta)$, the solution of the η-approximation, satisfy (2.56) and (2.57). Our final goal is to take the limit $\eta \to 0$ in (2.56) and in

(2.57), and to establish the existence of a suitable weak solution to our original problem.

Using weak lower semicontinuity of appropriate norms, we find that (2.32)–(2.39), (2.41)$_2$ and (2.42)$_2$ hold. These estimates together with the Aubin–Lions Lemma (Lemma 2.5.5) and Corollary 2.5.2 are sufficient to find a (labelled again by ε) subsequence of (v^η, c^η, p^η) such that

$$v^\eta_{,t} \rightharpoonup v_{,t} \quad \text{weakly in } L^m(0,T; W_n^{-1,m}) \cap (X^{r,\frac{5r}{5r-8}})^*, \quad (2.58)$$

$$v^\eta \rightharpoonup v \quad \text{weakly in } L^r(0,T; W_n^{1,r}), \quad (2.59)$$

$$v^\eta \to v \quad \text{strongly in } L^h(0,T; L^h(\Omega)^3) \text{ for all } h < \tfrac{5r}{3}, \quad (2.60)$$

$$\text{tr } v^\eta \to \text{tr } v \quad \text{strongly in } L^2(0,T; L^2(\partial\Omega)^3), \quad (2.61)$$

$$c^\eta_{,t} \rightharpoonup c_{,t} \quad \text{weakly in } L^{s'}(0,T; W^{-1,s'}(\Omega)), \quad (2.62)$$

$$c^\eta \rightharpoonup c \quad \text{weakly in } L^q(0,T; W^{1,q}(\Omega)), \quad (2.63)$$

$$c^\eta \to c \quad \text{strongly in } L^h(0,T; L^h(\Omega)) \text{ for all } h < \infty, \quad (2.64)$$

$$p^\eta \rightharpoonup p \quad \text{weakly in } L^m(0,T; L^m(\Omega)), \quad (2.65)$$

$$\mathbf{S}^\eta \rightharpoonup \overline{\mathbf{S}} \quad \text{weakly in } L^{r'}(0,T; L^{r'}(\Omega)^{3\times3}), \text{ and} \quad (2.66)$$

$$q^\eta_c \to \overline{q_c} \quad \text{weakly in } L^s(0,T; L^s(\Omega)^3). \quad (2.67)$$

Assume for a moment that $\mathbf{D}(v^\eta)$ converges to $\mathbf{D}(v)$ almost everywhere in Q. Then using the same procedure as in Subsection 2.3.2 we get $\overline{q_c} = q_c(c, \mathbf{D}(v)\nabla c)$. Then we are able to take the limit in (2.57) and to achieve (2.22).

Thus it remains to show that ∇v^η converges a.e. in Q. To do this, we follow the approach described in Boccardo & Murat (1992), Frehse, Málek, & Steinhauer (2000), or Bulíček et al. (2008). We define

$$g^\eta := |\nabla v^\eta|^r + |\nabla v|^r + (|\mathbf{S}^\eta| + |\mathbf{S}|)\,(|\mathbf{D}(v^\eta)| + |\mathbf{D}(v)|).$$

It follows from (2.59) and (2.66) that there is $K \in [1,\infty)$ so that for all η

$$0 \le \int_0^T \int_\Omega g^\eta \, dx \, dt \le K.$$

Let $\varepsilon^* > 0$ be arbitrary. Then the following statement is proved in Bulíček et al. (2008):

> There exist $L \leq \dfrac{\varepsilon^*}{K}$, a subsequence $\{\boldsymbol{v}^j\}_{j=1}^\infty \subset \{\boldsymbol{v}^\eta\}_{\eta > 0}$,
>
> and sets $E^j := \{(t,x) \in Q; L^2 \leq |\boldsymbol{v}^j(t,x) - \boldsymbol{v}(t,x)| < L\}$ $\qquad$ (2.68)
>
> such that $\displaystyle\int_{E^j} g^j \, dx \, dt \leq \varepsilon^*.$

For the obtained sequence $\{\boldsymbol{v}^j\}_{j=1}^\infty$ and L we define $\boldsymbol{u}^j$ and the sets Q^j as

$$\boldsymbol{u}^j := (\boldsymbol{v}^j - \boldsymbol{v})\left(1 - \min\left\{\tfrac{|\boldsymbol{v} - \boldsymbol{v}^j|}{L}, 1\right\}\right);$$
$$Q^j := \{(t,x) \in Q; |\boldsymbol{v} - \boldsymbol{v}^j| < L\}.$$

By using (2.59), (2.60), and the fact that $|\boldsymbol{u}^j| \leq L$ in Q we have (as $j \to \infty$)

$$\boldsymbol{u}^j \rightharpoonup \boldsymbol{0} \qquad \text{weakly in } L^r(0,T; W_{\boldsymbol{n}}^{1,r}), \qquad\qquad (2.69)$$
$$\boldsymbol{u}^j \to \boldsymbol{0} \qquad \text{strongly in } L^s(0,T; L^s(\Omega)^3) \quad \forall s < \infty, \qquad (2.70)$$
$$\operatorname{tr} \boldsymbol{u}^j \to \boldsymbol{0} \qquad \text{strongly in } L^2(0,T; L^2(\partial\Omega)^3). \qquad\qquad (2.71)$$

Since (see equation (2.60) in Bulíček et al., 2008, for details)

$$\int_0^T \|\operatorname{div} \boldsymbol{u}^j\|_r^r \, dt \leq C\varepsilon^*,$$

the Helmholtz decomposition $\boldsymbol{u}^j = \boldsymbol{u}_{\text{div}}^j + \nabla g^{\boldsymbol{u}^j}$ then implies that

$$\int_0^T \|g^{\boldsymbol{u}^j}\|_{2,r}^r \, dt \leq C\varepsilon^* \qquad\qquad (2.72)$$

and

$$\boldsymbol{u}_{\text{div}}^j \to \boldsymbol{0} \qquad \text{strongly in } L^s(0,T; L^s(\Omega)^3) \quad \text{for all } s < \infty.$$

Using the assumption (2.9) we have

$$0 \leq (\mathbf{S}(c^j, \mathbf{D}(\boldsymbol{v}^j)) - \mathbf{S}(c^j, \mathbf{D}(\boldsymbol{v})), \mathbf{D}(\boldsymbol{v}^j - \boldsymbol{v}))_{Q^j}$$
$$= -(\mathbf{S}(c^j, \mathbf{D}(\boldsymbol{v})), \mathbf{D}(\boldsymbol{v}^j - \boldsymbol{v}))_{Q^j} + (\mathbf{S}(c^j, \mathbf{D}(\boldsymbol{v}^j)), \mathbf{D}(\boldsymbol{v}^j - \boldsymbol{v}))_{Q^j}$$
$$=: Y_1 + Y_2.$$

By virtue of (2.64), the Lebesgue Dominated Convergence Theorem and (2.7), we observe that

$$\mathbf{S}(c^j, \mathbf{D}(\boldsymbol{v})) \to \mathbf{S}(c, \mathbf{D}(\boldsymbol{v})) \qquad \text{strongly in } L^{r'}(0,T; L^{r'}(\Omega)^{3\times 3}).$$

Therefore (as $j \to \infty$)

$$Y_1 := (\mathbf{S}(c^j, \mathbf{D}(v)), \mathbf{D}(v - v^j))_{Q^j}$$

$$= (\mathbf{S}(c^j, \mathbf{D}(v)), \mathbf{D}(u^j))_Q + (\mathbf{S}(c^j, \mathbf{D}(v)), \mathbf{D}((v - v^j)\tfrac{|v - v^j|}{L}))_{Q^j}$$

$$\overset{(2.3.3)}{\underset{(2.69)}{\leq}} o(1) + (\mathbf{S}(c^j, \mathbf{D}(v)), \mathbf{D}((v - v^j)\tfrac{|v - v^j|}{L}))_{Q^j \setminus E^j}$$

$$+ (\mathbf{S}(c^j, \mathbf{D}(v)), \mathbf{D}((v - v^j)\tfrac{|v - v^j|}{L}))_{E^j}$$

$$\overset{(2.68)}{\leq} o(1) + CL + C\varepsilon^* \leq o(1) + C\varepsilon^*,$$

where $o(1) \to 0$ as $j \to \infty$. To estimate Y_2 we set $\varphi = u^j_{\mathrm{div}}$ in (2.56) and denoting $\mathbf{S}^j := \mathbf{S}(c^j, \mathbf{D}(v^j))$ obtain

$$Y_2 := (\mathbf{S}^j, \mathbf{D}(v^j - v))_{Q^j} = (\mathbf{S}^j, \mathbf{D}(u^j))_Q + (\mathbf{S}^j, \mathbf{D}((v - v^j)\tfrac{|v - v^j|}{L}))_{Q^j}$$

$$= (\mathbf{S}^j, \mathbf{D}(u^j_{\mathrm{div}}))_Q + (\mathbf{S}^j, \mathbf{D}(\nabla g^{u^j}))_Q + (\mathbf{S}^j, \mathbf{D}((v - v^j)\tfrac{|v - v^j|}{L}))_{Q^j}$$

$$\overset{(2.68)}{\underset{(2.72)}{\leq}} (\mathbf{S}^j, \mathbf{D}(u^j_{\mathrm{div}}))_Q + C\varepsilon^* \overset{(2.56)}{=}: \sum_{i=1}^{4} I_i + C\varepsilon^*,$$

where[2]

$$I_1 = -\langle v^j_{,t}, u^j_{\mathrm{div}} \rangle = -\langle v_{,t}, u^j_{\mathrm{div}} \rangle + \langle v_{,t} - v^j_{,t}, u^j_{\mathrm{div}} \rangle$$

$$\leq o(1) + \langle v_{,t} - v^j_{,t}, u^j_{\mathrm{div}} \rangle \overset{\mathrm{div}\, v - v^j = 0}{=} o(1) + \langle v_{,t} - v^j_{,t}, u^j \rangle \leq o(1),$$

$$I_2 = -\left([\nabla v^j] v^j_{\eta(j)}, u^j_{\mathrm{div}} \right)_Q \leq C\|u^j\|_{\frac{5r}{5r-8}, Q} \|v^j\|_{\frac{5r}{3}, Q} \|\nabla v^j\|_{r, Q} \overset{(2.70)}{=} o(1),$$

$$I_3 = -\alpha(v^j, u^j_{\mathrm{div}})_\Gamma \leq C\|u^j\|_{L^2(\Gamma)} \overset{(2.71)}{=} o(1), \quad \text{and}$$

$$I_4 = \langle f, u^j_{\mathrm{div}} \rangle = o(1).$$

Thus, we can conclude that for $\theta < 1$

$$\int_Q |(\mathbf{S}^j - \mathbf{S}(c^j, \mathbf{D}(v))) \cdot \mathbf{D}(v^j - v)|^\theta = \int_{Q_j} |\dots|^\theta + \int_{Q \setminus Q_j} |\dots|^\theta$$

$$\leq C(Y_1 + Y_2)^\theta + C|Q \setminus Q_j|^{1-\theta} \leq o(1)$$

and using strict monotonicity of $\mathbf{S}$, i.e. the assumption (2.9) we obtain that $\mathbf{D}(v^j)$ converges to $\mathbf{D}(v)$ a.e. in Q. Thus with the help of Vitali's Theorem, it is standard to take limit in (2.56) to obtain (2.21). The proof is complete.

---

[2]For details concerning the estimate of I_1 see Bulíček et al. (2008).

2.4 Extensions of Theorem 2.2.2

This section is devoted to the formulation of extensions that are not proved in detail anywhere but are direct consequences of the approaches described here and those presented in recent studies by Wolf (2007) and Diening et al. (2008). These extensions are focused on including general domains with no smoothness of the boundary, incorporating no-slip boundary conditions, and lowering the power-law index r. We wish to underline that to the best of our knowledge we can deal with globally integrable pressure for the low exponents of r only for domains of class $\mathcal{R}$ and for Navier's slip boundary conditions, the case that is studied in previous sections.

We first formulate the result for the same range of r as in Theorem 2.2.2 and for no-slip boundary conditions (corresponding to $\alpha = \infty$ in $(2.5)_1$) which also enables us to consider any bounded domain. The pressure is omitted from the weak formulation. As the pressure is not in general an integrable function it is also no longer clear how to fix its mean value over any spatial subdomain.

Theorem 2.4.1 *Let Ω be a bounded domain and let (2.13)–(2.14) hold (with the exception of the condition fixing the pressure). Assume that S satisfies (2.7)–(2.8) with $r > \frac{8}{5}$, and $\boldsymbol{q}_c$ satisfies (2.9) with $-r < 2\beta < r$. Let m, q and s be defined as in (2.15). Then there is a weak solution $(\boldsymbol{v}, c)$ of (2.3)–(2.6) such that*

$$\boldsymbol{v} \in \mathcal{C}([0,T]; L_w^2(\Omega)^3) \cap L^r(0,T; W_{0,\mathrm{div}}^{1,r}), \quad \boldsymbol{v}_{,t} \in L^m(0,T; W_{0,\mathrm{div}}^{-1,m}),$$

$$c - \widetilde{c}_b \in L^q(0,T; W_{\Gamma_D}^{1,q}(\Omega)), \quad c_{,t} \in L^{s'}(0,T; W_{\Gamma_D}^{-1,s'}(\Omega)),$$

$$0 \le c \le 1 \ a.e. \ in \ Q,$$

$$(1 + |\mathsf{D}(\boldsymbol{v})|^2)^{\frac{\beta}{2}} \nabla c \in L^2(0,T; L^2(\Omega)^3),$$

$(\boldsymbol{v}, c)$ satisfy the following weak formulation

$$\langle \boldsymbol{v}_{,t}, \boldsymbol{\varphi} \rangle - (\boldsymbol{v} \otimes \boldsymbol{v}, \nabla \boldsymbol{\varphi})_Q + \alpha(\boldsymbol{v}, \boldsymbol{\varphi})_\Gamma + (\mathsf{S}(c, \mathsf{D}(\boldsymbol{v})), \mathsf{D}(\boldsymbol{\varphi}))_Q = \langle \boldsymbol{f}, \boldsymbol{\varphi} \rangle$$
$$\textit{for all } \boldsymbol{\varphi} \in L^{m'}(0,T; W_{0,\mathrm{div}}^{1,m'}),$$

$$(c_{,t}, \varphi)_Q - (c\boldsymbol{v}, \nabla\varphi)_Q + (\mathsf{K}(c, |\mathsf{D}(\boldsymbol{v})|^2)\nabla c, \nabla\varphi)_Q = 0$$
$$\textit{for all } \varphi \in L^s(0,T; W_{\Gamma_D}^{1,s}(\Omega)),$$

and $(\boldsymbol{v}, c)$ attain the initial conditions as in (2.23).

The next theorem extends the result stated in Theorem 2.2.2 also to the case $r \in (6/5, 8/5]$.

Theorem 2.4.2 *Let $\frac{6}{5} < r \le \frac{8}{5}$. Then for any data fulfilling (2.13)–(2.14) and for any $T \in (0,\infty)$ there exists a weak solution to (2.3)–(2.6) in the sense of Definition 2.2.1.*

The extension of the existence result in the sense of Theorem 2.4.1 to no-slip boundary conditions for $r \in (6/5, 8/5]$ can be also proved. We skip the formulation.

Our final remark concerns the possible extension of the result established here and formulated in the above theorem to models describing unsteady flows of incompressible chemically reacting and heat conducting fluids. Such a problem requires one to consider the full thermodynamical system including the balance of energy and possibly also the second law of thermodynamics in combination with the convection-diffusion equation for c. An interested reader could combine the results established here with those proved in Bulíček et al. (2009) and Bulíček et al. (2008) where incompressible Navier–Stokes–Fourier-like systems are analysed.

2.5 Appendix

The following lemmas summarize helpful properties of functions belonging to certain Sobolev spaces.

Lemma 2.5.1 *Let $1 < q_1, q_2 < \infty$. Set*

$$\mathcal{S} := \{v; v \in L^\infty(0,T; L^2(\Omega)^3) \cap L^r(0,T; W_{\boldsymbol{n}}^{1,r}), v_{,t} \in L^{q_1}(0,T; W_{\boldsymbol{n},\mathrm{div}}^{-1,q_2})\}.$$

If $r > \frac{3}{2}$ and $\{v^i\}_{i=1}^\infty$ is bounded in $\mathcal{S}$, then $\{\mathrm{tr}\, v^i\}_{i=1}^\infty$ is precompact in $L^p(0,T; L^s(\partial\Omega)^3)$ for all $p, s \in (1,\infty)$ satisfying

$$p < s\frac{5r-6}{3s-4}, \quad \max(2,r) \le s \le \frac{2r}{3-r}. \tag{2.73}$$

Proof See Bulíček et al. (2008, Lemma 1.4) where an even more general result is proved. □

Corollary 2.5.2 *Let $r > \frac{8}{5}$. Let $\{v^i\}_{i=1}^\infty$ be bounded in $\mathcal{S}$. Then $\{\mathrm{tr}\, v^i\}_{i=1}^\infty$ is precompact in $L^2(0,T; L^2(\partial\Omega)^3)$.*

Lemma 2.5.3 (Korn's inequality) *Let $q \in (1,\infty)$. Then there exists a positive constant C depending only on Ω and q such that for all $v \in W^{1,q}(\Omega)^3$ that has trace $\mathrm{tr}\, v \in L^2(\partial\Omega)^3$ the following inequality holds*

$$C\|v\|_{1,q} \le \|\mathbf{D}(v)\|_q + \|v\|_{L^2(\partial\Omega)}. \tag{2.74}$$

Proof For the proof we refer the reader to a modification of the result in Nečas (1966) and Bulíček et al. (2008). □

Lemma 2.5.4 (Interpolation inequalities) *For* $2 \leq q \leq \frac{3r}{3-r}$ *(if* $6/5 < r < 3$*) and for any* $q \in [1, \infty)$ *if* $r \geq 3$ *the following inequality holds*

$$\|z\|_q \leq \|z\|_2^{\frac{6r-6q+2qr}{q(5r-6)}} \|z\|_{1,r}^{\frac{3r(q-2)}{q(5r-6)}}. \tag{2.75}$$

Proof See Nirenberg (1966). □

Lemma 2.5.5 (Aubin–Lions) *Let* V_1, V_2, V_3 *be reflexive separable Banach spaces such that*

$$V_1 \hookrightarrow \hookrightarrow V_2 \text{ and } V_2 \hookrightarrow V_3.$$

Let $1 < p < \infty$, $1 \leq q \leq +\infty$ *and* $0 < T < \infty$. *Then*

$$\{v; v \in L^p(0, T; V_1), v_{,t} \in L^q(0, T; V_3)\}$$

is compactly embedded into $L^p(0, T; V_2)$.

Proof See for example Simon (1987) or Feireisl (2004, Lemma 6.3). □

Acknowledgements

The contribution of M. Bulíček to this work was supported by the Jindřich Nečas Center for Mathematical Modeling, the project LC06052 financed by MSMT. J. Málek's contribution is a part of the research project MSM 0021620839 financed by MSMT; the support of GACR 201/08/0315 and GACR 201/06/0352 is also acknowledged. K.R. Rajagopal thanks the National Science Foundation for its support.

References

Adkins, J.E. (1963a) Non-linear diffusion. I. Diffusion and flow of mixtures of fluids. *Philos. Trans. Roy. Soc. London Ser. A* **255**, 607–633.

Adkins, J.E. (1963b) Non-linear diffusion. II. Constitutive equations for mixtures of isotropic fluids. *Philos. Trans. Roy. Soc. London Ser. A* **255**, 635–648.

Anand, M. & Rajagopal, K.R. (2004) A shear-thinning viscoelastic fluid model for describing the flow of blood. *International Journal of Cardiovascular Medicine and Science* **4**, 59–68.

Anand, M., Rajagopal, K., & Rajagopal, K.R. (2003) A model incorporating some of the mechanical and biochemical factors underlying clot formation and dissolution in flowing blood. *J. Theor. Med.* **5**, no. 3-4, 183–218.

Anand, M., Rajagopal, K., & Rajagopal, K.R. (2005) A model for the formation and lysis of blood clots. *Pathophysiology of Haemostasis and Thrombosis* **34**, 109–120.

Antontsev, S.N., Kazhikhov, A.V., & Monakhov, V.N. (1990) *Boundary value problems in mechanics of nonhomogeneous fluids.* Studies in Mathematics and its Applications **22**, North-Holland Publishing Co., Amsterdam.

Atkin, R.J. & Craine, R.E. (1976a) Continuum theories of mixtures: applications. *J. Inst. Math. Appl.* **17**, no. 2, 153–207.

Atkin, R.J. & Craine, R.E. (1976b) Continuum theories of mixtures: basic theory and historical development. *Quart. J. Mech. Appl. Math.* **29**, no. 2, 209–244.

Boccardo, L. & Murat, F. (1992) Almost everywhere convergence of the gradients of solutions to elliptic and parabolic equations. *Nonlinear Anal.* **19**, no. 6, 581–597.

Bowen, R.M. (1975) *Continuum physics* III, Academic Press, New York.

Bridges, C. & Rajagopal, K.R. (2006) Pulsatile flow of a chemically-reacting nonlinear fluid. *Comput. Math. Appl.* **52**, no. 6-7, 1131–1144.

Brinkman, H.C. (1947a) A calculation of the viscous force exerted by a flowing fluid on a dense swarm of particles. *Applied Scientific Research* **A1**, 27–34.

Brinkman, H.C. (1947b) On the permeability of media consisting of closely packed porous particles. *Applied Scientific Research* **A1**, 81–86.

Bulíček, M., Málek, J., & Rajagopal, K.R. (2008) Mathematical analysis of unsteady flows of fluids with pressure, shear-rate and temperature dependant material moduli, that slip at solid boundaries. *SIAM J. Math. Anal*, to appear.

Bulíček, M., Feireisl, E., & Málek, J. (2009) Navier-Stokes-Fourier system for incompressible fluids with temperature dependent material coefficients. to appear in *Nonlinear Analysis: Real World Applications*, **10**, 992–1015.

Darcy, H. (1856) *Les Fontaines Publiques de La Ville de Dijon.* Victor Dalmont, Paris.

Diening, L., Růžička, M., & Wolf, J. (2008) Existence of weak solutions to the equations of non-stationary motion of non-Newtonian fluids with shear rate dependent viscosity. *Ann. Scuola Norm. Sup. Pisa Cl. Sci*, to appear.

Fernández-Cara, E., Guillén, F., & Ortega, R.R. (1997) Some theoretical results for viscoplastic and dilatant fluids with variable density. *Nonlinear Anal.* **28**, no. 6, 1079–1100.

Feireisl, E. (2004) *Dynamics of viscous compressible fluids.* Oxford Lecture Series in Mathematics and its Applications **26**, Oxford University Press, Oxford.

Forchheimer, P. (1901) Wasserbewegung Durch Boden. *Zeits. V. Deutsch. Ing.* **45**, 1781–1788.

Frehse, J., Málek, J., & Steinhauer, M. (2000) On existence results for fluids with shear dependent viscosity - unsteady flows. Partial differential equations (Praha, 1998). *Res. Notes Math.* **406**, Chapman & Hall/CRC, Boca Raton, FL.

Frehse, J. & Růžička, M. (2008) Non-homogeneous generalized Newtonian fluids. *Math. Z.* **260**, no. 2, 355–375.

Green, A.E. & Naghdi, P.M. (1969) On Basic Equations for Mixtures. *Quart. J. Mech. Appl. Math.* **22**, 427–438.

Gray, W.G. (1983) General conservation equations for multiphase systems 4. Constitutive theory including phase change. *Adv. Water Resources* **6**, 130–140.

Grisvard, P. (1985) *Elliptic problems in nonsmooth domains.* Monographs and Studies in Mathematics **24**, Pitman (Advanced Publishing Program), Boston, MA.

Guillén-González, F. (2004) Density-dependent incompressible fluids with non-Newtonian viscosity. *Czechoslovak Math. J.* **54(129)**, no. 3, 637–656.

Hassanizadeh, M. (1986) Derivation of basic equations of mass transport in porous media, Part 1. Macroscopic balance laws. *Adv. Water Resources* **9**, 196–206.

Lions, P.L. (1996) *Mathematical topics in fluid mechanics.* Vol. 1. Oxford Lecture Series in Mathematics and its Applications **3**, The Clarendon Press Oxford University Press, New York.

Málek, J., Nečas, J., Rokyta, M., & Růžička, M. (1996) *Weak and measure-valued solutions to evolutionary PDEs.* Chapman & Hall, London.

Nečas, J. (1966) *Séminaire Equations aux Dérivées Partielles.* Les Presses de l'Université.

Nirenberg, L. (1966) An extended interpolation inequality. *Ann. Scuola Norm. Sup. Pisa (3)* **20**, 733–737.

Novotný, A. & Straškraba, I. (2004) *Introduction to the mathematical theory of compressible flow.* Oxford Lecture Series in Mathematics and its Applications **27**, Oxford University Press, Oxford.

Rajagopal, K.R. (2007) On a hierarchy of approximate models for flows of incompressible fluids through porous solids. *Math. Models Methods Appl. Sci.* **17**, no. 2, 215–252.

Rajagopal, K.R. & Tao, L. (1995) *Mechanics of mixtures.* World Scientific Publishing Co. Inc., River Edge, NJ.

Rayleigh, Lord (1883) Investigation of the character of the equilibrium of an incompressible heavy fluids of variable density. *Proceedings of the London Mathematical Society* **14**, 170–177.

Roubíček, T. (2005) Incompressible ionized fluid mixtures: a non-Newtonian approach. *IASME Trans.* **2**, no. 7, 1190–1197.

Roubíček, T. (2007) Incompressible ionized non-Newtonean fluid mixtures. *SIAM J. Math. Anal.* **39**, no. 3, 863–890.

Simon, J. (1987) Compact sets in the space $L^p(0,T;B)$. *Ann. Mat. Pura Appl.* **146**, 65–96.

Thurston, G.B. (1972) Viscoelasticity of human blood. *Biophysical Journal* **12**, no. 9, 1205–1217.

Thurston, G.B. (1973) Frequency and shear rate dependence of viscoelasticity of blood. *Biorheology* **10**, 375–381.

Truesdell, C.A. (1991) *A first course in rational continuum mechanics* **1**, Academic Press, New York.

Wolf, J. (2007) Existence of weak solutions to the equations of non-stationary motion of non-Newtonian fluids with shear rate dependent viscosity. *J. Math. Fluid Mech.* **9**, no. 1, 104–138.

Yeleswarapu, K.K. (1996) *Evaluation of Continuum Models for Characterizing the Constitutive Behavior of Blood.* PhD Thesis, University of Pittsburgh.

Yeleswarapu, K.K., Kameneva, M. V., Rajagopal, K.R., & Antaki, J.F. (1998) The flow of blood in tubes: theory and experiment. *Mechanics Research Communication* **25**, 257–262.

Yih, C.S. (1965) *Dynamics of non-homogeneous fluids*. MacMillan, New York.

Yih, C.S. (1980) *Stratified flows*. Academic Press, New York.

3

The uniqueness of Lagrangian trajectories in Navier–Stokes flows

Masoumeh Dashti & James C. Robinson
Mathematics Institute, University of Warwick,
Coventry, CV4 7AL. UK
M.Dashti@warwick.ac.uk
J.C.Robinson@warwick.ac.uk

Abstract

Given an initial condition in $H^{(d/2)-1}$, the d-dimensional Navier–Stokes equations $(d = 2, 3)$ have a unique solution; when $d = 2$ this solution exists for all $t \geq 0$, while when $d = 3$ its existence can only be guaranteed on some time interval $[0, T)$. Given such a solution, we show by elementary methods (in the periodic case) that the Lagrangian particle trajectories are also unique. In fact, we prove a general result that solutions of the ordinary differential equation $\dot{X} = u(X, t)$ are unique if $u \in L^2(0, T; H^{(d/2)-1})$ and $\sqrt{t}\, u \in L^2(0, T; H^{(d/2)+1})$, and verify that these conditions hold for the Navier–Stokes equations via some straightforward energy estimates. We also show that the Lagrangian trajectories depend continuously on the Eulerian initial data $u_0 \in H^{(d/2)-1}$.

3.1 Introduction

We consider the Navier–Stokes equations

$$u_t - \nu \Delta u + (u \cdot \nabla)u + \nabla p = 0 \qquad \nabla \cdot u = 0. \tag{3.1}$$

In the two-dimensional case it is well known that an initial condition $u_0 \in L^2$ gives rise to a unique weak solution u with $u \in L^\infty(0, T; L^2) \cap L^2(0, T; H^1)$ for every $T > 0$ (see Constantin & Foias, 1988, for example). In the three-dimensional case, if $u_0 \in H^{1/2}$ then (Kato & Fujita, 1962; Chemin et al., 2006) there is a unique solution $u \in L^\infty(0, T; H^{1/2}) \cap L^2(0, T; H^{3/2})$ that exists for at least some small time interval $[0, T^*)$. In other words, for $u_0 \in H^{(d/2)-1}$ there exists a unique solution $u \in L^\infty(0, T; H^{(d/2)-1}) \cap L^2(0, T; H^{d/2})$, $d = 2, 3$.

Using Littlewood–Paley theory, Chemin & Lerner (1995) showed that for flows on the whole of $\mathbb{R}^d$ $(d = 2, 3)$ the associated Lagrangian

Published in *Partial Differential Equations and Fluid Mechanics*, edited by James C. Robinson and José L. Rodrigo. © Cambridge University Press 2009.

trajectories – the paths traced out by 'fluid particles' – inherit this uniqueness, despite the rough nature of the vector field u. Here we give a much more elementary proof of the same result, in a form that is largely independent of the structure of the Navier–Stokes equations themselves.

For the sake of simplicity we restrict here to periodic boundary conditions and initial conditions that have zero average (a condition which is preserved under the evolution). However, a similar analysis is also possible in the more technically challenging case of flows in bounded domains with no slip boundary conditions: for details see Dashti & Robinson (2009).

More precisely, the question is whether the solutions of the ordinary differential equation

$$\dot{X} = u(X, t) \qquad X(0) = X_0 \tag{3.2}$$

are unique, when $u(t)$ is a solution of the Navier–Stokes equations with $u_0 \in H^{(d/2)-1}$. An elementary computation – which uses the result

$$|u(X) - u(Y)| \le c\|u\|_{H^{(d/2)+1}} |X - Y|(-\log|X - Y|)^{1/2}$$

due to Zuazua (2002) – shows that u does not have enough regularity to proceed in an entirely straightforward manner.

To this end, let $X(t)$ and $Y(t)$ be two solutions of (3.2) with $X(0) = Y(0) = X_0$. Then with $W(t) = X(t) - Y(t)$ one has

$$\frac{\mathrm{d}}{\mathrm{d}t}|W| \le |u(X, t) - u(Y, t)| \le c\|u(t)\|_{H^{(d/2)+1}} |W|(-\log|W|)^{1/2}.$$

Integrating this differential inequality between $s > 0$ and $t > s$ one obtains

$$-(-\log|W(t)|)^{1/2} \le -(-\log|W(s)|)^{1/2} + c\int_s^t \|u(r)\|_{H^{(d/2)+1}} \, \mathrm{d}r. \tag{3.3}$$

If one had $u \in L^1(0, T; H^{(d/2)+1})$ then one could put $s = 0$ in (3.3) and immediately obtain uniqueness. However, such regularity for u is not known to be true for the Navier–Stokes equations, nor even for solutions of the heat equation.

3.2 An abstract result for ordinary differential equations

We now prove two abstract results. One guarantees uniqueness of solution of the ODE $\dot{X} = u(X, t)$ given sufficient regularity of u, the second is related to continuity of solutions with respect to perturbations of the vector field u. These results are valid in any dimension d; we show in Section 3.3 that they apply to solutions of the Navier–Stokes equations with initial conditions in $H^{(d/2)-1}$ for $d = 2, 3$ when the boundary conditions are periodic.

3.2.1 Existence and uniqueness

We now state and prove an abstract existence and uniqueness result. One can weaken the assumption on u to $u \in L^p(0, T; H^{(d/2)-1})$ for some $p > 1$ (Dashti & Robinson, 2009); and note that the assumption that $\sqrt{t}u \in L^2(0, T; H^{(d/2)+1})$ implies that $u \in L^r(0, T; H^{(d/2)+1})$ for any $r < 1$, since

$$\int_0^T \|u(t)\|_{H^{(d/2)+1}}^r \, dt = \int_0^T t^{-r/2} t^{r/2} \|u(t)\|_{H^{(d/2)+1}}^r \, dt$$

$$\leq \left(\int_0^T t^{-r/(2-r)} \, dt \right)^{1-(r/2)} \left(\int_0^T t \|u(t)\|_{H^{(d/2)+1}}^2 \, dt \right)^{r/2}.$$

Theorem 3.2.1 *Let Ω be $\mathbb{R}^d$, a periodic domain in $\mathbb{R}^d$, or a bounded domain in $\mathbb{R}^d$ whose boundary is sufficiently smooth. Then provided that*

$$u \in L^2(0, T; H^{(d/2)-1}(\Omega)) \quad and \quad \sqrt{t}u \in L^2(0, T; H^{(d/2)+1}(\Omega)) \quad (3.4)$$

along with the additional assumption that $u = 0$ on $\partial\Omega$ if Ω is bounded, the ordinary differential equation

$$\frac{dX}{dt} = u(X, t) \quad with \quad X(0) = X_0 \quad (3.5)$$

has a unique solution for every $X_0 \in \Omega$.

The argument of the proof depends only on certain Sobolev embedding results and Hölder's inequality; bounded domains are required to be 'sufficiently smooth' that these results are valid, see Adams (1975) for details.

Proof We need to guarantee first that (3.5) does indeed have at least one solution, which we do, following Foias, Guillopé, & Temam (1985), by considering the equation in its equivalent integral form

$$X(t) = X_0 + \int_0^t u(X(s), s) \, ds. \quad (3.6)$$

Note that (3.4) implies that $u \in L^1(0, T; L^\infty)$; indeed, using Agmon's inequality $\|u\|_\infty \leq c\|u\|_{H^{(d/2)-1}}^{1/2} \|u\|_{H^{(d/2)+1}}^{1/2}$ (see Lemma 4.9

in Constantin & Foias, 1988, for example), and the extended Hölder inequality with exponents $(2, 4, 4)$,

$$\int_0^s \|u(t)\|_\infty \, dt \le c \int_0^s \|u(t)\|_{H^{(d/2)-1}}^{1/2} \|u(t)\|_{H^{(d/2)+1}}^{1/2} \, dt$$

$$= c \int_0^s t^{-1/4} \|u(t)\|_{H^{(d/2)-1}}^{1/2} t^{1/4} \|u(t)\|_{H^{(d/2)+1}}^{1/2} \, dt$$

$$\le c \left(\int_0^s t^{-1/2} \, dt \right)^{1/2} \left(\int_0^s \|u(t)\|_{H^{(d/2)-1}}^2 \, dt \right)^{1/4}$$

$$\times \left(\int_0^s t \|u(t)\|_{H^{(d/2)+1}}^2 \, dt \right)^{1/4}$$

$$\le c s^{1/4} \|u\|_{L^2(0,T;H^{(d/2)-1})}^{1/2} \|\sqrt{t} u\|_{L^2(0,T;H^{(d/2)+1})}^{1/2}. \tag{3.7}$$

In the case of a periodic or bounded domain let u_n be a sequence of Galerkin approximations of u based on eigenfunctions of the Laplacian on Ω with appropriate boundary conditions, while in the case of $\mathbb{R}^d$ let u_n be a sequence of mollified versions of u. Thus u_n has the same regularity as that of u given in (3.4), with the bounds for u_n being dominated by those for u. Let X_n be the solution of the integral equation

$$X_n(t) = X_0 + \int_0^t u_n(X_n(s), s) \, ds,$$

which by standard results exists and is unique on $[0, T]$; when Ω is bounded the eigenfunctions used in the Galerkin expansion preserve the condition that $u|_{\partial\Omega} = 0$, which ensures that $X_n(t)$ cannot leave Ω. (Of course, X_n is precisely the solution of the smooth ordinary differential equation $\dot{X}_n = u_n(X_n, t)$ with $X_n(0) = X_0$.) Note that (3.7) in fact shows that

$$|X_n(t) - X_n(s)| \le K|t - s|^{1/4},$$

and so the X_n form a bounded equicontinuous family. It follows using the Arzelà–Ascoli Theorem that there is a subsequence that converges uniformly on $[0, T]$ to some $X(\cdot)$. To guarantee that X solves (3.6), it suffices to prove that $u_n(X_n(\cdot), \cdot) \to u(X(\cdot), \cdot)$ in $L^1(0, T)$. To this end, we write

$$|u_n(X_n(t), t) - u(X(t), t)|$$

$$\le |u_n(X_n(t), t) - u(X_n(t), t)| + |u(X_n(t), t) - u(X(t), t)|$$

$$\le \|u_n(t) - u(t)\|_{L^\infty} + |u(X_n(t), t) - u(X(t), t)|.$$

We already know that $u_n \to u$ in $L^1(0, T; L^\infty)$, so it only remains to show that

$$u(X_n(\cdot), \cdot) \to u(X(\cdot), \cdot) \quad \text{in} \quad L^1(0, T). \tag{3.8}$$

But for almost every $t > 0$, $u(\cdot, t) \in H^{(d/2)+1}$, and therefore $u(t)$ is continuous, from whence $u(X_n(t), t) \to u(X(t), t)$ for almost every t. Since we also have $|u(X_n(t), t)| \leq \|u(t)\|_\infty \leq c\|u\|_{H^{(d/2)-1}}^{1/2}\|u\|_{H^{(d/2)+1}}^{1/2}$ and the right-hand side of this inequality is integrable by (3.7), we obtain (3.8) using Lebesgue's Dominated Convergence Theorem.

We now prove uniqueness. To do this we want to fix $t > 0$ and let $s \to 0$ on the right-hand side of (3.3) in the hope of deducing that $W(t) = 0$. Note that this need only hold for all $t \leq t_0$ for some $t_0 > 0$, since $\sqrt{t}\,u \in L^2(0, T; H^{(d/2)+1})$ implies that $u \in L^1(t_0, T; H^{(d/2)+1})$ for any $t_0 > 0$, and so uniqueness for $t \in [t_0, T]$ is easy to obtain.

We therefore require

$$\lim_{s \to 0} \left[(-\log |W(s)|)^{1/2} - \int_s^t \|u(r)\|_{H^{(d/2)+1}} \, dr \right] = +\infty.$$

In order to prove this we find an upper bound on both $|W(s)|$ and $\int_s^t \|u(r)\|_{H^{(d/2)+1}} \, dr$.

For an upper bound on $|W(s)|$, we have

$$\frac{d}{dt}|W| \leq |u(X, t) - u(Y, t)| \leq 2\|u(t)\|_\infty,$$

and so integrating from 0 to s, using the fact that $|W(0)| = 0$,

$$|W(s)| \leq 2 \int_0^s \|u(t)\|_\infty \, dt.$$

It follows from (3.7) that $|W(s)| \leq Ks^{1/4}$, and hence

$$(-\log |W(s)|)^{1/2} \geq \left(-\log K - \frac{1}{4}\log s\right)^{1/2} \geq \alpha(-\log s)^{1/2},$$

for some appropriate fixed $\alpha > 0$ for all $s \leq s_0$ for some s_0 sufficiently small.

An upper bound on the integral follows easily, since

$$\int_s^t \|u(r)\|_{H^{(d/2)+1}} \, \mathrm{d}r = \int_s^t r^{-1/2} r^{1/2} \|u(r)\|_{H^{(d/2)+1}} \, \mathrm{d}r$$

$$\leq \left(\int_s^t r^{-1} \, \mathrm{d}r \right)^{1/2} \left(\int_s^t r \|u(r)\|^2_{H^{(d/2)+1}} \, \mathrm{d}r \right)^{1/2}$$

$$\leq (\log t - \log s)^{1/2} \left(\int_0^t r \|u(r)\|^2_{H^{(d/2)+1}} \, \mathrm{d}r \right)^{1/2} .$$

So for all $s \leq s_0$ the right-hand side of (3.3) is bounded below by

$$\alpha(-\log s)^{1/2} - (\log t - \log s)^{1/2} \left(\int_0^t r \|u(r)\|^2_{H^{(d/2)+1}} \, \mathrm{d}r \right)^{1/2} .$$

Since $r\|u(r)\|^2_{H^{(d/2)+1}}$ is integrable, one can choose t^* small enough that

$$\int_0^{t^*} r \|u(r)\|^2_{H^{(d/2)+1}} \, \mathrm{d}r < \frac{\alpha^2}{4}.$$

Fixing such a t^*, for any $t \leq t^*$ one can then take $s \to 0$ to deduce that $X(t) = 0$, and hence the solutions of $\dot{X} = u(X, t)$ are unique. $\square$

3.2.2 Continuity with respect initial data

We now prove another abstract result which, for the case of the Navier–Stokes equations, will guarantee continuity of the Lagrangian trajectories $X(t)$ with respect to the Eulerian initial data u_0.

Theorem 3.2.2 *Let Ω be as in Theorem 3.2.1. Suppose that $u_n \to u$ strongly in $L^2(0, T; H^{(d/2)-1}(\Omega))$ and that $\sqrt{t} u_n$ is uniformly bounded in $L^2(0, T; H^{(d/2)+1}(\Omega))$. For some $X_0 \in \Omega$ let $X_n(t)$ be the unique solution of*

$$\dot{X}_n = u_n(X_n, t) \qquad X_n(0) = X_0.$$

Then $X_n(t) \to X(t)$ uniformly on $[0, T]$, where $X(t)$ solves

$$\dot{X} = u(X, t) \qquad X(0) = X_0. \tag{3.9}$$

Proof Applying the argument of (3.7) to $u_n - u$ yields

$$\int_0^s \|u_n(t) - u(t)\|_\infty \, \mathrm{d}s$$

$$\leq cs^{1/4} \|u_n - u\|^{1/2}_{L^2(0, T; H^{(d/2)-1})} \|\sqrt{t}(u_n - u)\|^{1/2}_{L^2(0, T; H^{(d/2)+1})},$$

and so $u_n \to u$ strongly in $L^1(0,T;L^\infty)$. One can now show that a subsequence of the $X_n(t)$ converges to a solution $X(t)$ of (3.9), essentially by repeating the argument used to prove existence of solutions in Theorem 3.2.1: one simply notes that the uniform bounds on u_n and u in $L^2(0,T;H^{(d/2)-1})$ and on $\sqrt{t}u_n$ and $\sqrt{t}u$ in $L^2(0,T;H^{(d/2)+1})$ are sufficient to guarantee the equicontinuity of $X_n(\cdot)$ via (3.7).

Since $X(t)$ is a solution to (3.9), and this solution is unique, any convergent subsequence of the $\{X_n\}$ must have the same limit. A contradiction argument then guarantees that the original sequence $X_n(t)$ itself must converge to $X(t)$, and the proof is complete.   □

3.3 A priori estimates for the Navier–Stokes equations

We now verify that the conditions of Theorems 3.2.1 and 3.2.2 are satisfied by the solutions of the d-dimensional Navier–Stokes equations on periodic domains with initial conditions in $H^{(d/2)-1}$.

To be more precise, let $\Omega = [0,L]^d$, and denote by $\mathcal{V}$ the space of smooth (C^∞) divergence-free, periodic functions on Ω that have zero average ($\int_\Omega u = 0$). We denote by H the completion of $\mathcal{V}$ with respect to the L^2 norm on Ω, and by V the completion of $\mathcal{V}$ with respect to the H^1 norm on Ω.

We let Π denote the orthogonal projection of $L^2(\Omega)$ onto H, and define the Stokes operator $A = -\Pi\Delta$. In the periodic case, which we are considering here, $Au = -\Delta u$ for $u \in D(A) = H^2(\Omega) \cap H$, and $D(A^{m/2}) = H^m(\Omega) \cap H$. For simplicity we denote $D(A^{m/2})$ by V^m, and equip it with the natural norm

$$\|u\|_m = |A^{m/2}u|;$$

on V^m this is equivalent to the standard H^m Sobolev norm,

$$c_m^{-1}\|u\|_{H^m} \le \|u\|_m \le c_m\|u\|_{H^m} \qquad \text{for all} \qquad u \in V^m$$

(this is easy to see using the Fourier expansion of u, appealing to the fact that u has zero average). We identify V^0 with H.

Theorem 3.3.1 *If $u_0 \in V^{(d/2)-1}$ and $u(t)$ is the corresponding solution of the d-dimensional Navier–Stokes equations on $[0,T]$, then the solution $X(\cdot)$ of $\dot{X} = u(X,t)$ with $X(0) = X_0$ is unique. Furthermore, for each fixed $X_0 \in \Omega$ the map $u_0 \mapsto X(\cdot)$ is continuous from $H^{(d/2)-1}$ into $C^0([0,T],\mathbb{R}^d)$.*

As remarked in the introduction, a similar result is true for bounded domains (Dashti & Robinson, 2009). This requires a more careful analysis to obtain the *a priori* estimates we prove below for the periodic case, but is broadly along the lines of what follows. In particular the three-dimensional case requires some consideration of the fractional powers of the Laplacian (rather than the Stokes operator), and estimates for the term involving the pressure, which does not vanish in the case of bounded domains.

3.3.1 The two-dimensional case

In the two-dimensional case, it is well known (e.g. Constantin & Foias, 1988) that if $u_0 \in H$ then there exists a unique weak solution u that for each $T > 0$ satisfies $u \in L^\infty(0, T; L^2) \cap L^2(0, T; H^1)$, with

$$\frac{1}{2}\|u(t)\|^2 + \int_0^t \|Du(s)\|^2 \, ds = \frac{1}{2}\|u_0\|^2. \tag{3.10}$$

Given this, the additional estimate required to apply Theorem 3.2.1 is straightforward to obtain. Indeed, taking the inner product of (3.1) with $t\Delta u$ one can use the orthogonality property

$$\int (u \cdot \nabla u) \cdot \Delta u \, dx = 0$$

(only available for periodic boundary conditions in the two-dimensional case) to deduce that

$$\frac{1}{2}\frac{d}{dt}(t\|Du\|^2) - \frac{1}{2}\|Du\|^2 + \nu t\|\Delta u\|^2 = 0.$$

An integration yields

$$t\|Du(t)\|^2 + 2\nu \int_0^t s\|\Delta u(s)\|^2 \, ds \le \int_0^t \|Du(s)\|^2 \, ds; \tag{3.11}$$

since $\|\Delta u\| = \|u\|_2 \ge c_2^{-1}\|u\|_{H^2}$, it follows that $\sqrt{t}u \in L^2(0, T; H^2)$. Theorem 3.2.1 now guarantees uniqueness of the Lagrangian trajectories corresponding to weak solutions of the equations in two dimensions.

Theorem 3.2.2 requires uniform estimates for $\sqrt{t}u_n$ in $L^2(0, T; H^2)$ when $u_n(0) \to u_0$ strongly in L^2. These follow immediately from (3.10) and (3.11) since $u_n(0)$ must be uniformly bounded in L^2. The strong convergence of u_n to u in $L^2(0, T; L^2)$ follows – for some appropriate subsequence – from uniform bounds on u_n in $L^2(0, T; H^1)$ and on du_n/dt in $L^2(0, T; H^{-1})$ via the Aubin–Lions compactness theorem (see Theorem 8.1 in Robinson, 2001, for example).

3.3.2 The three-dimensional case

In the three-dimensional case we take $u_0 \in V^{1/2}$, and obtain a locally unique solution $u \in L^\infty(0, T; H^{1/2}) \cap L^2(0, T; H^{3/2})$ for some $T > 0$ (Kato & Fujita, 1962; Chemin et al., 2006).

Taking the inner product of (3.1) with $tA^{3/2}u$ yields

$$\frac{1}{2}\frac{d}{dt}\left(t\|u\|_{3/2}^2\right) - \frac{1}{2}\|u\|_{3/2}^2 + \nu t\|u\|_{5/2}^2 \leq t\|(u \cdot \nabla)u \cdot A^{3/2}u\|_{L^1}.$$

In the periodic case $A = -\Delta$ for $u \in D(A)$, and so we can integrate by parts on the right-hand side,

$$
\begin{aligned}
\|(u \cdot \nabla u) \cdot A^{3/2}u\|_{L^1} &\leq \||Du|^2|D^2u|\|_{L^1} + \||u||D^2u|^2\|_{L^1} \\
&\leq \|Du\|_{L^3}^2\|D^2u\|_{L^3} + \|u\|_{L^6}\|D^2u\|_{L^3}\|D^2u\|_{L^2} \\
&\leq \|u\|_{H^{3/2}}^2\|u\|_{H^{5/2}} + \|u\|_{H^{1/2}}^{1/2}\|u\|_{H^{3/2}}\|u\|_{H^{5/2}}^{3/2} \\
&\leq c\|u\|_{3/2}^4 + c\|u\|_{1/2}^2\|u\|_{3/2}^4 + \frac{\nu}{2}\|u\|_{5/2}^2.
\end{aligned}
$$

Thus one can obtain

$$\frac{d}{dt}\left(t\|u\|_{3/2}^2\right) - \|u\|_{3/2}^2 + \nu t\|u\|_{5/2}^2 \leq ct\|u\|_{3/2}^4.$$

Multiplying by $E(t) = \exp(-c\int_0^t \|u(s)\|_{3/2}^2\,ds)$ gives

$$\frac{d}{dt}\left(t\|u(t)\|_{3/2}^2 E(t)\right) + \nu t E(t)\|u\|_{5/2}^2 \leq E(t)\|u\|_{3/2}^2,$$

and integrating between 0 and t one obtains

$$t\|u(t)\|_{3/2}^2 E(t) + \nu \int_0^t sE(s)\|u(s)\|_{5/2}^2\,ds \leq \int_0^t \|u(s)\|_{3/2}^2\,ds.$$

It follows that

$$\nu \int_0^t s\|u(s)\|_{5/2}^2\,ds \leq \left(\int_0^t \|u(s)\|_{3/2}^2\,ds\right)\exp\left(c\int_0^t \|u(s)\|_{3/2}^2\,ds\right) < \infty.$$

Standard energy estimates show that u_n is uniformly bounded in $L^2(0, T; H^1)$, and du_n/dt is uniformly bounded in $L^{4/3}(0, T; H^{-1})$ (this is true even if $u_0 \in L^2$). It follows from the Aubin–Lions compactness theorem that $u_n \to u$ strongly in $L^2(0, T; H^s)$ for any $s < 1$; in particular $u_n \to u$ strongly in $L^2(0, T; H^{1/2})$.

3.4 Acknowledgements

MD was partially supported by a Postgraduate Scholarship from the University of Warwick, and by EPSRC Grant ER/F050798/1. JCR was partially supported by a Royal Society University Research Fellowship, the Leverhulme Trust, and an EPSRC Leadership Fellowship EP/G007470/1. We would both like to thank Andrew Stuart, who originally suggested the problem of uniqueness of Lagrangian trajectories for weak solutions of the two-dimensional equations, and Isabelle Gallagher who asked us about the dependence of these trajectories on the Navier–Stokes initial conditions.

References

Adams, R.A. (1975) *Sobolev spaces.* Academic Press, New York.

Chemin, J.Y., Desjardins, B., Gallagher, I., & Grenier, E. (2006) *Mathematical Geophysics: An introduction to rotating fluids and the Navier–Stokes equations.* Oxford University Press, Oxford.

Chemin, J.Y. & Lerner, N. (1995) Flot de champs de vecteurs non lipschitziens et équations de Navier–Stokes. *J. Differential Equations* **121**, 314–328.

Constantin, P. & Foias, C. (1988) *Navier–Stokes Equations.* University of Chicago Press, Chicago.

Dashti, M. & Robinson, J.C. (2009) A simple proof of uniqueness of the particle trajectories for solutions of the Navier–Stokes equations. *Nonlinearity* **22**, 735–746.

Foias, C., Guillopé, C., & Temam, R. (1985) Lagrangian representation of a flow. *J. Differential Eq.* **57**, 440–449.

Kato, T. & Fujita, H. (1962) On the nonstationary Navier–Stokes system. *Rend. Sem. Mat. Univ. Padova* **32**, 243–260.

Robinson, J.C. (2001) *Infinite-dimensional dynamical systems.* Cambridge Texts in Applied Mathematics, Cambridge University Press, Cambridge.

Zuazua, E. (2002) Log-Lipschitz regularity and uniqueness of the flow for a field in $(W_{\mathrm{loc}}^{n/p+1,p}(\mathbb{R}^n))^n$. *C. R. Math. Acad. Sci. Paris* **335**, 17–22.

4

Some controllability results in fluid mechanics

Enrique Fernández-Cara

Departamento de Ecuaciones Diferenciales y Análisis Numérico,
Facultad de Matemáticas, Universidad de Sevilla,
Apartado 1160, 41080 Sevilla. Spain.
cara@us.es

Abstract

This contribution is devoted to recalling several recent results concerning the controllability of some linear and nonlinear equations from fluid mechanics. More precisely, the local and global exact controllability to bounded trajectories will be analyzed.

4.1 Introduction: the null controllability problem

Let us first recall some general ideas that can be applied to a large family of linear and nonlinear evolution problems.

Suppose that we are considering an abstract *state equation* of the form

$$\begin{cases} y_t - A(y) = Bv, & t \in (0,T), \\ y(0) = y^0, \end{cases} \tag{4.1}$$

which governs the behaviour of a physical system (y_t denotes the time derivative of y). For simplicity, in the analysis of the system above we will restrict to the context of Hilbert spaces, and so we assume that the spaces U and H introduced below are Hilbert spaces. We make the following additional assumptions about (4.1):

- $A : D(A) \subset H \mapsto H$ is a (generally nonlinear) operator,
- $y : [0,T] \mapsto H$ is the *state*, i.e. the variable that serves to identify the physical properties of the system,
- $v : [0,T] \mapsto U$ is the *control*, i.e. the variable that we can choose,
- $B \in \mathcal{L}(U;H)$, and
- $y^0 \in H$.

Suppose that the state equation is well posed in the sense that, for any $y^0 \in H$ and $v \in L^2(0,T;U)$, it possesses exactly one solution. Then the *null controllability* problem for (4.1) can be stated as follows:

Published in *Partial Differential Equations and Fluid Mechanics*, edited by James C. Robinson and José L. Rodrigo. © Cambridge University Press 2009.

For each $y^0 \in H$, find $v \in L^2(0,T;U)$ such that the associated solution satisfies $y(T) = 0$.

For each system of the form (4.1), the null controllability problem leads to many interesting questions. Let us mention several of them:

- First, are there controls v such that $y(T) = 0$?
- Then, if this is the case, what is the *cost* we have to pay to drive y to zero? In other words, what is the minimal norm of a control $v \in L^2(0,T;U)$ satisfying this property?
- How can these controls be computed?

From the practical viewpoint, controllability results are crucial since, roughly speaking, they allow one to drive the solution to rest and, consequently, are associated with finite-time work.

The controllability of differential systems is an important area of research and has been the subject of many papers in recent years. In particular, in the context of partial differential equations, some relevant references for the null controllability problem are Russell (1973, 1978), Lions (1988a,b), Imanuvilov (1995), and Lebeau & Robbiano (1995). For semilinear systems of this kind, the first contributions have been given in Zuazua (1991), Fabre, Puel, & Zuazua (1995), and Fursikov & Imanuvilov (1996).

In these notes, I will try to recall some of the results that are known concerning the controllability of systems stemming from fluid mechanics.

4.2 The classical heat equation. Carleman estimates

In this section, in order to fix the main ideas, we will consider the controlled heat equation, complemented with initial and Dirichlet boundary conditions:

$$
\begin{cases}
y_t - \Delta y = v 1_\omega, & (x,t) \in \Omega \times (0,T), \\
y(x,t) = 0, & (x,t) \in \partial\Omega \times (0,T), \\
y(x,0) = y^0(x), & x \in \Omega.
\end{cases} \tag{4.2}
$$

Here (and also in the following sections), $\Omega \subset \mathbb{R}^N$ is a nonempty bounded domain, $\omega \subset\subset \Omega$ is a (small) nonempty open subset, 1_ω is the characteristic function of ω, and $y^0 \in L^2(\Omega)$.

It is well known that for every $y^0 \in L^2(\Omega)$ and every $v \in L^2(\omega \times (0,T))$, there exists a unique solution y of (4.2), with

$$y \in L^2(0,T; H_0^1(\Omega)) \cap C^0([0,T]; L^2(\Omega)).$$

In this context, the null controllability problem reads:

For each $y^0 \in L^2(\Omega)$, find $v \in L^2(\omega \times (0,T))$ such that the corresponding solution of (4.2) satisfies

$$y(x,T) = 0 \quad a.e. \ in \ \Omega. \tag{4.3}$$

Together with (4.2), for each $\varphi^1 \in L^2(\Omega)$ we can introduce the associated adjoint system

$$\begin{cases} -\varphi_t - \Delta\varphi = 0, & (x,t) \in \Omega \times (0,T), \\ \varphi(x,t) = 0, & (x,t) \in \partial\Omega \times (0,T), \\ \varphi(x,T) = \varphi^1(x), & x \in \Omega. \end{cases} \tag{4.4}$$

Then, it is well known, see for instance Coron (2007), that the null controllability of (4.2) is equivalent to the following property of (4.4):

There exists $C > 0$ such that

$$\|\varphi(\cdot,0)\|_{L^2}^2 \le C \iint_{\omega \times (0,T)} |\varphi|^2 \, dx \, dt \qquad \forall \, \varphi^1 \in L^2(\Omega). \tag{4.5}$$

This is called an *observability* estimate. We thus find that, in order to solve the null controllability problem for (4.2), it suffices to prove (4.5).

The estimate (4.5) is implied by the so-called global Carleman inequalities. These have been introduced in the context of the controllability of PDEs by Imanuvilov (1995) and Fursikov & Imanuvilov (1996). When they are applied to the solutions of the adjoint systems (4.4), they take the form[1]

$$\iint_{\Omega \times (0,T)} \rho^2 \, |\varphi|^2 \, dx \, dt \le K \iint_{\omega \times (0,T)} \rho^2 \, |\varphi|^2 \, dx \, dt \qquad \forall \, \varphi^1 \in L^2(\Omega), \tag{4.6}$$

where $\rho = \rho(x,t)$ is an appropriate weight, depending on Ω, ω and T and the constant K only depends on Ω and ω.

---

[1] In order to prove (4.6), as a first step we have to use a weight ρ decreasing to zero, as $t \to 0$ and also as $t \to T$, for instance exponentially. Then, the parabolicity of (4.4) makes it possible to choose ρ only decreasing to zero as $t \to 0$.

Combining (4.6) and the dissipativity properties of the backwards heat equation (4.4), it is not difficult to deduce (4.5) for some C only depending on Ω, ω and T.

As a consequence, we have:

Theorem 4.2.1 *The linear system (4.2) is null controllable. In other words, for each $y^0 \in L^2(\Omega)$, there exists $v \in L^2(\omega \times (0,T))$ such that the corresponding solution of (4.2) satisfies (4.3).*

There are many generalizations and variants of this result that provide the null controllability of other similar linear state equations:

- Time-space dependent (and sufficiently regular) coefficients can appear in the equation, other boundary conditions can be used, boundary control (instead of distributed control) can be imposed, etc. For a review of recent applications of Carleman inequalities to the controllability of parabolic systems, see Fernández-Cara & Guerrero (2006).
- The controllability of Stokes-like systems can also be analyzed with these techniques. This includes systems of the form

$$y_t - \Delta y + (a \cdot \nabla)y + (y \cdot \nabla)b + \nabla p = v 1_\omega, \quad \nabla \cdot y = 0,$$

where a and b are regular enough; see for instance Fernández-Cara et al. (2004).
- Other linear parabolic (non-scalar) systems can also be considered.

As mentioned above, an interesting question related to Theorem 4.2.1 concerns the cost of null controllability. One has the following result from Fernández-Cara & Zuazua (2000a):

Theorem 4.2.2 *For each $y^0 \in L^2(\Omega)$, let us set*

$$C(y^0) = \inf\{ \|v\|_{L^2(\omega \times (0,T))} : \text{solution of (4.2) with } y(x,T) = 0 \text{ in } \Omega \}.$$

Then we have the following estimate

$$C(y^0) \le e^{C\left(1 + \frac{1}{T}\right)} \|y^0\|_{L^2},$$

where the constant C only depends on Ω and ω.

Remark 4.2.3 Notice that Theorem 4.2.1 ensures the null controllability of (4.2) for any ω and T. Of course, this is related to the fact that, in a parabolic equation, the information is transmitted at infinite speed. For instance, this is not the case for the wave equation[2].

4.3 Positive and negative controllability results for the one-dimensional Burgers equation

Let us now consider the following system for the viscous Burgers equation:

$$\begin{cases} y_t - y_{xx} + yy_x = v1_\omega, & (x,t) \in (0,1) \times (0,T), \\ y(0,t) = y(1,t) = 0, & t \in (0,T), \\ y(x,0) = y^0(x), & x \in (0,1). \end{cases} \quad (4.7)$$

Some controllability properties of (4.7) have been studied by Fursikov & Imanuvilov (1996, Chapter 1, Theorems 6.3 and 6.4). It is shown there that, in general, a stationary solution of (4.7) with large L^2-norm cannot be reached (not even approximately) at any time T. In other words, with the help of one control, the solutions of the Burgers equation cannot be driven to an arbitrary state in finite time.

For each $y^0 \in L^2(0,1)$, let us introduce

$$T(y^0) = \inf\{\, T > 0 : (4.7) \text{ is null controllable at time } T \,\}.$$

Then, for each $r > 0$, let us define the quantity

$$T^*(r) = \sup\{\, T(y^0) : \|y^0\|_{L^2} \le r \,\}.$$

Let us show that $T^*(r) > 0$, with explicit sharp estimates from above and from below. In particular, this will imply that (global) null controllability at any positive time does not hold for (4.7). We have the following result from Fernández-Cara & Guerrero (2007b):

Theorem 4.3.1 *Let $\phi(r) = (\log \frac{1}{r})^{-1}$. We have*

$$C_0\phi(r) \le T^*(r) \le C_1\phi(r) \quad as \quad r \to 0, \quad (4.8)$$

for some positive constants C_0 and C_1 not depending on r.

---

[2] For the linear wave equation, null controllability is equivalent to exact controllability, but does not always hold. On the contrary, the couple (ω, T) has to satisfy appropriate geometrical assumptions; see Lions (1988b) and Bardos, Lebeau, & Rauch (1992) for more details.

Remark 4.3.2 The same estimates hold when the control v acts on the system (4.7) through the boundary *only* at $x = 1$ (or only at $x = 0$). Indeed, it is easy to transform the boundary controlled system

$$\begin{cases} y_t - y_{xx} + yy_x = 0, & (x,t) \in (0,1) \times (0,T), \\ y(0,t) = 0, \quad y(1,t) = w(t), & t \in (0,T), \\ y(x,0) = y^0(x), & x \in (0,1) \end{cases}$$

into a system of the kind (4.7). The boundary controllability of the Burgers equation with *two* controls (at $x = 0$ and $x = 1$) has been analyzed in Guerrero & Imanuvilov (2007); see also Coron (2007). There, it is shown that even in this more favourable situation null controllability does not hold for small time. It is also proved in that paper that exact controllability does not hold for large time[3].

Proof (Theorem 4.3.1) The proof of the estimate from above in (4.8) can be obtained by solving the null controllability problem for (4.7) via a (more or less) standard fixed point argument, using global Carleman inequalities to estimate the control and energy inequalities to estimate the state and being very careful with the role of T in these inequalities.

Let us give more details. We will denote by Q the domain $(0,1) \times (0,T)$. First we recall that, as long as $a \in L^\infty(Q)$, we can find controls v such that the solution of the linear system

$$\begin{cases} y_t - y_{xx} + a(x,t)y_x = v1_\omega, & (x,t) \in (0,1) \times (0,T), \\ y(0,t) = y(1,t) = 0, & t \in (0,T), \\ y(x,0) = y^0(x), & x \in (0,1) \end{cases}$$

satisfies

$$y(x,T) = 0 \text{ in } \Omega. \tag{4.9}$$

This is implied by the observability of the associated adjoint system

$$\begin{cases} -\varphi_t - \varphi_{xx} + a(x,t)\varphi_x = 0, & (x,t) \in (0,1) \times (0,T), \\ \varphi(0,t) = \varphi(1,t) = 0, & t \in (0,T), \\ \varphi(x,T) = \varphi^0(x), & x \in (0,1), \end{cases}$$

which, in turn, is implied by an appropriate Carleman estimate, see for example Fursikov & Imanuvilov (1996).

---

[3] However, the results in Guerrero & Imanuvilov (2007) do not allow one to estimate $T(r)$; in fact, the proofs are based on contradiction arguments.

Moreover, the controls v can be found in $L^\infty(\omega \times (0,T))$ and such that

$$\|v\|_\infty \le e^{C^*/T}\|y^0\|_{L^2(0,1)} \qquad (4.10)$$

for some $C^* = C^*(\omega, \|a\|_\infty)$; see Fernández-Cara & Zuazua (2000b).

Let us show that this provides a local controllability result for the nonlinear system (4.7):

Lemma 4.3.3 *Assume that*

$$y^0 \in H_0^1(0,1), \quad \|y^0\|_\infty \le \frac{1}{2} \quad and \quad \|y^0\|_{L^2(\Omega)} \le \frac{1}{2T}e^{-C_1^*/T}, \qquad (4.11)$$

where C_1^* corresponds to the constant C^* in (4.10) for $\|a\|_\infty = 1$. Then there exist controls $v \in L^\infty(\omega \times (0,T))$ such that the associated solutions to (4.7) satisfy (4.9).

Proof (Sketch) The proof of this lemma relies on well known arguments, but we present a sketch for the sake of completeness.

For s in $(\frac{1}{2}, 1)$ we define the set-valued mapping $\mathcal{A} : H^s(Q) \to H^s(Q)$ as follows: for each $z \in H^s(Q)$, we first denote by $\mathcal{A}_0(z)$ the set of all controls $v \in L^\infty(\omega \times (0,T))$ such that (4.10) is satisfied and the associated solution of

$$\begin{cases} y_t - y_{xx} + z(x,t)y_x = v1_\omega & (x,t) \in Q, \\ y(0,t) = y(1,t) = 0 & t \in (0,T), \qquad (4.12) \\ y(x,0) = y^0(x) & x \in (0,1) \end{cases}$$

fulfills (4.9); then, $\mathcal{A}(z)$ is by definition the family of these associated solutions.

Let $\mathcal{K}$ be the closed convex set $\mathcal{K} = \{z \in H^s(Q) : \|z\|_\infty \le 1\}$. Let us check that the hypotheses of Kakutani's Fixed-Point Theorem are satisfied by $\mathcal{A}$ in $\mathcal{K}$ (for the statement of this theorem, see for instance Aubin, 1984):

- First, we note that the solution of (4.12) belongs to the space

$$X := L^2(0,T; H^2(0,1)) \cap H^1(0,T; L^2(0,1)).$$

In particular, $y \in H^s(Q)$. Then, an application of the classical maximum principle yields

$$\|y\|_\infty \le T\|v\|_\infty + \|y^0\|_\infty.$$

Now, from (4.10) and (4.11), we deduce that

$$\|y\|_\infty \le 1.$$

Consequently, $\mathcal{A}$ maps $\mathcal{K}$ into $\mathcal{K}$.

• It is not difficult to prove that, for each $z \in \mathcal{K}$, $\mathcal{A}(z)$ is a nonempty compact convex subset of $H^s(Q)$, in view of the compactness of the embedding $X \hookrightarrow H^s(Q)$.

• Furthermore, $\mathcal{A}$ is upper hemicontinuous in $H^s(Q)$, that is for each $\mu \in (H^s(Q))'$, the single-valued mapping $z \mapsto \sup_{y \in \mathcal{A}(z)} \langle \mu, y \rangle$ is *upper semi-continuous*. Indeed, let us assume that $z_n \in K$ for all n and $z_n \to z_0$ in $H^s(Q)$. For each n, there exists $y_n \in \mathcal{A}(z_n)$ such that

$$\sup_{y \in \mathcal{A}(z_n)} \langle \mu, y \rangle = \langle \mu, y_n \rangle.$$

Then, from classical regularity estimates for the linear heat equation (see for instance Ladyzenskaya, Solonnikov, & Uraltzeva, 1967), we see that, at least for a subsequence, one has

$$y_n \to y^* \text{ weakly in } L^2(0, T; H^2(0, 1))$$

and

$$y_{n,t} \to y_t^* \text{ weakly in } L^2(Q),$$

whence

$$y_n \to y^* \text{ strongly in } L^2(0, T; H_0^1(0, 1)).$$

Consequently, $z_n y_{n,x}$ converges weakly in $L^2(Q)$ to $z_0 y_x^*$ and the limit function y^* satisfies (4.12) and $y^* \in \mathcal{A}(z_0)$. This shows that

$$\limsup_{n \to \infty} \sup_{y \in \mathcal{A}(z_n)} \langle \mu, y \rangle \leq \sup_{y \in \mathcal{A}(z_0)} \langle \mu, y \rangle,$$

as desired.

In view of Kakutani's Theorem, there exists $\hat{y} \in \mathcal{K}$ such that $\hat{y} \in \mathcal{A}(\hat{y})$. This ends the proof of Lemma 4.3.3. □

Let us now finish the proof of the right inequality in (4.8). Assume that $y^0 \in L^2(0, 1)$ and $\|y^0\|_{L^2(0,1)} \leq r$. In a first step, we take $v(x, t) \equiv 0$. Then, from classical parabolic regularity results, see for instance Ladyzenskaya et al. (1967), we know that $y(\cdot, t) \in H_0^1(0, 1)$ for any $t > 0$ and there exist constants τ and M such that the solution of (4.7) satisfies

$$\|y(\cdot, t)\|_\infty \leq M t^{-1/4} \|y^0\|_{L^2(0,1)} \quad \forall t \in (0, \tau).$$

We allow the solution to evolve freely, until it reaches a set of the form

$$\left\{ w \in L^\infty(0,1) : \|w\|_\infty \leq \frac{1}{2}, \ \|w\|_{L^2(0,1)} \leq r \right\}.$$

More precisely, we take $v(x,t) \equiv 0$ for $t \in (0,T_0)$, where $T_0 = (2M)^4 r^4$.
Let us set $\overline{y}^0 = y^0(\cdot, T_0)$. Then $\overline{y}^0 \in H_0^1(0,1)$,

$$\|\overline{y}^0\|_\infty \leq \frac{1}{2}, \quad \text{and} \quad \|\overline{y}^0\|_{L^2(0,1)} \leq r.$$

Let us now consider the system

$$\begin{cases} y_t - y_{xx} + yy_x = v1_\omega, & (x,t) \in (0,1) \times (T_0, T_0 + T_1), \\ y(0,t) = y(1,t) = 0, & t \in (T_0, T_0 + T_1), \\ y(x,T_0) = \overline{y}^0(x), & x \in (0,1), \end{cases} \quad (4.13)$$

where

$$T_1 = \frac{C_1^*}{\log \frac{1}{r}}.$$

Since one can assume that $\frac{1}{2C^*} \log \frac{1}{r} \geq 1$, in view of Lemma 4.3.3
there exist controls $\hat{v} \in L^\infty(\omega \times (T_0, T_0 + T_1))$ such that the associated
solutions of (4.13) satisfy

$$y(x, T_0 + T_1) = 0 \quad \text{in} \quad (0,1).$$

We then set $v(x,t) \equiv \hat{v}(x,t)$ for $t \in (T_0, T_0 + T_1)$. In this way, we have
shown that we can drive the solution of (4.7) exactly to zero in a time
interval of length

$$T_0 + T_1 = (2M)^4 r^4 + \frac{C_1^*}{\log \frac{1}{r}}.$$

Hence, the second inequality in (4.8) is proved.

Let us now give the proof of the estimate from below in (4.8). This is
inspired by the arguments in Anita & Tataru (2002).

We will prove that there exist positive constants C_0 and C_0' such that,
for any sufficiently small $r > 0$, we can find initial data y^0 satisfying
$\|y^0\|_{L^2} \leq r$ with the following property: for any state y associated to y^0,
one has

$$|y(x,t)| \geq C_0' r \quad \text{for some} \ \ x \in (0,1) \ \ \text{and any} \ \ t : 0 < t < C_0 \phi(r).$$

Thus, let us set $T = \phi(r)$ and let $\rho_0 \in (0,1)$ be such that $(0, \rho_0) \cap \omega = \emptyset$.
Notice that this is not restrictive, since it is always possible to work in
a suitable open subset $\widetilde{\omega} \subset \omega$.

We can suppose that $0 < r < \rho_0$. Let us choose $y^0 \in L^2(0,1)$ such that $y^0(x) = -r$ for all $x \in (0, \rho_0)$ and let us denote by y an associated solution of (4.7).

Let us introduce the function $Z = Z(x,t)$, with

$$Z(x,t) = \exp\left\{-\frac{2}{t}\left(1 - e^{-\rho_0^2(\rho_0-x)^3/(\rho_0/2-x)^2}\right) + \frac{1}{\rho_0 - x}\right\}.$$

Then one has $Z_t - Z_{xx} + ZZ_x \geq 0$.

Let us now set $w(x,t) = Z(x,t) - y(x,t)$. It is immediate that

$$\begin{cases} w_t - w_{xx} + ZZ_x - yy_x \geq 0, & (x,t) \in (0, \rho_0) \times (0,T), \\ w(0,t) \geq 0, \quad w(\rho_0, t) = +\infty, & t \in (0,T), \\ w(x,0) = r, & x \in (0, \rho_0). \end{cases} \quad (4.14)$$

Consequently, $w^-(x,t) \equiv 0$, where w^- denotes the negative part of w. Indeed, let us multiply the differential equation in (4.14) by $-w^-$ and let us integrate in $(0, \rho_0)$. Since w^- vanishes at $x = 0$ and $x = \rho_0$, after some manipulation we find that

$$\frac{1}{2}\frac{\mathrm{d}}{\mathrm{d}t}\int_0^{\rho_0} |w^-|^2\,\mathrm{d}x + \int_0^{\rho_0} |w_x^-|^2\,\mathrm{d}x$$
$$= \int_0^{\rho_0} w^-(ZZ_x - yy_x)\,\mathrm{d}x \leq C\int_0^{\rho_0} |w^-|^2\,\mathrm{d}x.$$

Hence,

$$y \leq Z \quad \text{in } (0, \rho_0) \times (0,T). \quad (4.15)$$

Let us set $\rho_1 = \rho_0/2$ and let $\tilde{r}$ be a regular function satisfying the following: $\tilde{r}(0) = \tilde{r}(\rho_1) = 0$; $\tilde{r}(x) = r$ for all $x \in (\delta\rho_1, (1-\delta)\rho_1)$ and some $\delta \in (0, 1/4)$; $-r \leq -\tilde{r}(x) \leq 0$; and

$$|\tilde{r}_x| \leq Cr \quad \text{and} \quad |\tilde{r}_{xx}| \leq C \quad \text{in } (0, \rho_1), \quad (4.16)$$

where $C = C(\rho_1)$ is independent of r.

Let us introduce the solution u of the auxiliary system

$$\begin{cases} u_t - u_{xx} + uu_x = 0, & (x,t) \in (0, \rho_1) \times (0,T), \\ u(0,t) = Z(\rho_1, t), \quad u(\rho_1, t) = Z(\rho_1, t), & t \in (0,T), \\ u(x,0) = -\tilde{r}(x), & x \in (0, \rho_1). \end{cases}$$

We will need the following lemma from Fernández-Cara & Guerrero (2007b), whose proof will be given below:

Lemma 4.3.4 *One has*

$$|u| \leq Cr \quad and \quad |u_x| \leq Cr^{1/2} \quad in \ (0, \rho_1) \times (0, \phi(r)), \tag{4.17}$$

where C is independent of r.

Taking into account (4.15) and that $u_x, y \in L^\infty((0, \rho_1) \times (0, T))$ (see Lemma 4.3.4 above), a standard application of Gronwall's Lemma shows that

$$y \leq u \quad in \ (0, \rho_1) \times (0, T).$$

On the other hand, we see from (4.17) that $u_t - u_{xx} \leq C^* r^{3/2}$ in $(0, \rho_1) \times (0, \phi(r))$ for some $C^* > 0$. Let us consider the functions p and q, given by $p(t) = C^* r^{3/2} t - r$ and

$$q(x, t) = c(e^{-(x - (\rho_1/4))^2/4t} + e^{-(x - 3(\rho_1/4))^2/4t}).$$

It is then clear that $b = u - p - q$ satisfies

$$b_t - b_{xx} \leq 0 \quad in \ (\rho_1/4, 3\rho_1/4) \times (0, \phi(r)),$$

$$b(\rho_1/4, t) \leq Z(\rho_1, t) - C^* r^{3/2} t + r - c(1 + e^{-\rho_1^2/(16t)}) \ \forall \ t \in (0, \phi(r)),$$

$$b(3\rho_1/4, t) \leq Z(\rho_1, t) - C^* r^{3/2} t + r - c(1 + e^{-\rho_1^2/(16t)}) \ \forall \ t \in (0, \phi(r)),$$

$$b(x, 0) = 0 \quad for \ x \in (\rho_1/4, 3\rho_1/4).$$

Obviously, in the definition of q the constant c can be chosen large enough to have $Z(\rho_1, t) - C^* r^{3/2} t + r - c(1 + e^{-\rho_1^2/(16t)}) < 0$ for any $t \in (0, \phi(r))$. If this is the case, we get $u \leq p + q$ and, in particular,

$$u(\rho_1/2, t) \leq (p + q)(\rho_1/2, t) = 2ce^{-\rho_1^2/(64t)} + C^* r^{3/2} t - r.$$

Therefore, we see that there exist C_0 and C_0' such that $u(\rho_1/2, t) < -C_0' r$ for any $t \in (0, C_0 \phi(r))$.

This proves (4.8) and, consequently, ends the proof of Theorem 4.3.1. $\qquad \square$

We now give the promised proof of Lemma 4.3.4.

Proof (Lemma 4.3.4) The first estimate in (4.17) can be obtained in a classical way, using arguments based on the maximum principle for the heat equation and the facts that

$$|\tilde{r}(x)| \leq r, \quad Z(\rho_1, t) \leq Cr^2, \quad and \quad Z_t(\rho_1, t) \leq Cr^2 \phi(r)^{-2}$$

for $x \in (0, \rho_1)$ and $t \in (0, C_0 \phi(r))$.

Let us explain how the second estimate in (4.17) can be deduced. Let us set $\widetilde{u}(x,t) = u(x,t) - Z(\rho_1, t)$. This function satisfies

$$\begin{cases} \widetilde{u}_t - \widetilde{u}_{xx} + (\widetilde{u} + Z(\rho_1, t))\widetilde{u}_x = -Z_t(\rho_1, t), & (x,t) \in (0, \rho_1) \times (0, C_0\phi(r)), \\ \widetilde{u}(0,t) = 0, \quad \widetilde{u}(\rho_1, t) = 0, & t \in (0, C_0\phi(r)), \\ \widetilde{u}(x,0) = -\widetilde{r}(x), & x \in (0, \rho_1). \end{cases}$$

- In a standard way, we can deduce energy estimates for $\widetilde{u}$:

$$\|\widetilde{u}\|_{L^\infty(0,C_0\phi(r);L^2(0,\rho_1))}^2 + \|\widetilde{u}_x\|_{L^2((0,\rho_1)\times(0,C_0\phi(r)))}^2$$

$$\leq C\|\widetilde{r}\|_{L^2(0,\rho_1)}^2 + C \int_0^{\rho_1} \int_0^{C_0\phi(r)} |\widetilde{u} \, Z_t(\rho_1, t)| \, dt \, dx.$$

Since $|\widetilde{u}| \leq Cr$, we obtain

$$\|\widetilde{u}\|_{L^\infty(0,C_0\phi(r);L^2(0,\rho_1))}^2 + \|\widetilde{u}_x\|_{L^2((0,\rho_1)\times(0,\phi(r)))}^2 \leq Cr^2. \qquad (4.18)$$

From the definition of $\widetilde{u}$, a similar estimate holds for u. Multiplying the equation for $\widetilde{u}$ by $\widetilde{u}_t$, we also get $\widetilde{u}_t \in L^2((0,\rho_1) \times (0, C_0\phi(r)))$, $\widetilde{u}_x \in C^0([0, C_0\phi(r)]; L^2(0, \rho_1))$ and

$$\|\widetilde{u}_t\|_{L^2((0,\rho_1)\times(0,C_0\phi(r)))}^2 + \|\widetilde{u}_x\|_{L^\infty(0,C_0\phi(r);L^2(0,\rho_1))}^2$$

$$\leq \frac{1}{2}\|\widetilde{u}_t\|_{L^2((0,\rho_1)\times(0,C_0\phi(r)))}^2 + C\left(\|(\widetilde{u} + Z(\rho_1,t))\widetilde{u}_x\|_{L^2((0,\rho_1)\times(0,C_0\phi(r)))}^2\right.$$

$$\left. + \|Z_t(\rho_1, \cdot)\|_{L^2(0,C_0\phi(r))}^2 + \|\widetilde{r}_x\|_{L^2(0,\rho_1)}^2\right).$$

Taking into account (4.16), (4.18) and the fact that $|\widetilde{u}| \leq Cr$, we deduce that

$$\|\widetilde{u}_t\|_{L^2((0,\rho_1)\times(0,C_0\phi(r)))}^2 + \|\widetilde{u}_x\|_{L^\infty(0,C_0\phi(r);L^2(0,\rho_1))}^2 \leq Cr^2. \qquad (4.19)$$

Obviously, this also holds for the norm of $\widetilde{u}_{xx}$ in $L^2((0,\rho_1) \times (0, C_0\phi(r)))$. Again, these estimates are satisfied by u.

- Next, by multiplying the equation for $\widetilde{u}$ by $-\widetilde{u}_{txx}$ and then integrating with respect to x in $(0, \rho_1)$, we have

$$\int_0^{\rho_1} |\widetilde{u}_{tx}|^2 \, dx + \frac{1}{2}\frac{d}{dt}\int_0^{\rho_1} |\widetilde{u}_{xx}|^2 \, dx$$

$$= \int_0^{\rho_1} \widetilde{u}_{txx}(\widetilde{u} + Z(\rho_1, t))\widetilde{u}_x \, dx + \int_0^{\rho_1} \widetilde{u}_{txx} Z_t(\rho_1, t) \, dx.$$

Integrating with respect to the time variable in $(0, t)$, we obtain the following after several integrations by parts:

$$\int_0^t \int_0^{\rho_1} |\tilde{u}_{sx}|^2 \, \mathrm{d}x \, \mathrm{d}s + \left(\int_0^{\rho_1} |\tilde{u}_{xx}|^2 \, \mathrm{d}x \right)(t) \leq \frac{1}{2} \left(\int_0^{\rho_1} |\tilde{u}_{xx}|^2 \, \mathrm{d}x \right)(t)$$

$$+ \frac{1}{2} \int_0^t \int_0^{\rho_1} |\tilde{u}_{sx}|^2 \, \mathrm{d}x \, \mathrm{d}s + C \left[\left(\int_0^{\rho_1} |\tilde{u} + Z(\rho_1, s)|^2 |\tilde{u}_x|^2 \, \mathrm{d}x \right)(t) \right.$$

$$+ \int_0^{\rho_1} \tilde{r} \tilde{r}_x \tilde{r}_{xx} \, \mathrm{d}x + \int_0^{\rho_1} |\tilde{r}_{xx}|^2 \, \mathrm{d}x + \int_0^t \int_0^{\rho_1} |\tilde{u}_{xx}|^2 |\tilde{u} + Z(\rho_1, s)|^2 \, \mathrm{d}x \, \mathrm{d}s$$

$$+ r^2 + \int_0^t \int_0^{\rho_1} ((|\tilde{u}_s|^2 + |Z_s(\rho_1, s)|^2) + |\tilde{u}_{xx}|^2) |\tilde{u}_x|^2 \, \mathrm{d}x \, \mathrm{d}s$$

$$+ |Z_t(\rho_1, t)|^2 + \int_0^t |Z_{ss}(\rho_1, s)|^2 \, \mathrm{d}s \right].$$

Using again that $|\tilde{u}| \leq Cr$ and (4.19), we deduce that

$$\|\tilde{u}_{tx}\|^2_{L^2((0,\rho_1)\times(0,C_0\phi(r)))} + \|\tilde{u}_{xx}\|^2_{L^\infty(0,C_0\phi(r);L^2(0,\rho_1))}$$
$$\leq C(r^4 + r^2 + 1 + r^4\phi(r)^{-4} + r^4\phi(r)^{-8}). \tag{4.20}$$

As a consequence, (4.20) implies that

$$\|\tilde{u}_{tx}\|^2_{L^2((0,\rho_1)\times(0,C_0\phi(r)))} + \|\tilde{u}_{xx}\|^2_{L^\infty(0,C_0\phi(r);L^2(0,\rho_1))} \leq C. \tag{4.21}$$

• Finally, in order to estimate $\tilde{u}_x$ in $L^\infty((0, \rho_1) \times (0, C_0\phi(r)))$, we observe that for each $t \in (0, C_0\phi(r))$ there exists $a(t) \in (0, \rho_1)$ such that $\tilde{u}_x(a(t), t) = 0$. Using this fact, we obtain:

$$|\tilde{u}_x(x, t)|^2 = \frac{1}{2} \int_{a(t)}^x \tilde{u}_x(\xi, t)\tilde{u}_{xx}(\xi, t) \, \mathrm{d}\xi.$$

Applying the estimates (4.19) and (4.21) to the functions $\tilde{u}_x$ and $\tilde{u}_{xx}$, which belong to $L^\infty(0, C_0\phi(r); L^2(0, \rho_1))$ and $L^\infty(0, C_0\phi(r); L^2(0, \rho_1))$ respectively, we deduce at once that

$$\|\tilde{u}_x\|^2_{L^\infty((0,\rho_1)\times(0,C_0\phi(r)))} \leq Cr,$$

which in particular implies the second estimate in (4.17).   □

4.4 Other more realistic nonlinear equations and systems

There are many more realistic nonlinear equations and systems from fluid mechanics that can also be considered in this context. First, we have the well known Navier–Stokes equations:

$$\begin{cases} y_t + (y \cdot \nabla)y - \Delta y + \nabla p = v1_\omega, & \nabla \cdot y = 0, & (x,t) \in Q, \\ y = 0, & & (x,t) \in \Sigma, \\ y(x,0) = y^0(x), & & x \in \Omega. \end{cases}$$

$$(4.22)$$

Here and below, Q and Σ respectively stand for the sets

$$Q = \Omega \times (0,T) \quad \text{and} \quad \Sigma = \partial\Omega \times (0,T),$$

where $\Omega \subset \mathbb{R}^N$ is a nonempty bounded domain, $N = 2$ or $N = 3$ and (again) $\omega \subset\subset \Omega$ is a nonempty open set.

To my knowledge, the best results concerning the controllability of this system have been given in Fernández-Cara et al. (2004) and Fernández-Cara et al. (2006)[4]. Essentially, these results establish the local exact controllability of the solutions of (4.22) to uncontrolled trajectories (this is, more or less, the analog of the *positive* controllability result in Theorem 4.3.1).

More precisely, in these references it is proved that, for any solution $(\hat{y}, \hat{p})$ of the Navier–Stokes system (with $v = 0$) with $\hat{y} \in L^\infty(Q)^N$, there exists $\varepsilon > 0$ such that, if $\|y^0 - \hat{y}(\cdot,0)\|_{H^1} \le \varepsilon$, then there exist controls v and associated solutions (y,p) of (4.22) with the following property:

$$y(x,T) = \hat{y}(x,T) \quad \text{in} \quad \Omega.$$

Similar results have been given in Guerrero (2006) for the Boussinesq equations

$$\begin{cases} y_t + (y \cdot \nabla)y - \Delta y + \nabla p = \theta k + v1_\omega, & \nabla \cdot y = 0, \\ \theta_t + y \cdot \nabla\theta - \Delta\theta = u1_\omega \end{cases}$$

complemented with initial and Dirichlet boundary conditions for y and θ (see Fernández-Cara et al. (2006) for a controllability result with a reduced number of scalar controls).

Let us also mention Barbu et al. (2003) and Havarneanu, Popa, & Sritharan (2006), where the controllability of the MHD and other related equations has been analyzed.

Another system is considered in Fernández-Cara & Guerrero (2007a):

$$\begin{cases} y_t + (y \cdot \nabla)y - \Delta y + \nabla p = \nabla \times w + v1_\omega, & \nabla \cdot y = 0, \\ w_t + (y \cdot \nabla)w - \Delta w - \nabla(\nabla \cdot w) = \nabla \times y + u1_\omega. \end{cases}$$

$$(4.23)$$

---

[4] The main ideas come from Fursikov & Imanuvilov (1999) and Imanuvilov (2001); some additional results have appeared recently in Guerrero, Imanuvilov, & Puel (2006) and González-Burgos, Guerrero, & Puel (2009).

Here, $N = 3$. These equations govern the behavior of a micropolar fluid, see Łukaszewicz (1999). As usual, y and p stand for the velocity field and pressure and w is the microscopic velocity of rotation of the fluid particles. Again, the local exact controllability of the solutions to the trajectories is established.

In this case, it is shown that, for any triplet $(\hat{y}, \hat{p}, \hat{w})$ solving (4.23) with $v = 0$ and $u = 0$ and satisfying $\hat{y}, \hat{w} \in L^\infty(Q)^N$, there exists $\varepsilon > 0$ such that, if $\|y^0 - \hat{y}(\cdot, 0)\|_{H^1} + \|w^0 - \hat{w}(\cdot, 0)\|_{H^1} \leq \varepsilon$, we can find controls v and associated solutions (y, p, w) of (4.23) with the following property:

$$y(x, T) = \hat{y}(x, T) \quad \text{and} \quad w(x, T) = \hat{w}(x, T) \quad \text{in} \quad \Omega.$$

Notice that this case involves a non-trivial difficulty. The main reason is that w is not a scalar variable and the equations satisfied by its components w_i are coupled through the second-order terms $\partial_i(\nabla \cdot w)$. This is a serious obstacle and an appropriate strategy has to be developed in order to deduce the required Carleman estimates.

For all these systems, the proof of the local controllability results can be achieved arguing as in the first part of the proof of Theorem 4.3.1. This is the general structure of the argument:

- First, consider a linearized similar problem and the associated adjoint system and rewrite the original controllability problem in terms of a fixed point equation.
- Then, prove a global Carleman inequality and an observability estimate for the adjoint system. This provides a controllability result for the linearized problem.
- Prove appropriate estimates for the control and the state (this needs some kind of *smallness* of the data); prove an appropriate compactness property of the state and deduce that there exists at least one fixed point.

There is an alternative method that relies on the Implicit Function Theorem, which corresponds to another strategy introduced by Fursikov & Imanuvilov (1996):

- First, rewrite the original controllability problem as a nonlinear equation in a space of admissible "state-control" pairs.
- Then, prove an appropriate global Carleman inequality and a regularity result and deduce that the linearized equation possesses at least one solution. Again, this provides a controllability result for a related linear problem.

- Check that the hypotheses of a suitable implicit function Theorem are satisfied and deduce a local result.

At present, no negative result is known to hold for these nonlinear systems (apart from the one-dimensional Burgers equation).

References

Anita, S. & Tataru, D. (2002) Null controllability for the dissipative semilinear heat equation. *Appl. Math. Optim.* **46**, 97–105.

Aubin, J.P. (1984) *L'Analyse non Linéaire et ses Motivations Économiques.* Masson, Paris.

Barbu, V., Havarneanu, T., Popa, C., & Sritharan, S.S. (2003) Exact controllability for the magnetohydrodynamic equations. *Comm. Pure Appl. Math.* **56**, 732–783.

Bardos, C., Lebeau, G., & Rauch, J. (1992) Sharp sufficient conditions for the observation, control and stabilization of waves from the boundary. *SIAM J. Cont. Optim.* **30**, 1024–1065.

Coron, J.M. (2007) *Control and nonlinearity.* Mathematical Surveys and Monographs, 136. American Mathematical Society, Providence.

Fabre, C., Puel, J.P., & Zuazua, E. (1995) Approximate controllability of the semilinear heat equation. *Proc. Royal Soc. Edinburgh* **125 A**, 31–61.

Fernández-Cara, E. & Guerrero, S. (2006) Global Carleman inequalities for parabolic systems and applications to controllability. *SIAM J. Control Optim.* **45**, no. 4, 1399–1446.

Fernández-Cara, E. & Guerrero, S. (2007a) Local exact controllability of micropolar fluids. *J. Math. Fluid Mech.* **9**, no. 3, 419–453.

Fernández-Cara, E. & Guerrero, S. (2007b) Null controllability of the Burgers system with distributed controls. *Systems & Control Letters* **56**, 366–372.

Fernández-Cara, E., Guerrero, S., Imanuvilov, O.Yu., & Puel, J.P. (2004) Local exact controllability to the trajectories of the Navier–Stokes equations. *J. Math. Pures Appl.* **83**, no. 12, 1501–1542.

Fernández-Cara, E., Guerrero, S., Imanuvilov, O.Yu., & Puel, J.P. (2006) Some controllability results for the N-dimensional Navier–Stokes and Boussinesq systems with $N-1$ scalar controls. *SIAM J. Control Optim.* **45**, no. 1, 146–173.

Fernández-Cara, E. & Zuazua, E. (2000a) The cost of approximate controllability for heat equations: The linear case. *Advances Diff. Eqs.* **5** no. (4–6), 465–514.

Fernández-Cara, E. & Zuazua, E. (2000b) Null and approximate controllability for weakly blowing up semilinear heat equations. *Ann. Inst. H. Poincaré Anal. Non Linéaire* **17**, no. 5, 583–616.

Fursikov, A.V. & Imanuvilov, O.Yu. (1996) *Controllability of Evolution Equations.* Lecture Notes 34, Seoul National University, Korea.

Fursikov, A.V. & Imanuvilov, O.Yu. (1999) Exact controllability of the Navier–Stokes and Boussinesq equations (Russian), *Uspekhi Mat. Nauk* **54**. no. 3 (327), 93–146; translation in *Russian Math. Surveys* **54**, no. 3, 565–618.

González-Burgos, M., Guerrero, S., & Puel, J.P. (2009) Local exact controllability to the trajectories of the Boussinesq system via a fictitious control on the divergence equation. *Commun. Pure Appl. Anal.* **8**, no.1, 311–333.

Guerrero, S. (2006) Local exact controllability to the trajectories of the Boussinesq system. *Annales IHP, Anal. non linéaire* **23**, 29–61.

Guerrero, S. & Imanuvilov, O. Yu. (2007) Remarks on global controllability for the Burgers equation with two control forces. *Annal. de l'Inst. Henri Poincaré, Analyse Non Linéaire* **24**, no. 6, 897–906.

Guerrero, S., Imanuvilov, O.Yu., & Puel, J.P. (2006) Remarks on global approximate controllability for the 2-D Navier–Stokes system with Dirichlet boundary conditions. *C. R. Math. Acad. Sci. Paris* **343**, no. 9, 573–577.

Havarneanu, T., Popa, C., & Sritharan, S.S. (2006) Exact internal controllability for the magnetohydrodynamic equations in multi-connected domains. *Adv. Differential Equations* **11**, no. 8, 893–929.

Imanuvilov, O.Yu. (1995) Boundary controllability of parabolic equations. *Russian Acad. Sci. Sb. Math.* **186**, 109–132 (in Russian).

Imanuvilov, O.Yu. (2001) Remarks on exact controllability for the Navier–Stokes equations. *ESAIM Control Optim. Calc. Var.* **6**, 39–72.

Ladyzenskaya, O.A., Solonnikov, V.A., & Uraltzeva, N.N. (1967) *Linear and Quasilinear Equations of Parabolic Type*. Trans. Math. Monograph, Moscow.

Lebeau, G. & Robbiano, L. (1995) Contrôle exact de l'équation de la chaleur. *Comm. P.D.E.* **20**, 335–356.

Lions, J.L. (1988) Exact controllability, stabilizability and perturbations for distributed systems. *SIAM Review* **30**, 1–68.

Lions, J.L. (1988) *Contrôlabilité Exacte, Stabilisation et Perturbations de Systèmes Distribués, Tomes 1 & 2*. Masson, RMA **8** & **9**, Paris.

Łukaszewicz, G. (1999) *Micropolar fluids. Theory and applications. Modeling and Simulation in Science, Engineering and Technology*. Birkhäuser Boston, Inc., Boston, MA.

Russell, D.L. (1973) A unified boundary controllability theory for hyperbolic and parabolic partial differential equations. *Studies in Appl. Math.* **52**, 189–221.

Russell, D.L. (1978) Controllability and stabilizability theory for linear partial differential equations. Recent progress and open questions. *SIAM Review* **20**, 639–739.

Zuazua, E. (1991) *Exact boundary controllability for the semilinear wave equation*, in Brezis, H. & Lions, J.L. (eds.) *Nonlinear Partial Differential Equations and their Applications*. Vol. **X**. Pitman, New York.

5

Singularity formation and separation phenomena in boundary layer theory

Francesco Gargano

Department of Mathematics, Via Archirafi 34, 90123 Palermo. Italy.
`gargano@math.unipa.it`

Maria Carmela Lombardo

Department of Mathematics, Via Archirafi 34, 90123 Palermo. Italy.
`lombardo@math.unipa.it`

Marco Sammartino

Department of Mathematics, Via Archirafi 34, 90123 Palermo. Italy.
`marco@math.unipa.it`

Vincenzo Sciacca

Department of Mathematics, Via Archirafi 34, 90123 Palermo. Italy.
`sciacca@math.unipa.it`

Abstract

In this paper we review some results concerning the behaviour of the incompressible Navier–Stokes solutions in the zero viscosity limit. Most of the emphasis is put on the phenomena occurring in the boundary layer created when the no-slip condition is imposed. Numerical simulations are used to explore the limits of the theory. We also consider the case of 2D vortex layers, i.e. flows with internal layers in the form of a rapid variation, across a curve, of the tangential velocity.

5.1 Introduction

The aim of this paper is to give a short review of some recent results in the study of the behaviour of a high-Reynolds-number fluid that has developed an internal scale due to the interaction with a physical boundary. Our starting point is the incompressible Navier–Stokes (NS) equations:

$$\partial_t \boldsymbol{u}_\nu + \boldsymbol{u}_\nu \cdot \boldsymbol{\nabla} \boldsymbol{u}_\nu + \boldsymbol{\nabla} p = \nu \Delta \boldsymbol{u}_\nu \qquad (5.1)$$

$$\boldsymbol{\nabla} \cdot \boldsymbol{u}_\nu = 0 \qquad (5.2)$$

$$\boldsymbol{u}_\nu(\boldsymbol{x}, t = 0) = \boldsymbol{u}_0(\boldsymbol{x}), \qquad (5.3)$$

Published in *Partial Differential Equations and Fluid Mechanics*, edited by James C. Robinson and José L. Rodrigo. © Cambridge University Press 2009.

where, with the coefficient ν, which we shall call viscosity, we denote the reciprocal of the Reynolds number Re.

Understanding the behaviour of a high Reynolds number flow is a fundamental and challenging problem which has received great attention from the research community. This is due to the relevance of this problem from the practical point of view and to the fact that the system (5.1)–(5.3) is one of the paradigms of a complex system. It is in fact well known that when the Reynolds number is increased above a critical value, internal flows and boundary layers undergo a remarkable transition from the laminar to the turbulent regime. The orderly pattern of laminar flow ceases to exist and the velocity and pressure fields exhibit very irregular, high-frequency fluctuations distributed over a large range of time and length scales. The mechanism responsible for the development of turbulence is still a matter of debate. A discussion of the various scenarios that have been proposed in the literature is beyond the scope of this paper. We shall restrict our attention to the question of whether, in the zero viscosity limit, the solutions of the NS system converge to the solutions of the Euler equations.

If the fluid does not interact with a physical boundary (i.e. when the fluid fills the whole $\mathbb{R}^n$ or for periodic domains) it is known that the answer is positive, see e.g. Swann (1971).

In this paper we shall focus on the case when the fluid interacts with a solid boundary. In this case the NS equations must be supplemented with the appropriate boundary condition, and most of our attention will be devoted to the no-slip boundary condition:

$$\boldsymbol{u}_\nu(\boldsymbol{x}, t) = 0 \qquad \text{for} \quad \boldsymbol{x} \in \partial\Omega\,, \tag{5.4}$$

where Ω is the region where the fluid is confined. The above boundary condition can be derived from a kinetic description based on the Boltzmann equation in the zero mean free-path limit. For example, suppose that the interaction of the molecules with the wall is described by the Maxwell kernel. Then, if a non negligible fraction of the molecules hitting the wall accommodates, i.e. is re-emitted with the Gaussian distribution of the wall, then in the zero mean free–path limit one derives (5.4). The Navier boundary condition (the main competitor of the no-slip BC) is derived when all the particles hitting the wall, except a fraction of the same order of magnitude as the mean free-path, are specularly reflected. The consensus is that (5.4) is the relevant BC for ordinary flows, while the Navier BC are a useful tool in the modelling of micro-fluids or of geo-fluids.

The reason why the analysis of a wall-bounded flow is challenging, is that this is a true singular perturbation problem. The fluid, close to the boundary, experiences a sharp transition to pass from rest, as prescribed by (5.4), to the free-stream non-viscous regime. Therefore a strong gradient in the direction normal to the wall appears and a large amount of vorticity is created. This is the first stage of a complicated process (not yet fully understood, as will be illustrated in more detail in Section 4) whose ultimate results are the injection of vorticity in the free-stream inviscid flow, the break-up of Prandtl's scenario and, probably, the failure of the convergence of the NS solutions to the Euler solutions (at least in the classical sense). An understanding of this process would be a major advance because it would shed light on the mechanisms of the onset of turbulence. From the mathematical point of view among the questions one would like to answer are the following. First, whether NS solutions converge to Prandtl's boundary layer solutions (and if so under which conditions and up to what time). If the answer to this question is positive this would imply that NS solutions converge to Euler solutions away from boundaries, thanks to Kato's criterion (Kato, 1984). Second, given that Prandtl's solutions in general develop a singularity, does this singularity have anything to do with the onset of turbulence, or is it a mere mathematical curiosity? Third, supposing that Prandtl's solution does not correctly describe the NS solutions close to the boundary, can one still hope to have the convergence of the NS solutions to the Euler solutions away from boundaries? And in this case, is it possible to describe the generation of vorticity at the boundary and the shedding of vorticity away from the boundary, without resorting to the full NS equations?

None of the above questions has so far received a fully satisfactory answer.

The plan of the paper is the following. In the next section we shall briefly review some of the most important results concerning the convergence of the Navier–Stokes solutions to the Euler solutions. We shall see that the theory is particularly unsatisfactory when the fluid interacts with a boundary at which no-slip is enforced. This leads to Section 3 where Prandtl's equations are derived and some of the available well-posedness theorems are reviewed. In Section 4 we report some results, based on numerics as well as on analysis, on the blow up of the solutions of Prandtl's equations. In Section 5 we focus on a planar inviscid flow with vorticity concentrated on a curve (vortex sheets). After a brief review of some of the results that have appeared in the last three decades, we consider the role of the viscosity and derive the vortex layer equations.

5.2 Convergence of Navier–Stokes solutions to Euler solutions

5.2.1 *Domains without boundaries*

In the absence of boundaries (periodic solutions or solutions decaying at infinity), the convergence of viscous planar flow to ideal planar flow with smooth initial data has been shown independently by McGrath (1968) and Golovkin (1966) with no restriction on the time interval of solution. The inviscid Eulerian dynamics is approached at a rate that is $O(\nu)$. Swann (1971) proved the existence of a unique classical solution to the Navier–Stokes equations in $\mathbb{R}^3$ for a small time interval independent of the viscosity and that, for vanishing viscosity, the solutions converge uniformly to a function that is solution to the Euler equations in $\mathbb{R}^3$.

Other results of this type (convergence at rate $O(\nu)$ for smooth enough initial data in 2D or 3D) can also be found in Marsden (1970), Kato (1972), and Beale & Majda (1981). Constantin & Wu (1995) considered the case of vortex patches (2D flows with vorticity supported on bounded sets) with smooth boundaries. They proved the following:

Theorem 5.2.1 (Constantin & Wu, 1995) *Consider the velocity difference*

$$w(x,t) = u_\nu(x,t) - u(x,t)$$

between a solution of the Navier–Stokes equation and a solution of the Euler equation on the plane. Assume that these have the same initial datum, corresponding to a vortex patch with smooth boundary. Then the difference $w(x,t)$ is square integrable and obeys the estimate:

$$\|w(\cdot,t)\|_{L^2}^2 \le (2\nu t)\|\omega_0\|_{L^2}^2 \exp\left(\int_0^t 2\|\nabla u(\cdot,\tau)\|_{L^\infty}\,d\tau\right).$$

Given that for vortex patches the quantity $\int_0^t 2\|\nabla u(\cdot,\tau)\|_{L^\infty}\,d\tau$ is globally bounded, convergence in L^2 at rate $(\nu t)^{1/2}$ follows. More recently this rate of convergence has been made sharp by Abidi & Danchin (2004) who proved that the optimal rate is $(\nu t)^{3/4}$.

Regarding the case of less regular data we mention the result of Chemin (1996), where an initial vorticity ω_0 in $L^2 \bigcap L^\infty$ is considered. In this case one still has convergence in L^2. However the rate is not $(\nu t)^{1/2}$ but it is $\exp\left(-Ct\|\omega_0\|_{L^2 \bigcap L^\infty}\right)/2$, thus degrading exponentially in time. See also Cozzi & Kelliher (2007).

5.2.2 Navier slip boundary conditions

The presence of boundaries complicates the situation significantly, since a boundary acts as a source of vorticity and it is precisely the lack of control on the production of vorticity that makes the problem difficult. This can be avoided by imposing the so-called free-boundary conditions: $\omega = 0$ and $u_\nu \cdot n = 0$ at the boundary, as was proposed by Lions (1969). However, in many situations, setting $\omega = 0$ at the boundary is not realistic.

A generalization of the free boundary condition is the Navier friction or slip boundary condition which still allows for vorticity production at the boundary but in a controlled way. Defining n and τ to be the normal and the tangent unit vector to the boundary $\partial\Omega$, and the rate-of-strain tensor $D(v) = \left(\nabla v + (\nabla v)^T\right)/2$, the Navier (or 'slip') boundary condition can be expressed in the form:

$$v \cdot n = 0 \qquad 2\left(n \cdot D(v)\right) \cdot \tau + \alpha v \cdot \tau = 0 \quad \text{on } \partial\Omega,$$

where $\alpha > 0$ is a fixed constant. It states that the tangential component of the viscous stress at the boundary has to be proportional to the tangential velocity. This boundary condition approximates the interaction of the flow with an infinitely rough boundary. In fact Jager & Mikelic (2001) derived, rigorously, the Navier friction boundary conditions as the limit of the no-slip boundary condition when a laminar flow interacts with a boundary whose profile oscillates. The limit is taken in the sense that the size and the amplitude of the oscillations go to zero.

The inviscid limit of the Navier–Stokes equations with Navier slip boundary conditions was studied in Clopeau, Mikelic, & Robert (1998) and Lopes Filho, Nussenzveig Lopes, & Planas (2005). These works show that the boundary layer arising from the inviscid limit can be controlled in 2D, thus proving the convergence to solutions of the Euler equations in $L^\infty([0,T]; L^2(\Omega))$. The rate of convergence with respect to the viscosity in the case of the Navier slip boundary conditions and bounded initial vorticity is studied by Kelliher (2006). The rate of convergence is essentially the same as that found by Chemin (1996) (in a different context) as mentioned in the previous section.

Finally we mention Chemetov & Antontsev (2008) where the vanishing viscosity limit with a permeable boundary is studied. In dimension three the problem with the Navier slip boundary condition has been studied by Iftimie & Planas (2006), where also the case of anisotropic viscosity is treated (see also Masmoudi, 1998, and Paicu, 2005).

More information on this line of research can be found in the nice review paper by Lopes Filho (2007).

5.2.3 No-slip boundary conditions: Kato's type criteria

Regarding the no-slip boundary condition, an important contribution to the mathematical understanding of the inviscid limit was given by Kato (1984). He proved a necessary and sufficient condition for the convergence of the weak solutions of the NS equations to the solution of Euler equations.

Theorem 5.2.2 (Kato, 1984) *Suppose that in a bounded domain Ω and on the time interval $[0, T]$, u is a smooth solution of the Euler system with initial velocity u_0. Let u_ν be a Leray–Hopf solution to the Navier–Stokes system with the same initial velocity. Then the following two conditions are equivalent:*

(i)

$$u_\nu \to u, \quad strongly \ in \quad L^\infty([0,T]; L^2(\Omega))$$

(ii)

$$\nu \int_0^T \|\nabla u_\nu(\cdot, t)\|_{L^2(\Gamma_{c\nu})}^2 \, dt \to 0 \qquad as \quad \nu \to 0.$$

where $\Gamma_{c\nu}$ denotes a boundary strip of width $c\nu$ with $c > 0$ fixed but arbitrary.

Kato's condition requires that, for the convergence to take place, the energy dissipation in a strip of width $c\nu$ close to the boundary must vanish as the viscosity tends to zero. The idea of the proof is to construct a time-dependent boundary layer velocity v that is different from zero only within a distance $O(\nu)$ from the boundary $\partial\Omega$ and equal to u on $\partial\Omega$ (thus v has no direct relation with the true boundary layer belonging to u_ν) and then use v to estimate the L^2 norm of $u_\nu - u$. The result is important as it places the condition for the convergence to hold in a sublayer $O(\nu)$, which is much smaller than the $O(\sqrt{\nu})$ boundary layer prescribed by Prandtl's theory.

There have been several results extending and improving Kato's criterion. Temam & Wang (1998) proved that Kato's condition can be replaced by a condition on the integrability of the pressure on the boundary. Their main result reads:

Theorem 5.2.3 (Temam & Wang, 1998) *Let (u_ν, p_ν) and (u, p) be the classical solutions of the Navier–Stokes equations and Euler equations respectively in the 2D strip $\Omega_\infty = \mathbb{R} \times (0, 1)$. Fix $T > 0$ and assume that there exist two constants κ and δ, not depending on ν, with $0 \le \delta \le 1/2$ such that*

$$\|\nabla p_\nu\|_{L^2([0,T];L^2(\partial\Omega))} \le \kappa \nu^{-\delta - \frac{1}{4}}.$$

Then there exists a constant c, not depending on ν such that:

$$\|u_\nu - u\|_{L^\infty([0,T];L^2(\Omega))} \le c \nu^{(1-2\delta)/15}.$$

Wang (2001) showed that, at the expense of a slight increase of the size of the boundary layer, one needs to consider only the tangential derivatives of the velocity. Namely he proved the following:

Theorem 5.2.4 (Wang, 2001) *Under the hypotheses of Theorem 5.2.2, conditions (i) and (ii) from that theorem are equivalent to the following:*

(iii) There exists $\delta(\nu)$ such that

$$\lim_{\nu \to 0} \frac{\nu}{\delta(\nu)} = 0$$

$$\lim_{\nu \to 0} \nu \int_0^T \int_{\Gamma_\delta} |\nabla_\tau u_{\nu\tau}|^2 = 0.$$

(iv) There exists $\delta(\nu)$ such that

$$\lim_{\nu \to 0} \frac{\nu}{\delta(\nu)} = 0$$

$$\lim_{\nu \to 0} \nu \int_0^T \int_{\Gamma_\delta} |\nabla_\tau u_{\nu n}|^2 = 0,$$

where ∇_τ denotes tangential (to the boundary) derivatives, $u_{\nu\tau}$ denotes the tangential components of the velocity, $u_{\nu n}$ denotes the normal components of the velocity, and Γ_δ is the $\delta(\nu)$ neighbourhood of the boundary $\partial\Omega$. One possible choice for $\delta(\nu)$ is $\nu \log \nu$.

The basic idea of the proof is to construct explicitly a corrector θ_ν (a background flow) supported in the ν/α neighbourhood of the boundary, where α is a free parameter which interpolates between the viscous sublayer used by Kato and the laminar boundary layer prescribed by Prandtl's theory. The explicit construction of the corrector and an upper bound on the energy dissipation rate independent of the viscosity then allow the estimate in $L^\infty([0,T]; L^2(\Omega))$ of the adjusted difference

$w = u_\nu - u - \theta_\nu$. One consequence of this result is that, in 2D, if the pressure gradient along the boundary does not grow too fast (less than $\nu^{-3/2}$), then convergence must occur. Thus, even in the presence of an adverse pressure gradient, when Prandtl's equations can develop a singularity, the viscous solution might converge to the Euler solution. This is a further indication of the fact that the problem of the inviscid limit in the presence of boundaries may not be related to the validity of the boundary layer equations. Moreover, another interesting consequence of Wang's work indicates the numerical difficulty in verifying the inviscid limit: essentially one has to resolve a small scale of the order $1/\nu$ in the direction parallel to the wall in order to be able to say anything about the inviscid limit (see also Cheng & Wang, 2007, for a discrete version of Kato's result).

Finally, leaving the size of Kato's boundary layer unchanged, Kelliher (2007) showed that the gradient of the velocity appearing in Kato's condition *(ii)* can be replaced by the vorticity.

However, all these results give no ultimate solution to the problem because of the unverified energy estimates on the Navier–Stokes solution.

5.2.4 Convergence with analytic initial data

When the flow satisfies the no–slip boundary conditions, one can prove the convergence of the Navier–Stokes solutions to the Euler solution under the assumption that the initial condition is analytic. This has been done for the half space $(x, y) \in \mathbb{R}^2 \times \mathbb{R}^+$ by Sammartino & Caflisch (1998b), following an earlier unpublished analysis of Asano (1988). Sammartino & Caflisch (1998b) also proved that, close to the boundary, the Navier–Stokes solutions converge to the solution of Prandtl's equations. Here we give only an informal statement of this result.

Theorem 5.2.5 (Sammartino & Caflisch, 1998b) *Suppose that the initial condition for the NS equations $u_0(x, y)$ is analytic with respect to x and y. Then, up to a time T_1 that does not depend on the viscosity, the solution of the Navier–Stokes equations has the following structure:*

$$u_\nu = u^P + \sqrt{\nu}\left(u_1^P + w\right) \qquad \text{close to the boundary}$$
$$u_\nu = u^E + \sqrt{\nu}\left(u_1^E + w\right) \qquad \text{away from the boundary}.$$

In the above theorem u^P denotes the solution of Prandtl's equations (which will be introduced in the next section and which, with analytic data, admit a unique solution, as proved by Sammartino & Caflisch,

1998a). With $\boldsymbol{u}_1^P$ and $\boldsymbol{u}_1^E$ we have denoted first order correctors to the Prandtl and Euler solutions. These correctors satisfy linear equations. Finally $\boldsymbol{w}$ is an overall correction term.

The meaning of "close to the boundary" or "away from the boundary" has to be understood in the following sense. The solutions $\boldsymbol{u}^P$ and $\boldsymbol{u}_1^P$ depend on the rescaled normal variable $Y = y/\sqrt{\nu}$. The first order corrector satisfies $\boldsymbol{u}_1^P(x, Y \to \infty) \to 0$, while $\boldsymbol{u}^P(x, Y \to \infty) \to \boldsymbol{u}^E(x, y = 0)$. In other words, in Sammartino & Caflisch (1998b), the solution of the Navier–Stokes equation is constructed as a matched asymptotic expansion involving the Euler (outer solution) and the Prandtl (inner solution) flows. The time of existence of this NS solution does not depend on the viscosity.

Remark 5.2.6 *The overall corrector $\boldsymbol{w}$ can be made $O(\nu^{N/2})$ for any N. This can be accomplished by considering the higher order terms in the matched asymptotic expansion (Van Dyke, 1975). In this case the NS solution would be of the form*

$$\boldsymbol{u}_\nu = \boldsymbol{u}^P + \sqrt{\nu}\boldsymbol{u}_1^P + \nu\boldsymbol{u}_2^P + \cdots + \nu^{N/2}\left(\boldsymbol{u}_N^P + \boldsymbol{w}\right)$$

close to the boundary, and

$$\boldsymbol{u}_\nu = \boldsymbol{u}^E + \sqrt{\nu}\boldsymbol{u}_1^E + \nu\boldsymbol{u}_2^E + \cdots + \nu^{N/2}\left(\boldsymbol{u}_N^E + \boldsymbol{w}\right)$$

away from the boundary.

Remark 5.2.7 *In Sammartino & Caflisch (1998b) the initial condition is more general than is considered in Theorem 5.2.5. In fact there $u_0(x, y)$ is also allowed to have a boundary-layer global corrector structure. In other words, one can allow an initial condition that, close to the boundary has an $O(1)$ dependence on the rescaled variable Y, and that away from the boundary has an $O(\sqrt{\nu})$ dependence on the rescaled variable Y. Denoting by u_0^{NS} the initial condition for the NS equations, one can allow the following structure:*

$$\boldsymbol{u}_0^{NS} = \boldsymbol{u}_0^P(x, Y) + \sqrt{\nu}\left(\boldsymbol{u}_{10}^P(x, Y) + \boldsymbol{w}_0(x, Y)\right)$$

close to the boundary, and

$$\boldsymbol{u}_0^{NS} = \boldsymbol{u}_0^E(x, y) + \sqrt{\nu}\left(\boldsymbol{u}_{10}^E(x, y) + \boldsymbol{w}_0(x, Y)\right)$$

away from the boundary, where the various terms in the two expansions must match smoothly.

As we mentioned, a byproduct of the above theorem is the existence of the NS solution (in a 3D domain with boundaries and with the no-slip boundary condition) for a time which does not depend on the viscosity. This result, for analytic data, can be obtained, as done by Lombardo (2001), directly without the analysis of the boundary layer structure of the solution. We now briefly sketch how this can be accomplished. First define the heat operator $\mathcal{E}(w)$ as the operator solving the heat equation in the half space with source term w and with zero boundary data and initial condition:

$$
\begin{aligned}
(\partial_t - \nu\Delta)\, u &= w \\
u(x, y, t = 0) &= 0 \\
u(x, y = 0, t) &= 0.
\end{aligned}
$$

Define the Stokes operator $\mathcal{S}^B(g)$ as the operator solving the Stokes equation in the half space with boundary data g and with zero source term and initial condition:

$$
\begin{aligned}
(\partial_t - \nu\Delta)\, \boldsymbol{u} + \boldsymbol{\nabla} p &= 0 \\
\boldsymbol{\nabla} \cdot \boldsymbol{u} &= 0 \\
\boldsymbol{u}(x, y, t = 0) &= 0 \\
\boldsymbol{u}(x, y = 0, t) &= g.
\end{aligned}
$$

One can define analogously the operator $\mathcal{S}^{IC}(\boldsymbol{u}_0)$ solving the Stokes equation in the half space with initial condition $\boldsymbol{u}_0$ and with zero source term and zero boundary data. Then one can define the Navier–Stokes operator as:

$$
\mathcal{N} = \mathcal{P}\mathcal{E} - \mathcal{S}\gamma\mathcal{P}\mathcal{E},
$$

where $\mathcal{P}$ is the Leray projection onto the divergence-free vector fields, which can be defined in such a way it commutes with the operator $(\partial_t - \nu\Delta)$, and where γ is the trace operator on the boundary $y = 0$. The solution of the Navier–Stokes equations (5.1)–(5.4), can be written as:

$$
\boldsymbol{u}_\nu = -\mathcal{N}\left(\boldsymbol{u}_\nu \cdot \boldsymbol{\nabla} \boldsymbol{u}_\nu\right) + \mathcal{S}^{IC}(\boldsymbol{u}_0). \tag{5.5}
$$

The difficulty in applying a fixed point theorem to the above expression lies in the fact that if one tries to use the regularizing properties of the heat kernel to compensate the derivatives, one gets estimates that are not uniform in the viscosity. If, on the other hand, the data are analytic, one can use the Cauchy estimate which, for a function f that is analytic in a strip of the complex plane of width ρ_0, can be written as:

$$|\partial_x f|_{\rho'} \leq \frac{|f|_\rho}{\rho - \rho'} \qquad \text{with} \quad \rho' < \rho < \rho_0.$$

Theorem 5.2.8 (Lombardo, 2001) *If $u_0(x, y)$ is analytic with respect to x in a strip of width ρ_0, and analytic with respect to y in an angular sector of the complex plane with angle θ_0, then there exists a $\beta > 0$ such that the Navier–Stokes equations (5.5) admit a unique solution $u_\nu(x, y, t)$ which is analytic with respect to x in a strip of width $\rho_0 - \beta t$, and analytic with respect to y in an angular sector of the complex plane with angle $\theta_0 - \beta t$.*

Clearly the solution exists up to a time $T_2 < \min(\rho_0/\beta, \theta_0/\beta)$. The quantity β is the speed at which the strip (or the angular sector) of analyticity shrinks with time.

The proof is based on the abstract Cauchy–Kowalewski Theorem (Safonov, 1995), which is a fixed point theorem on a scale of Banach spaces B_δ such that $B_{\delta'} \supset B_\delta$ when $\delta' < \delta$. In our case the index δ is the vector index $\delta = (\rho, \theta)$ expressing the width of the strip and of the angular sector of analyticity. Various estimates, uniform in ν, for the Navier–Stokes operator are also a key ingredients for the proof of Theorem 5.2.8.

Remark 5.2.9 *Theorem 5.2.5 ensures that up to a time T_1 the NS solution has a boundary layer structure (made up of analytic solutions of the Prandtl and Euler equations). Theorem 5.2.8 ensures the existence, up to a time T_2, of an analytic solution of the NS solution. Both T_1 and T_2 do not depend on the viscosity. Clearly $T_1 \leq T_2$.*

It would be interesting to know whether $T_1 = T_2$, i.e. if the break up of the boundary layer structure (for example because Prandtl's solution becomes singular) necessarily leads to the break up of the analytic solution of the NS solution.

Remark 5.2.10 *Grenier (2000) proved that there exist solutions of the NS equations that initially have a boundary layer structure and that, in zero time, violate the expansion. The building block for this construction is a shear layer profile that is unstable for the Euler equations.*

5.3 Prandtl's equations

The origin of the theory of boundary layers can be traced back to the beginning of the 20th century when Prandtl established the mathematical basis of flows for very large Reynolds number. The basic observation of Prandtl was that the Navier–Stokes equations are a singular perturbation of the Euler equations. This means that, in the zero viscosity limit, one derives the Euler equations because the terms with higher derivatives, i.e. the Laplacian in (5.1), are dropped. When the fluid interacts with a physical boundary, the change in the order of the equations is reflected in the boundary conditions to be imposed. In fact the Euler equations admit, as a boundary condition, only a condition on the velocity normal to the boundary, while the tangential velocity remains unknown and must be recovered through the solution of the equations. On the other hand, for the Navier–Stokes equations, one must impose the value of the velocity tangential to the boundary. Therefore the NS solution and the Euler solution take different values at the boundary. If one still wants to keep the hypothesis of the convergence of the NS solutions to the Euler solutions one must allow a layer, close to the boundary, where the NS solutions adjust to the Euler solution. In real wall-bounded laminar flows one observes that this adjustment is very rapid, and the higher the Reynolds number, the smaller the boundary layer is. There is also a *mathematical* reason why the thickness of the boundary layer should scale with some power of the Reynolds number. If there is a discrepancy between the Euler and the NS flows, this is necessarily due to viscous forces, i.e. to the Laplacian of the velocity times the viscosity. To have the viscous term $(\nu \Delta u_\nu)$ comparable with the inertial term $(u_\nu \cdot \nabla u_\nu)$, one must have the derivatives of u_ν to be of the order of some inverse power of the viscosity.

The simplest asymptotic hypothesis based on the experiments on high Reynolds number laminar flows (rapid variation close to the boundary), and on the above heuristic considerations, is that close to the boundary

$$\partial_y u_\nu = O(\nu^{-1/2}) \quad \text{and} \quad \partial_x u_\nu = O(1).$$

This implies that the boundary layer thickness (necessary for the adjustment of the NS solution to the inviscid regime) is $O(\nu^{1/2})$. Another key observation is that at the boundary the normal velocity is zero (both for the NS and the Euler flows). Therefore the normal velocity, inside the boundary layer, must be $O(\nu^{1/2})$.

All these observations can be implemented in a formal setting by introducing a rescaled normal variable $Y = y/\varepsilon$, with $\varepsilon = \sqrt{\nu}$, and assuming that close to the boundary,

$$\boldsymbol{u}_\nu = (u^P, \varepsilon v^P) \qquad \text{with} \quad \partial_Y u^P \sim \partial_Y v^P \sim \partial_x u^P \sim \partial_x v^P \sim O(1) \,.$$

In the above expressions u^P denotes the tangential components of the velocity, while ∂_x denotes the tangential gradient.

If one introduces the above asymptotic ansatz into the NS equations, at the leading order one derives the Prandtl equations:

$$\partial_t u^P + u^P \partial_x u^P + v^P \partial_Y u^P \quad = \quad \partial_t U + U \partial_x U + \partial_{YY} u^P \quad (5.6)$$

$$\partial_x u^P + \partial_Y v^P \quad = \quad 0 \qquad\qquad\qquad (5.7)$$

$$u^P|_{t=0} \quad = \quad u_0 \qquad\qquad\qquad (5.8)$$

$$u^P|_{Y=0} \quad = \quad 0 \qquad\qquad\qquad (5.9)$$

$$u^P \quad \longrightarrow \quad U \quad \text{as} \quad Y \to \infty \,. \quad (5.10)$$

In the above equations $U(x,t)$ denotes the Euler solution calculated at the boundary.

Notice that there is no evolution equation for the normal component v^P. In fact, the conservation of momentum in the normal direction, at the leading order $O(\varepsilon^{-1})$, simply gives $\partial_Y p^P = 0$, i.e. the pressure is constant inside the boundary layer. Imposing that this value matches with the pressure predicted by the Euler equations at the boundary, one gets $\partial_x p^P = -\partial_t U - U \partial_x U$. This identity has been used to write (5.6).

The normal velocity v^P can be recovered using (5.7), which can be written as

$$v^P = -\int_0^Y \partial_x u^P(x, Y', t) \, dY' \,. \qquad (5.11)$$

Equation (5.9) is the no–slip condition, while (5.10) is the matching condition with the outer flow. The above equations are written supposing that the domain in the tangential variable x is either periodic or the real line (the plane in 3D). If the domain is finite in the x direction (say $[0, L]$) then the above equation must be supplemented with some inflow boundary conditions.

Prandtl's equations, from the mathematical point of view, pose a difficult problem. There are several technical points that have, so far, prevented researchers from proving a well-posedness theorem under general hypotheses. We mention two of them. The lack of the streamwise viscosity, i.e. the term $\partial_{xx} u$, is one difficulty because one does not have a regularizing effect which can prevent the energy transfer to high Fourier modes. The second difficulty is given by the peculiar expression for the normal velocity (5.11), which involves the derivative of the stream-wise velocity (therefore with a loss of regularity), and an integration in the normal velocity. This integration, in general, leads to a linear growth in the normal direction Y.

In the sixties Oleinik (1967a,b) obtained several important existence results. These results are based on monotonicity hypotheses for the data: one has to assume that

$$U(x,t) > 0 \quad \text{and} \quad \partial_Y u_0(x,Y) > 0 \,,$$

together with the compatibility between initial and boundary data. The monotonicity allows one to use the Crocco transformation that recasts Prandtl's equations as a degenerate parabolic equation. For a review of the various results of Oleinik and coworkers, see the book she wrote together with Samokhin (Oleinik & Samokhin, 1999), and also the review papers by Caflisch & Sammartino (2000) and E (2000). Oleinik's results are for short time, or for long time when one considers Prandtl's equations for $x \in [0, L]$ with prescribed monotone inflow at $x = 0$ and for sufficiently small L. The requirement of small L was removed by Xin & Zhang (2004) where global weak solutions are constructed by imposing a favourable pressure gradient, i.e.

$$-\partial_x p = \partial_t U + U \partial_x U \geq 0 \qquad t > 0 \quad 0 < x < L \,.$$

This theory is particularly interesting because it corresponds to what one would expect on physical grounds. In fact a precursor of the sin-gularity is the formation of a back flow. Oleinik's hypotheses have the physical meaning that initially no back flow is present. The hypothesis of Xin & Zhang means that the external forcing (the outer Euler flow) does not create back flow.

A different point of view was adopted by Sammartino & Caflisch (1998a). No restriction was assumed on the sign of the data, which were however supposed to be analytic. They proved short time exis-tence and uniqueness of analytic solutions. This result was improved in Cannone, Lombardo, & Sammartino (2001) and Lombardo, Cannone, & Sammartino (2003), where the data were supposed to be analytic in the streamwise variable, but only C^2 in the normal variable. Lombardo et al. (2003) introduced the variable $\tilde{u} = u^P - U$ and wrote Prandtl's equations in the following form:

$$\left(\partial_t - \partial_{YY} + Y\partial_x U \partial_Y\right)\tilde{u} + 2\tilde{u}\partial_x\tilde{u} - \partial_Y\left(\tilde{u}\int_0^Y dY'\partial_x\tilde{u}\right)$$
$$+ U\partial_x\tilde{u} + \tilde{u}\partial_x U = 0.$$

An inversion of the parabolic operator $(\partial_t - \partial_{YY} + Y\partial_x U\,\partial_Y)$ through an operator E, which can be written explicitly in terms of the Gaussian, gives

$$\tilde{u} = E[\tilde{u}\partial_x\tilde{u}] + E\left[\partial_Y\left(\tilde{u}\int_0^Y dY'\partial_x\tilde{u}\right)\right] + E[\text{LT}] + \text{IC} + \text{BC}, \quad (5.12)$$

where by LT we denote the linear terms and by IC and BC the initial and boundary data.

Equation (5.12) is in a form suitable for application of the abstract Cauchy-Kowalewski Theorem (Safonov, 1995) in a time-integrated version. The term $\tilde{u}\partial_x\tilde{u}$ is bounded using the Cauchy estimate for the derivative of an analytic function. The term involving the normal derivative ∂_Y requires an integration by parts so that one can use the regularizing properties of the diffusion in the normal direction.

What is missing in the theory of Prandtl's equations is a well-posedness result for data that are not monotone and with a weaker restriction on the regularity than the analyticity assumption. Many evolution problems that are known to be ill-posed for Sobolev data because of the exponential growth of high Fourier modes (the typical example of this situation is the Kelvin–Helmholtz problem), turn out to be well posed for analytic data. In fact the exponential decay of the Fourier spectrum of an analytic datum can compensate, at least for a short time for the exponential growth of the higher modes.

The longstanding lack of a general well-posedness result has led to the conjecture that Prandtl's equations could be ill-posed, but no proof of this is available. The only result on this direction is numerical evidence of the ill-posedness in H^1 (Gargano, Sammartino, & Sciacca, 2007). However this is far from being a full answer to the question.

An important phenomenon that Prandtl's equations show is the appearance, in finite time, of a singularity, typically in the form of the blow up of the tangential derivative of the solution. This phenomenon, besides being intriguing from the mathematical point of view, is also interesting from a physical perspective. The formation of a singularity could in fact be related to the separation of the boundary layer, which is one of the most striking phenomena observed in wall-bounded flows, and which is one of the most important mechanisms of transition to turbulence.

The finite-time blow-up of solutions to Prandtl's equations is the subject of the next section.

5.4 Singularity and separation

The possibility of the spontaneous generation of a singularity in solutions of Prandtl's equations was a long standing conjecture. During the seventies many authors had tackled this problem from the numerical point of view, but the difficulties in the numerics did not allow for any definite conclusion.

It was in the seminal work of Van Dommelen & Shen (1980) that a new approach, the use of Lagrangian coordinates, finally provided strong numerical evidence that solutions of Prandtl's equations can blow up in a finite time.

E & Engquist (1997) finally gave a rigorous proof of this fact. The singularity construction is based on an initial datum with a stagnation line throughout the boundary layer and on recasting Prandtl's equation, along the stagnation line, in the form of a nonlinear heat equation for $\partial_x u$. E & Engquist were able to prove that the solution of this nonlinear heat equation blows up in finite time. The construction of E & Engquist is based on physical grounds, because it is known that the formation of back flow is the precursor of a possible singularity and that stagnation points are the preferred location for the blow up of solutions of Prandtl's equation.

5.4.1 Van Dommelen & Shen's singularity

All boundary layer calculations before Van Dommelen & Shen (1980) were carried out using mesh grids with fixed partial mesh in the conventional Eulerian description, see for example Collins & Dennis (1973), Walker (1978), and Brickman & Walker (2001). It was evident that numerical difficulties arose in these calculations, and it was impossible to resolve the problem because of the low computational resources available at the time. A loss of accuracy was evident mainly because of the development of large normal velocity and large streamwise gradients. The flow assumed an eruptive character and the computation could not be continued reliably.

The main idea of Van Dommelen & Shen was to treat the problem using a Lagrangian description, which turned out to be the right choice to follow the growth of the boundary layer.

Van Dommelen & Shen considered the case of the two-dimensional flow past an impulsively started disc in a uniform background flow. This means that they solved (5.6)–(5.10) in a domain periodic in $x \in [0, 2\pi]$, with $U = 2\sin(x)$ and with $u_0(x, Y) = 2\sin(x)$. The incompatibility between the initial datum and the boundary datum is immediately smoothed out by the normal viscosity which makes the flow regular in zero time.

The flow past an impulsively started disc shows several interesting physical phenomena such as the appearance of reverse flow, recirculation, separation, and vorticity shedding. In Van Dommelen & Shen (1980) the boundary layer terminal state was reached via an accurate numerical integration, revealing, at time $t = 1.5$, the formation of a singularity which starts the interaction of the viscous boundary layer with the inviscid outer flow. The singularity observed in the Lagrangian setting of Van Dommelen & Shen, consisted in the focusing of the boundary layer into a narrow region that forms in the upstream side of the recirculation zone; as a fluid particle is rapidly compressed in the streamwise location ($\nabla x = 0$), an eruption in the normal direction occurs, with a sharp spike forming in the displacement thickness.

The important result obtained by Van Dommelen & Shen (VDS) seemed to confirm the ideas reported by Sears & Telionis (1975) who proposed to define "unsteady separation" as the phenomenon that occurs when the velocity and the shear stress become singular. This is generally called the MRS (Moore, Rott, and Sears) model of separation, according to which the unsteady separation is connected with a breakdown of the boundary layer assumptions. Before this, the classical definition of unsteady separation was connected with the formation of reversed flow and the vanishing of wall shear. However, Sears & Telionis (1975) observed that the vanishing of the wall shear and the presence of reversed flow is not in itself sufficient to lead to unsteady separation; they quoted several examples of flows with vanishing wall shear and for which a breakaway is never expected to occur. An interesting and provocative review about boundary layer theory and the many numerical experiments exploring the problem of unsteady separation which followed Van Dommelen & Shen's work is given by Cowley (2001).

5.4.2 Tracking Van Dommelen & Shen's singularity

The singularity tracking method was initiated by Sulem, Sulem, & Frisch (1983) and the long-term goal of this research is the investigation of the possibility of a finite time singularity for the 3D Euler solutions. The idea behind the method is that a singularity does not come *out of the blue* (Frisch, Matsumoto, & Bec, 2003), but stays in the complex plane at some distance from the real axis. Its presence before the real blow up of the solution can be detected through a careful study of the asymptotic properties of the Fourier spectrum. When the singularity hits the real axis one has the real blow up of the solution. It is clear that if the issue is to know whether an equation develops a singularity, the method can *run out of steam* (Frisch et al., 2003) when the singularity is closer to the real axis than (let us say) three or four mesh sizes, particularly when the rate at which the singularity approaches the real axis assumes an exponential character. If instead there is a reasonable confidence that a particular equation develops a singularity, the tracking method has been revealed as a powerful tool to follow and characterize the whole process. Prandtl's equations are therefore the perfect candidate for the use of the singularity tracking method.

Suppose that an analytic function $f(z)$ has an algebraic singularity at $z^* = x^* + i\delta$, i.e. that close to z^* one has $f(z) \sim (z - z^*)^\alpha$. Then the Fourier coefficients have the following asymptotic expression (Carrier, Krook, & Pearson, 1966; Boyd, 2000):

$$f_k = Ck^{-(\alpha+1)}e^{-\delta k}e^{ikx^*}. \tag{5.13}$$

The singularity tracking technique consists of fitting the Fourier coefficients of $f(z)$ using the asymptotic formula (5.13) to determine the distance δ of the singularity that is closest to the real axis and the algebraic character α of this complex singularity.

It is possible to extend the technique of singularity tracking to two-dimensional functions. Let the analytic function $u(z, w)$ admit the following Fourier expansion:

$$u(z, w) = \sum_{h,k} a_{hk} \, e^{-ihz}e^{-ikw} \ .$$

One can consider the coefficients a_{hk} along the direction specified by $(h, k) = \kappa(\cos\theta, \sin\theta)$ in the space of Fourier modes. Denoting the rate of exponential decay in κ as $\delta(\theta)$, the width of the analyticity strip is

$$\delta = \min_\theta \delta(\theta)$$

(Matsumoto, Bec, & Frisch, 2005). The angular direction $\theta_\star$ along which the minimum is achieved is called the most singular direction.

We now discuss some results obtained applying this methodology to the case of the VDS singularity (Della Rocca et al., 2006; Gargano et al., 2007). Although the initial condition is not analytic in the normal direction one can show that the presence of the diffusion in the normal direction makes the solution analytic in Y when $t > 0$. In Figure 5.1 the angular dependence of δ is shown. One can see that the most singular direction is $\theta_\star = 0$, i.e. the direction parallel to the x variable. The

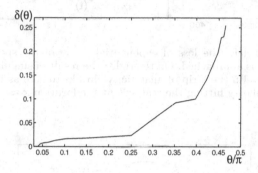

Fig. 5.1. The angular dependence of the rate of exponential decay of the Fourier coefficients a_{hk}. The most singular direction is $\theta_\star = 0$.

VDS singularity is therefore the result of a complex singularity hitting the real x-axis at $t = 1.5$; this is also confirmed by the study of the spectrum in the x variable of the solution $u(x, Y)$ keeping Y as parameter. The singularity occurs at $Y = 5$ (in agreement with the results of Van Dommelen & Shen) as a loss of exponential decay of the spectrum at time $t = 1.5$. The real location of the singularity $x^\star \approx 1.94$ also agrees with the findings of Van Dommelen & Shen. These results are shown in Figure 5.2.

Let us now discuss our result in terms of the physical phenomena occurring within the boundary layer that lead to the singularity formation. Vorticity is generated at the boundary because of the no-slip condition (5.9) imposed to the velocity at the wall. The solution exhibits a recirculation region with back-flow formation at approximately $t \approx 0.35$ in response to the adverse pressure gradient imposed by the outer flow. The formation of the recirculation region can be inferred, in this case, from the vanishing of the wall shear (see Figure 5.3), defined

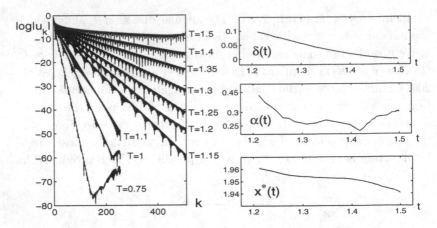

Fig. 5.2. On the left: the loss of exponential decay of the spectrum as the singularity time is approached. On the right: the result of singularity tracking at $Y = 5$. At $t \approx 1.5$ the strip of analyticity shrinks to zero as the result of a cubic-root singularity hitting the real axis at the location $x \approx 1.94$.

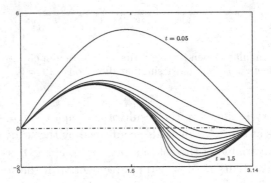

Fig. 5.3. The evolution of wall shear from $t = 0.05$ up to $t = 1.4$ with a time step increment of 0.15, and at time $t = 1.5$. At time $t = 0.35$ the wall shear has just vanished in correspondence with the formation of a recirculation region.

here as $\tau_w(x, t) = \frac{\partial u}{\partial Y}|_{Y=0}$. For a flow that, at the initial time, has an everywhere positive wall shear, the condition $\tau_w(x, t) = 0$ signals the onset of reversed flow within the boundary layer. The first point of zero wall shear appears at the rear stagnation point, and moves rapidly upstream along the cylinder surface, defining the upstream location of a growing recirculation region attached to the back of the cylinder.

As time passes, the recirculating eddy grows in both the streamwise and normal direction within the boundary layer, until a kink forms in the vorticity contours on the upstream side of the recirculation region at approximately $t \approx 1.35$ (see Figure 5.4). This is the precursor of the interaction of the boundary layer flow with the inviscid outer flow, which is revealed by the formation of a sharp spike as consequence of the singularity formation for the Prandtl solution at time $t = 1.5$.

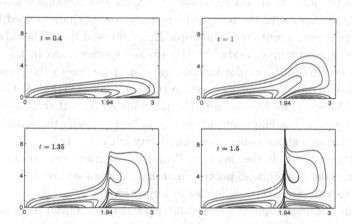

Fig. 5.4. Time evolution of the vorticity. At time $t = 0.4$ a little recirculation region attached to the wall has been formed. At $t \approx 1.35$ a kink is formed and evolves in a sharp spike at the singularity time when the vorticity shows an erupting behaviour.

5.4.3 Zero-time blow-up in Prandtl's equations

Gargano et al. (2007) consider the Van Dommelen & Shen solution at time $t = 0.75$, and perturb this solution with an analytic function having a dipole singularity in the complex plane at distance δ. The dipole singularity is constructed in such a way that this initial condition has norm bounded in H^1 by a constant K, no matter how close the singularity is to the real axis (i.e. how small δ is). The presence of this singularity speeds up the process of singularity formation. Moreover, it is shown that the singularity time seems to tend to zero when $\delta \to 0$.

This provides numerical evidence that Prandtl's equations are ill-posed in H^1.

5.4.4 A rigorous result: E & Engquist's singularity

All the above is based on numerical approximations of Prandtl's equations. As already remarked, after the breakthrough of Van Dommelen & Shen, the existence of a finite time singularity for the solutions of the Prandtl equations with physically relevant initial data is based on solid numerical evidence. The boundary-layer unsteady separation is caused by the presence of an adverse streamwise pressure gradient, which involves the formation of a recirculation region and eventually leads to an intense eruption of the near-wall vorticity, and is characterized by a sharp spike that erupts into the outer Euler inviscid flow. The adverse pressure gradient may be due to the surface geometry, as in the case of the flow around a circular cylinder or it may be due to the presence of a vortex convecting near a surface (Cassel, 2000; Obabko & Cassel, 2002). Moreover, we have seen in the previous sections that using complex singularity tracking methods it is possible to classify the separation singularity as a shock cubic–root singularity with $\alpha = 1/3$.

Another advance in the theory of Prandtl's equations was obtained by E & Engquist (1997), who proved rigorously that solutions of Prandtl's equations can develop singularities in finite time.

Let us consider Prandtl's equations with periodic initial data in the domain $D = [-\pi, \pi] \times [0, \infty)$ and with zero Euler forcing at infinity:

$$\partial_t u^P + u^P \partial_x u^P + v^P \partial_Y u^P = \partial_{YY} u^P \tag{5.14}$$

$$\partial_x u^P + \partial_Y v^P = 0 \tag{5.15}$$

$$u^P|_{t=0} = u_0 \tag{5.16}$$

$$u^P|_{Y=0} = v^P|_{Y=0} = 0 \tag{5.17}$$

$$u^P|_{x=-\pi} = u^P|_{x=\pi} \tag{5.18}$$

$$u^P \longrightarrow 0 \qquad \text{when} \quad Y \to \infty . \tag{5.19}$$

We impose an initial condition of the following form:

$$u_0 = -\sin(x) b_0(x, Y), \tag{5.20}$$

with $b_0 \geq 0$ a regular function of x and Y in D, that is zero at $Y = 0$ and decays to zero as $Y \to \infty$. It is clear that equation (5.14) preserves the odd symmetry of the initial condition, so that we may assume the solution to have the following form:

$$u^P(x, Y, t) = -\sin(x) b(x, Y, t), \tag{5.21}$$

$$v^P(x, Y, t) = \int_0^Y \left(\cos(x) b(x, y', t) + \sin(x) \partial_x b(x, y', t) \right) dy'.$$

Substituting the expressions (5.21) into (5.14), we can write

$$\partial_t b = \partial_{YY} b + \cos(x)\, b^2 + \sin(x)\, b\, \partial_x b - v^P\, \partial_Y b. \qquad (5.22)$$

If we evaluate equation (5.22) at $x = 0$, and define $a(Y,t) = b(0,Y,t)$, we get

$$\partial_t a = \partial_{YY} a + a^2 - (\partial_Y a)\int_0^Y a(y',t)\, dy', \qquad (5.23)$$

with the following boundary and initial conditions

$$a(0,t) = 0; \qquad \lim_{Y\to\infty} a(Y,t) = 0; \qquad a(Y,0) = a_0(Y) = b_0(0,Y). \qquad (5.24)$$

Equation (5.23) is a nonlinear heat equation with quadratic nonlinearity (which is well known to admit finite time blow-up), plus an integral term, which E & Engquist were able to prove does not prevent singularity formation.

Lemma 5.4.1 (E & Engquist, 1997) *Let the initial condition be* $a_0 \geq 0$, *and assume that*

$$E(a_0) = \int_0^\infty \left(\frac{1}{2}(\partial_Y a_0)^2 - \frac{1}{4}a_0^3\right) dY < 0. \qquad (5.25)$$

Then there exists a finite time T *such that*

$$either \quad \lim_{t\to T} \max_Y |a| = +\infty \quad or \quad \lim_{t\to T} \partial_Y a(0,t) = +\infty,$$

where $a(Y,t)$ *is the solution of equation (5.23) with boundary and initial conditions given in (5.24).*

This means that we have $\lim_{t\to T} \sup_{Y>0} \left|\frac{u(x,Y,t)}{\sin x}\right|_{x=0} = +\infty$ or

$$\lim_{t\to T} \sup_{Y>0} |\partial_x u(0,Y,t)| = +\infty.$$

One can therefore conclude the following:

Theorem 5.4.2 (E & Engquist, 1997, periodic version) *Assume that the initial condition for the periodic Prandtl equations (5.14)–(5.19) has the form (5.20) and suppose that it satisfies the condition (5.25), where* $a_0 = b|_{x=0}$. *Then there exists a finite time* T *such that the* x-*derivative of the solution blows up.*

In the next section we shall investigate E & Engquist's singularity numerically. What seems to emerge is the fact that the structure of this singularity is different from Van Dommelen & Shen's singularity (although the algebraic character as given by the exponent α is again $1/3$).

5.4.5 A numerical investigation of E & Engquist's singularity

We impose an initial condition of the form (5.20) with

$$b_0(x, Y) = a_0(Y) = \frac{1}{4}Y^2 e^{-(Y-2)}, \qquad (5.26)$$

which satisfies the condition (5.25).

As was done in the previous Section for the VDS initial condition, we shall solve Prandtl's equations (5.14)–(5.19) using a spectral method in the streamwise x–direction and a finite difference method in the normal Y-direction.

In Figure 5.5 we show the behaviour in time of the solutions for Prandtl's equations with initial condition given by (5.26). In this case the configuration corresponds to flow impinging from the left and right at the line $x = 0$, which is the rear stagnation point. At time $t \lesssim 1.20$ the

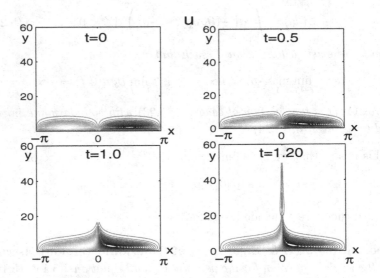

Fig. 5.5. The streamwise velocity u of Prandtl's equations with initial datum given by (5.26). One can follow the formation of the shock at $t \approx 1.20$.

Gibbs phenomenon can be observed for the velocity u, so we argue that at this time Prandtl's equations have already developed a singularity.

As the singularity forms, the singular structure seems to be convected at infinity in the normal direction, thus making the numerical computations difficult. The computational domain used is $[-\pi, \pi] \times [0, Y_{\mathrm{MAX}}]$, and we chose $Y_{\mathrm{MAX}} = 60$ where the velocity u is of the order of 10^{-18} at time $t \approx 1.20$.

On the left in Figure 5.6 we show the behaviour of $a(Y, t) = \partial_x u(0, Y, t)$. One can see that at time $t \approx 1.20$ the maximum grows around the location $Y \approx 20$. On the right in the same figure we show, at different times, the profile of u at the location $Y = 20$ where the maximum of a seems to be reached.

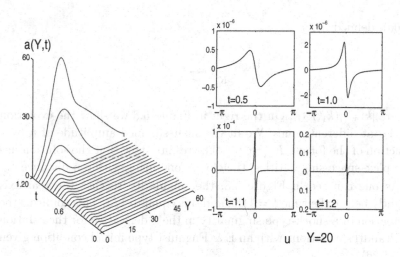

Fig. 5.6. On the right the behaviour of the function $a(Y, t) = \partial_x u(0, Y, t)$. On the left the profiles of u at the cut $Y = 20$ and at different times.

The structure of the complex singularity found by E & Engquist seems to be more complicated than the VDS singularity. If one investigates the angular dependence $\delta(\theta)$, reported in Figure 5.7, one sees that the most singular direction is $\theta_* \approx \pi/4$, which differs from the VDS case for which the most singular direction is parallel to the x-axis.

Hence we consider the full (both in x then in Y) spectrum $\widehat{u}_{k_1 k_2}$ of the velocity u and analyse the behaviour in time of the corresponding shell–summed Fourier amplitude (Matsumoto et al., 2005; Pauls et al.,

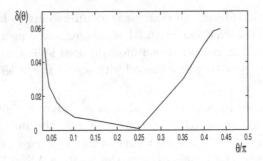

Fig. 5.7. The angular dependence of $\delta(\theta)$ at the singularity time $t = 1.20$. The distance δ seems to reach its minimum at $\theta \approx \pi/4$.

2006), defined by:

$$A_K \equiv \sum_{K \le |k| < K+1} \widehat{u}_k,$$

where $|k| = |(k_1, k_2)|$. On the right in Figure 5.8 we show the evolution of A_K at different times. We fit the shell–summed amplitude of u by a function of the form $CK^{-\alpha}e^{-\delta K}$. The results show that the solution has complex–space singularities, the closest one being within a distance δ.

As one can see in Figure 5.8, the singularity reaches the real axis slightly before the time $t = 1.20$, and α is equal to $1/3$. This confirms the formation of a shock-type singularity in the x derivative for the solution of Prandtl's equations with an E & Engquist type initial condition given by (5.26).

5.5 Vortex layers

In this section we shall consider the physically interesting case when the vorticity is highly concentrated. Vortex patches or vortex sheets are two examples of this situation. A plane vortex sheet is a curve across which the tangential component of the fluid velocity has a discontinuity, while the normal component is continuous.

Suppose that at time $t = 0$ in the plane (x, y) the flow has a tangential discontinuity across the curve $y = \phi_0(x)$ and that the flow outside the curve is irrotational. Then the initial vorticity ω_0 can be represented as a Dirac measure supported on the curve, see Marchioro & Pulvirenti

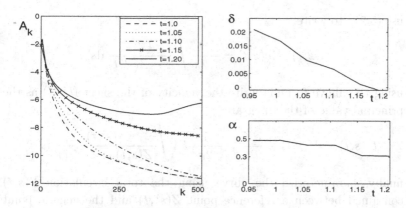

Fig. 5.8. The singularity tracking results for the solution of Prandtl's equation with initial datum given by (5.26). On the right is the time evolution of the shell–summed amplitude of the Fourier spectrum. On the left is the evolution in time of the analyticity strip δ and of the algebraic factor α.

(1994) or Saffman (1992):

$$\omega_0(x, y) = \gamma(x)\delta(y - \phi_0(x)),$$

where $\gamma(x)$ is the intensity of the tangential jump discontinuity at the point $(x, \phi(x))$. Supposing the fluid to be inviscid, the evolution of the vorticity is governed by

$$\partial_t \omega + \boldsymbol{u} \cdot \boldsymbol{\nabla}\omega = 0$$
$$\boldsymbol{u} = K \star \omega \qquad\qquad (5.27)$$
$$\omega(\boldsymbol{x}, t = 0) = \omega_0,$$

where the kernel K is given by $K(\boldsymbol{x}) = \boldsymbol{\nabla}^{\perp} \log|\boldsymbol{x}|/2\pi$.

If one makes the ansatz that, under the flow specified by the above equations, a vortex sheet remains a vortex sheet (i.e. that the vorticity will remain supported on a curve at a later time), one can write

$$\omega(x, y, t) = \gamma(x, t)\delta(y - \phi(x, t)),$$

and one is interested in finding $\phi(x, t)$. At this point the analysis proceeds as follows:

- characterize a point in the plane as a complex variable $z = x + iy$;
- denote by Z the points on the curve and parameterize the curve using the intrinsic arc length s:

$$x + i\phi(x, t) = Z(s, t);$$

- use (5.27) to write:

$$u(z,t) - \mathrm{i}v(z,t) = -\frac{\mathrm{i}}{2\pi} \int \frac{\gamma(s',t)}{z - Z(s',t)} \, \mathrm{d}s' \; ;$$

- using the above formula write the velocity of the sheet (U,V) as the principal value of the integral:

$$U(s,t) - \mathrm{i}V(s,t) \equiv \frac{\partial \bar{Z}}{\partial t} = -\frac{\mathrm{i}}{2\pi} \, PV\!\! \int \frac{\gamma(s',t)}{Z(s,t) - Z(s',t)} \, \mathrm{d}s' \; ;$$

- finally, parameterize the curve using the total circulation $\Gamma(s,t)$ contained between a reference point $Z(s^*,t)$ and the generic point $Z(s,t)$:

$$\frac{\partial \bar{Z}}{\partial t} = -\frac{\mathrm{i}}{2\pi} \, PV\!\! \int \frac{\mathrm{d}\Gamma'}{Z(\Gamma,t) - Z(\Gamma',t)} \; .$$

The above equation is the Birkhoff–Rott equation, which has been extensively studied from the mathematical point of view as well as from the numerical one.

The Birkhoff–Rott equation develops a singularity, which first manifests itself as an infinite curvature (a cusp) in the shape of the sheet. Moreover, the singularity time can be made short, thus making the Birkhoff–Rott equation an ill–posed problem. This was found using asymptotic methods by Moore (1979), verified numerically by Krasny (1986), and proved rigorously by Caflisch & Orellana (1986).

On the other hand it was proved by Sulem et al. (1981) that, if the initial condition has an exponentially decaying Fourier spectrum, then one has local-in-time existence and uniqueness of the solution. One can achieve long-time existence for small enough perturbations of the flat sheet (Caflisch & Orellana, 1986) or taking into account the regularizing effect of the surface tension (Hou, Lowengrub, & Shelley, 1997).

A different kind of regularization has been recently achieved in Bardos, Linshiz, & Titi (2008) using the 2D Euler-α equations instead of the Euler equations. The Euler-α equations are

$$\partial_t \omega + \boldsymbol{u} \cdot \boldsymbol{\nabla} \omega = 0$$
$$\boldsymbol{u} = K^\alpha \star \omega$$
$$\omega(\boldsymbol{x}, t = 0) = \omega_0,$$

where $K^\alpha = G^\alpha \star K$ with G^α a regularizing kernel that has symbol $(1 + \alpha^2 |k|^2)^{-1}$, and k is the dual Fourier variable of $\boldsymbol{x}$.

The fact that the Birkhoff–Rott–α model has a better behaviour than the usual Birkhoff–Rott equations can be understood by studying the linear stability of the flat sheet. In this case it is more convenient to start with the equations

$$\partial_t \phi = -u_1 \partial_x \phi + u_2$$
$$\partial_t \gamma = -\gamma \partial_x u_1 - u_1 \partial_x \gamma$$

(Marchioro & Pulvirenti, 1994). Denoting by $\widetilde{\phi}$, $(\widetilde{u}_1, \widetilde{u}_2)$, and $\widetilde{\gamma}$ the (small) deviations from a flat sheet with constant jump intensity γ_0, one derives the linear system

$$\partial_t \widetilde{\phi} = \widetilde{u}_2$$
$$\partial_t \widetilde{\gamma} = -\gamma_0 \partial_x \widetilde{u}_1 .$$

If one passes to the Fourier representation and uses the fact that the velocity is governed by the Euler–α equations, one can show that the above system admits a positive eigenvalue $\lambda(k)$,

$$\lambda(k) = \frac{1}{2}|\gamma_0||k| \left[1 - \left(\frac{\alpha^2 k^2}{1 + \alpha^2 k^2} \right)^{1/2} \right],$$

which however decays to zero like $1/(\alpha^2 k)$ as $k \to \infty$. This has to be contrasted with the usual Birkhoff–Rott behaviour (which can be recovered from the above equation when $\alpha = 0$) which gives a linear growth with k of the positive eigenvalue. It is this growth which, ultimately, is the origin of the ill–posed behaviour of the Birkhoff–Rott equations.

A different line of thought, which we shall not review here, is based on the analysis of the vortex sheet data in terms of measure solutions of the Euler equations. The reader can consult the papers by Delort (1991), Majda (1993), Evans & Muller (1994), Schochet (1996), and Lopes Filho, Nussenzveig Lopes, & Schochet (2007), and references therein.

5.5.1 Vortex layers

From the physical point of view, it is clear that the most natural kind of regularization to be considered is viscous regularization. If a small viscosity is present the vorticity would be spread out. Even initially if it is highly concentrated on a curve, after a small time τ the vorticity would stay on a layer of thickness $\sqrt{\nu\tau}$. A vortex sheet is therefore an approximation of a vortex layer and the main mathematical question regarding this model is to know if, when the viscosity is small, a vortex layer of thickness $O(\sqrt{\nu})$ would move as predicted by the Birkhoff–Rott equation (up to corrections that are small with the viscosity).

Moore (1978) considered the problem of the motion of a small layer of uniform vorticity. He identified the centre of the vortex layer with a moving curve $r = R(s,t)$ described parametrically with the arc length s. Denoting by $\widehat{s}$ and $\widehat{n}$ the unit tangent and normal vectors (as in Figure 5.9), one can write the position of a point r close to the curve as

$$r = R + n\widehat{n} \,.$$

The line element in the orthogonal frame $(\widehat{s}, \widehat{n})$ is

$$\mathrm{d}r = h_1 \widehat{s}\,\mathrm{d}s + \widehat{n}\,\mathrm{d}n \,,$$

where $h_1 = 1 - n/\rho$, with ρ the radius of curvature of the sheet, and where the Frenet–Serret formula have been used:

$$\frac{\partial \widehat{s}}{\partial s} = \frac{\widehat{n}}{\rho} \quad \text{and} \quad \frac{\partial \widehat{n}}{\partial s} = -\frac{\widehat{s}}{\rho} \,.$$

Denoting by $\partial/\partial\tau$ the time derivative in the comoving frame one can

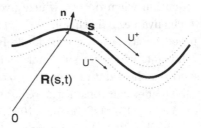

Fig. 5.9. Tangent and normal vectors near the centre $R(s,t)$ of the vortex layer.

also introduce the angular velocity Ω of the comoving frame $(\widehat{s}, \widehat{n})$ and write:

$$\frac{\partial \widehat{s}}{\partial \tau} = \Omega\widehat{n} \quad \text{and} \quad \frac{\partial \widehat{n}}{\partial \tau} = -\Omega\widehat{s} \,.$$

The fluid velocity of a particle close to the layer can be decomposed into a velocity relative to the centre of the sheet (which is $u\widehat{s} + v\widehat{n}$), and a part due to the motion of the sheet (which is $\partial_\tau \mathbf{R} + n\partial_\tau \hat{\mathbf{n}}$):

$$U = [u + \partial_\tau X - \Omega(Y + n)]\,\hat{\mathbf{s}} + [v + \partial_\tau Y + \Omega X]\,\hat{\mathbf{n}}\,. \qquad (5.28)$$

In the above equation, by X and Y we have denoted the components of $\mathbf{R}$ (the vector giving the centre of the sheet) in the intrinsic coordinate system, i.e. $\mathbf{R} = X\hat{\mathbf{s}} + Y\hat{\mathbf{n}}$.

Moore supposed that the vorticity is uniformly distributed in a small layer surrounding the curve of thickness $2H = O(\varepsilon)$, while it is zero outside this layer. Moreover he assumed that the radius of curvature is uniformly much larger than the thickness of the layer, specifically $H \leq \rho\varepsilon$. Introducing the rescaled normal variable

$$N = \frac{n}{\varepsilon}, \qquad (5.29)$$

and expanding the velocity field as:

$$u(s,n,t) = u_0(s,n,t) + \varepsilon u_1(s,n,t) + \varepsilon^2 u_2(s,n,t) + \dots \quad (5.30)$$
$$v(s,n,t) = \varepsilon v_0(s,n,t) + \varepsilon^2 v_1(s,n,t) + \dots, \qquad (5.31)$$

Moore performed an asymptotic matching procedure between the outer flow and the inner flow and obtained the following equation for the motion of the centre of the sheet:

$$\frac{\partial Z^*}{\partial t} = -\frac{\mathrm{i}}{2\pi}PV\!\int \frac{\mathrm{d}\Gamma'}{Z(\Gamma,t) - Z(\Gamma',t)} - \varepsilon\frac{\mathrm{i}}{6\omega}\frac{\partial}{\partial\Gamma}\left(\mathcal{G}^4\frac{\partial Z^*}{\partial\Gamma}\right) + O(\varepsilon^2)\,,$$
$$(5.32)$$

which contains an $O(\varepsilon)$ correction to the Birkhoff–Rott equation. In (5.32), $\mathcal{G}$ denotes the quantity $\gamma(s,t)$ expressed in terms of the variable Γ, and $\omega = O(1)$ denotes the rescaled constant vorticity of the layer.

Later, Benedetto & Pulvirenti (1992) proved rigorously, in an analytic function space, that the Euler dynamics of a layer of constant vorticity converges to the dynamics of the Birkhoff–Rott equation when the thickness of the layer goes to zero.

Clearly, assuming that vorticity is a constant inside the layer does not describe correctly the smoothing effects of the viscosity. Later Dhanak (1994a,b) derived an interesting generalization of Moore's equation. He took into account the role of the viscosity ν and assumed that the vorticity, not supposed uniform, decays exponentially fast away from the centre of the sheet,

For the motion of the sheet he derived the following equation, valid to $O(\varepsilon^2)$,

$$\frac{\partial Z^*}{\partial t} = -\frac{i}{2\pi} PV \int \frac{d\Gamma'}{Z(\Gamma,t) - Z(\Gamma',t)} - i\varepsilon \frac{\partial}{\partial \Gamma}\left(\delta_2 \mathcal{G}^3 \frac{\partial Z^*}{\partial \Gamma}\right) + \nu \mathcal{G}\frac{\partial \mathcal{G}}{\partial \Gamma}\frac{\partial Z^*}{\partial \Gamma},$$

(5.33)

where δ_2 is the momentum thickness of the layer:

$$\delta_2 = \int_{-\infty}^{\infty} \frac{(U^+ - u)(U^- - u)}{(U^+ - U^-)^2} \, dN,$$

with U^+ and U^- the free stream tangential velocities. One can recover Moore's equation (5.32) if the vorticity is constant and supported in a layer of thickness $2H = O(\varepsilon)$.

If one supposes that the thickness of the layer ε is related to the square root of the viscosity, as suggested by the diffusive scaling, then the third term in (5.33) is $O(\varepsilon^2)$ and should be consistently neglected.

Notice that, to get the momentum thickness of the layer, one has to specify the distribution of the velocity inside the layer. This is the problem we shall address in the following sections.

5.5.2 The vortex layer equations

If one inserts the expression (5.28) for the velocity U into the Navier–Stokes equations and writes the differential operators in the intrinsic coordinate frame comoving with the curve, one gets the following expressions for (u, v) which are the tangential and normal components of the relative velocity, see Caflisch & Sammartino (2006):

$$\partial_\tau u + \partial_{\tau\tau} X - 2\Omega(v + \partial_\tau Y) - \Omega^2 X - \partial_\tau \Omega(Y + n)$$
$$+ \frac{u}{h_1}\left[\partial_s u + \partial_s \partial_\tau X - \partial_s \Omega(Y + n) - \frac{1}{\rho}(v + \partial_\tau Y)\right] + v\partial_n u + \frac{\partial_s p}{h_1}$$
$$= \nu\,(\Delta U) \cdot \hat{s}$$

$$\partial_\tau v + \partial_{\tau\tau} Y + 2\Omega(u + \partial_\tau X) - \Omega^2(Y + n) + \partial_\tau \Omega X$$
$$+ \frac{u}{h_1}\left[\partial_s v + \partial_s \partial_\tau Y + \partial_s \Omega X + \frac{1}{\rho}(u + \partial_\tau X)\right] + v\partial_n v + \partial_n p$$
$$= \nu\,(\Delta U) \cdot \hat{n}$$

$$\frac{1}{h_1}\left\{\partial_s u + \partial_n\left[h_1 v\right]\right\} = 0.$$

In the above equations we have denoted by (ΔU) the Laplacian of U in the comoving frame; this is given by a complicated expression which we do not report here, but can be found in Caflisch & Sammartino (2006).

If one introduces the rescaled normal variable (5.29) and expands the velocity field as in (5.30)–(5.31), the leading order terms yield the following equations for u_0 and v_0:

$$\partial_\tau u + \partial_{\tau\tau} X - 2\Omega\partial_\tau Y - \Omega^2 X - \partial_\tau\Omega Y$$

$$+ u\left[\partial_s u + \partial_s\partial_\tau X - Y\partial_s\Omega - \frac{\partial_\tau Y}{\rho}\right] + v\partial_N u + \partial_s p^L = \partial_{NN} u \quad (5.34)$$

$$\partial_N p^L = 0 \quad (5.35)$$

$$\partial_s u + \partial_N v = 0 \quad (5.36)$$

$$u(s, N \to \pm\infty, t) \longrightarrow u^\pm(s, t) \quad (5.37)$$

$$u(s, N, t = 0) = u^{in}(s, N), \quad (5.38)$$

where we have renamed u_0 and v_0 as u and v to simplify the notation.

Equation (5.35) implies that the pressure is constant inside the vortex layer, which is consistent with the fact that the pressure is continuous across the layer. The fact that the pressure is continuous across the layer can be recovered through the same argument used to derive the continuity of the pressure across a vortex sheet, see pp. 28–29 in Saffman (1992). The value of the pressure must be recovered through a matching with the pressure of the outer flow. If one writes the Euler equation in the frame adapted to the centre of the layer, and calculates the tangential momentum equation at $n = 0$ (i.e. at the layer), one gets the following two expressions for the pressure gradient, the first taking the limit from above the sheet, the second from below:

$$-\partial_s p^L = \partial_\tau u^+ + \partial_{\tau\tau} X - 2\Omega\partial_\tau Y - \Omega^2 X - Y\partial_\tau\Omega$$
$$+ \left[\partial_s u^+ + \partial_s\partial_\tau X - Y\partial_s\Omega - \partial_\tau Y\rho^{-1}\right] u^+$$

$$-\partial_s p^L = \partial_\tau u^- + \partial_{\tau\tau} X - 2\Omega\partial_\tau Y - \Omega^2 X - Y\partial_\tau\Omega$$
$$+ \left[\partial_s u^- + \partial_s\partial_\tau X - Y\partial_s\Omega - \partial_\tau Y\rho^{-1}\right] u^- .$$

Here u^+ and u^- denote the matching values of the vortex layer tangential velocity with the outside Euler velocity, which are related to U^+ and U^- by the relations (see (5.28))

$$U^+ = u^+ + \partial_\tau X - \Omega Y , \qquad U^- = u^- + \partial_\tau X - \Omega Y .$$

Equation (5.36) is the incompressibility condition and allows one to recover the normal velocity from the tangential velocity through an integration. Equation (5.37) is the matching condition with the outer flow.

The similarities of the structure of the above equations with Prandtl's system are evident. In the next section we shall suppose that $X(s,t)$ and $Y(s,t)$ are given by the solution of the Birkhoff–Rott equation, and we shall briefly sketch a proof that the vortex layer equations are well posed.

5.5.3 Well posedness of the vortex layer equations

In this section we shall see that the vortex layer equations are well posed supposing that the data are analytic with respect to the tangential variable s. The proof is based on a version of the Cauchy–Kowalewski Theorem that allows a mild singularity in time, see Lombardo et al. (2003).

Define $H_{\delta,T}^m$ as the space of functions $f(s,t)$ that are analytic (together with the s derivatives up to order m) with respect to s on a strip of width δ and that are C^1 with respect to t in $[0,T]$.

Define $H_{\delta,\mu}^m$ as the space of functions $f(s,N)$ that are analytic (together with the s derivatives up to order m) with respect to s on a strip of width δ, that are C^2 with respect to Y and exponentially decaying to zero, at rate μ, for $N \to \pm\infty$.

The space $H_{\delta,\mu,T}^m$ is defined analogously.

Define the functions $\varphi^\pm(N)$ so that, at an exponential rate μ_0 one has $\varphi^\pm \to 1$ when $N \to \pm\infty$, and $\varphi^\pm \to 0$ when $N \to \mp\infty$. For example $\varphi^\pm = \exp(\pm\mu_0 N)/(\exp(\mu_0 N) + \exp(-\mu_0 N))$. Let us introduce the function $\varphi = u^+\varphi^+ + u^-\varphi^-$ and define the new variable $\widetilde{u}$:

$$\widetilde{u} = u - \varphi .$$

The $\widetilde{u}$ above is defined so that, if u satisfies (5.37) exponentially, then $\widetilde{u}$ decays to zero exponentially when $N \to \pm\infty$. We can now give a formal statement of the well posedness of the vortex layer equations:

Theorem 5.5.1 *Suppose that the free-streaming velocities u^+ and u^- are in $H_{\delta_0,T}^m$ and that the initial condition u^{in} is such that $u^{in} - \varphi$ is in H_{δ_0,μ_0}^m. Then, there exists $\beta > 0$ such that there exists a unique solution u of the vortex layer equations (5.34)–(5.38) with $u \in H_{\delta_0-\beta t,\mu_0-\beta t,T}^m$.*

We now briefly sketch the proof. Written in terms of the new variable $\tilde{u}$, the vortex layer equations read

$$\left[\partial_t - \partial_{NN} - (\Phi^+ \partial_s u^+ + \Phi^- \partial_s u^-)\partial_N\right] \tilde{u} = K_1[\tilde{u}] + \partial_N K_2[\tilde{u}] + L[\tilde{u}] + \mathcal{S} \tag{5.39}$$

with

$$\tilde{u}(s, N, t = 0) = u^{in} - \varphi,$$

where

$$
\begin{aligned}
\Phi^\pm(N) &= \int_0^N \varphi^\pm(N')\,\mathrm{d}N', \\
K_1[\tilde{u}] &= -2\tilde{u}\partial_s\tilde{u}, \\
K_2[\tilde{u}] &= \tilde{u}\int_0^N \partial_s\tilde{u}(s, N', t)\,\mathrm{d}N', \\
L[\tilde{u}] &= -\tilde{u}\partial_s\varphi, \quad \text{and} \\
\mathcal{S} &= -\varphi^+(\varphi^+ - 1)u^+\partial_s u^+ - \varphi^-(\varphi^- - 1)u^-\partial_s u^- \\
&\quad -(u^+\partial_s u^- + u^-\partial_s u^+)\varphi^+\varphi^- + u^+\partial_{NN}\varphi^+ + u^-\partial_{NN}\varphi^-.
\end{aligned}
$$

Notice that Φ^+ grows to infinity linearly fast when $N \to \infty$, and that Φ^- grows to infinity linearly fast when $N \to -\infty$. The term $L[\tilde{u}]$ is linear in $\tilde{u}$, while the source term $\mathcal{S}$ decays to zero exponentially fast when $N \to \pm\infty$.

The next step is to rewrite (5.39) in a form appropriate for the use of the Cauchy–Kowalewski Theorem.

To accomplish this we introduce some operators that, roughly speaking, invert the parabolic operator $[\partial_t - \partial_{NN} - (\Phi^+ \partial_s u^+ + \Phi^- \partial_s u^-)\partial_N]$.

The operator M_0 solves a homogeneous parabolic equation with non-zero initial conditions,

$$
\begin{aligned}
\left[\partial_t - \partial_{NN} - (\Phi^+ \partial_s u^+ + \Phi^- \partial_s u^-)\partial_N\right] M_0 u_0 &= 0, \\
M_0 u_0(s, N, t = 0) &= u_0,
\end{aligned}
$$

while the operator M_2 solves the same parabolic equation with a source term but zero initial condition,

$$
\begin{aligned}
\left[\partial_t - \partial_{NN} - (\Phi^+ \partial_s u^+ + \Phi^- \partial_s u^-)\partial_N\right] M_2 f &= f, \\
M_2 f(s, N, t = 0) &= 0.
\end{aligned}
$$

The operator M_3 has the property that $M_3 f = M_2 \partial_N f$.

With the use of these operators, one can recast (5.39) as

$$\tilde{u} = G(\tilde{u}, t),$$

where

$$G[\widetilde{u}, t] = M_2 \left(K_1[\widetilde{u}] + L[\widetilde{u}] + \mathcal{S} \right) + M_3 K_2[\widetilde{u}] + M_0 \left(u_0 - \varphi \right).$$

Using the Cauchy–Kowalewski Theorem, the Cauchy estimate on the derivative of an analytic function, and some estimates on the operators M_i, the proof of the well posedness is relatively straightforward.

5.6 Concluding remarks

In this paper we have reviewed some relevant results, both numerical and theoretical, regarding the well posedness of the boundary layer equations in relation to the zero-viscosity limit of the Navier–Stokes equations. In this respect a well-established mathematical framework is still lacking, the main open problems being the following:

- In presence of a boundary, both in two and three spatial dimensions, the convergence of the solutions of the Navier–Stokes equations to the solutions of the Euler equations for smooth initial data is still an open question. The problem is even more striking on plane domains, where regular global solutions of both the Navier–Stokes and the Euler equations are known to exist. If the data are analytic, then convergence occurs.

  The study of the inviscid limit may also be complicated, even in the absence of boundary effects, in the case of non-smooth initial data such as vortex patches or sheets. In this case the Birkhoff–Rott equation, which governs the evolution of the vortex sheet, is an ill-posed problem. In the case of a vortex layer of thickness $O(\sqrt{\nu})$ the equations of motion are known but their well-posedness has been proved only in the space of analytic functions.

- Prandtl's equations, which govern the behaviour of the fluid in a $O(\sqrt{\nu})$ neighbourhood of the boundary, are conjectured to be generally ill-posed in Sobolev spaces. In fact the existing well-posedness theorems require either monotone or analytic (with respect to the tangential variable) data. Related to the well-posedness of Prandtl's equations is the important mathematical problem of the finite time blow-up of its solutions: in fact the only known theoretical result in this direction holds for initial data with strong recirculation.

Careful numerical studies are required to investigate the above questions. We have presented some recent numerical results that confirm the development of a zero-time singularity formation (in H^1) for the boundary-layer equations.

We conclude this paper with a few words about the bibliography. The reference list, although of some length (as one would expect for a subject in which many different approaches have contributed to our current understanding), is far from being exhaustive. It should be considered only a guide for the interested reader who will find, in the cited papers and in the references therein, material for further investigation.

Acknowledgements

M. Sammartino acknowledges several interesting discussions on the vortex layer equations with Xinyu He. It is also a pleasure to thank James Robinson and José Rodrigo for the organization of a fantastic workshop.

This work has been supported by the INDAM and by the PRIN grant *Propagazione non lineare e stabilità nei processi termodinamici del continuo*.

References

Abidi, H. & Danchin, R. (2004) Optimal bounds for the inviscid limit of Navier–Stokes equations. *Asymptotic Analysis* **38**, no. 1, 35–46.

Asano, K. (1988) Zero viscosity limit of the incompressible Navier–Stokes equations *I* and *II*. Preprint.

Bardos, C., Linshiz, J.S., & Titi, E.S. (2008) Global regularity for a Birkhoff-Rott-α approximation of the dynamics of vortex sheets of the 2D Euler equations, *Physica D* **237**, 1905–1911.

Beale, J.T. & Majda, A. (1981) Rates of convergence for viscous splitting of the Navier–Stokes equations. *Math. Comp.* **37**, no. 156, 243–259.

Benedetto, D. & Pulvirenti, M. (1992) From vortex layers to vortex sheets. *SIAM J. Appl. Math.* **52**, 1041–1056.

Boyd, J.P. (2000) *Chebyshev and Fourier Spectral Methods*. Dover Publications.

Brinckman, K.W. & Walker, J.D.A. (2001) Instability in a viscous flow driven by streamwise vortices. *J. Fluid Mech.* **432**, 127–166.

Caflisch, R.E. & Orellana, O.F. (1986) Long time existence for a slightly perturbed vortex sheet. *Comm. Pure Appl. Math.* **39**, 807–838.

Caflisch, R.E. & Sammartino, M. (2000) Existence and singularities for the Prandtl boundary layer equations. *Z. Angew. Math. Mech.* **80**, 733-744.

Caflisch, R.E. & Sammartino, M. (2006) Vortex layers in the small viscosity limit, in Monaco, R., Mulone, G., Rionero, S., & Ruggeri T. (eds.) *Proceedings WASCOM 2005*. World Sci. Publ., Hackensack.

Cannone, M., Lombardo, M.C., & Sammartino, M. (2001) Existence and uniqueness for the Prandtl equations. *C. R. Acad. Sci. Paris Sér. I Math.* **332**, 277–282.

Carrier, F.G., Krook, M., & Pearson, C.E. (1966) *Functions of a complex variable. Theory and technique*. McGraw Hill.

Cassel, K.W. (2000) A comparison of Navier–Stokes solutions with the theoretical description of unsteady separation. *Phil. Trans. R. Soc. Lond. A* **358**, 3207–3227.

Chemetov, N.V. & Antontsev, S.N. (2008) Euler equations with non homogeneous Navier slip boundary conditions. *Physica D* **237**, 92–105.

Chemin, J.Y. (1996) A remark on the inviscid limit for two dimensional incompressible fluids. *Comm. Partial Differential Equations* **21**, 1771–1779.

Cheng, W. & Wang, X. (2007) Discrete Kato-type theorem on inviscid limit of Navier–Stokes flows. *J. Math. Phys.* **48**, 065303.

Clopeau, T., Mikelic, A., & Robert, R. (1998) On the vanishing viscosity limit for the 2D incompressible Navier–Stokes equations with the friction type boundary conditions. *Nonlinearity* **11**, 1625–1636.

Collins, W.M. & Dennis, S.C.R. (1973) Flow past an impulsively started circular cylinder. *J. Fluid Mech.* **60**, 105–127.

Constantin, P. & Wu, J. (1995) Inviscid limit for vortex patches. *Nonlinearity* **8**, 735–742.

Cowley, S.J. (2001) Laminar boundary-layer theory: a 20th century paradox? in Aref, H. & Phillips, J.W. (eds.) *Mechanics for a New Millennium, Proceedings of 20th ICTAM*. Springer, New York.

Cozzi, E. & Kelliher, J.P. (2007) Vanishing viscosity in the plane for vorticity in borderline spaces of Besov type. *J. Differential Equations* **235**, 647–657.

Della Rocca, G., Lombardo, M.C., Sammartino, M., & Sciacca, V. (2006) Singularity tracking for Camassa-Holm and Prandtl's equations. *Appl. Num. Math.* **56**, 1108–1122.

Delort, J.M. (1991) Existence de nappes de tourbillon en dimension deux. *J. Amer. Math. Soc.* **4**, 553–586.

Dhanak, M.R. (1994a) Equation of motion of a diffusing vortex sheet. *J. Fluid Mech.* **269**, 265–281.

Dhanak, M.R. (1994b) On the equation of motion of a thin layer of uniform vorticity. *Stud. Appl. Math.* **92**, 115–125.

E, W. (2000) Boundary layer theory and the zero-viscosity limit of the Navier–Stokes equation. *Acta Math. Sin.* **16**, no. 2, 207–218.

E, W. & Engquist, B. (1997) Blowup of solutions of the unsteady Prandtl's equation. *Comm. Pure Appl. Math.* **50**, no. 12, 1287–1293.

Evans, L.C. & Muller, S. (1994) Hardy spaces and the two-dimensional Euler equations with nonnegative vorticity. *J. Amer. Math. Soc.* **7**, 199–219.

Frisch, U., Matsumoto, T., & Bec, J. (2003) Singularities of Euler flow? Not out of the blue! *J. Stat. Phys.* **113**, 761–781.

Gargano, F., Sammartino, M., & Sciacca, V. (2007) Singularity formation for Prandtl's equations, submitted.

Golovkin, K.K. (1966) Vanishing viscosity in cauchy's problem for hydromechanics. *Trudy Mat. Inst. Steklov.* **92**, 31–49.

Grenier, E. (2000) On the nonlinear instability of Euler and Prandtl equations. *Comm. Pure Appl. Math.* **53**, no. 9, 1067–1091.

Hou, T.Y., Lowengrub, J.S., & Shelley, M.J. (1997) The long-time motion of vortex sheets with surface tension. *Phys. Fluids* **9**, 1933–1954.

Iftimie, D. & Planas, G. (2006) Inviscid limits for the Navier–Stokes equations with Navier friction boundary conditions. *Nonlinearity* **19**, 899–918.

Jager, W. & Mikelic, A. (2001) On the roughness induced effective boundary conditions for an incompressible viscous flow. *J. Differential Equations* **170**, 96–122.

Kato, T. (1972) Nonstationary flows of viscous and ideal fluids in $\mathbb{R}^3$. *J. Funct. Anal.* **9**, 296–305.

Kato, T. (1984) Remarks on zero viscosity limit for nonstationary Navier–Stokes flows with boundary. *Seminar on nonlinear partial differential equations* (Berkeley, Calif., 1983) 85–98. Springer, New York.

Kelliher, J.P. (2006) Navier–Stokes equations with Navier boundary conditions for a bounded domain in the plane. *SIAM J. Math. Anal.* **38**, 210–232.

Kelliher, J.P. (2007) On Kato's conditions for vanishing viscosity. *Indiana Univ. Math. J.* **56**, no. 4, 1711–1721.

Krasny, R. (1986) A study of singularity formation in a vortex sheet by the point-vortex approximation. *J. Fluid Mech.* **167**, 65–93.

Lions, J.L. (1969) *Quelques méthodes de résolutions des problemes aux limites non linéaires.* Dunod, Gauthier–Villars, Paris.

Lombardo, M.C. (2001) Analytic solutions of the Navier–Stokes equations. *Rend. Circ. Mat. Palermo* **50**, 299–311.

Lombardo, M.C., Cannone, M., & Sammartino, M. (2003) Well-posedness of the boundary layer equations. *SIAM J. Math. Anal.* **35**, no. 4, 987–1004.

Lopes Filho, M.C. (2007) Boundary layers and the vanishing viscosity limit for incompressible 2D flow. arXiv0712.0875v1.

Lopes Filho, M.C., Nussenzveig Lopes, H.J., & Planas, G. (2005) On the inviscid limit for 2D incompressible flow with Navier friction condition. *SIAM J. Math. Anal.* **36**, 1130–1141.

Lopes Filho, M.C., Nussenzveig Lopes, H.J., & Schochet, S. (2007) A criterion for the equivalence of the Birkhoff–Rott and Euler descriptions of vortex sheet evolution. *Trans. Amer. Math. Soc.* **359**, 4125–4142.

Majda, A. (1993) Remarks on weak solutions for vortex sheets with a distinguished sign. *Indiana Univ. Math. J.* **42**, 921–939.

Marchioro, C. & Pulvirenti, M. (1994) *Mathematical theory of incompressible nonviscous fluids.* Springer–Verlag, New York.

Marsden, J.E. (1970) *Nonlinear semigroups associated with the equations for a non-homogeneous fluid.* University of California, Berkeley.

Masmoudi, N. (1998) The Euler limit of the Navier–Stokes equations, and rotating fluids with boundary. *Arch. Rational Mech. Anal.* **142**, 375–394.

Matsumoto, T., Bec, J., & Frisch, U. (2005) The analytic structure of 2D Euler flow at short times. *Fluid Dyn. Res.* **36**, 221–237.

McGrath, F.J. (1968) Nonstationary plain flow of viscous and ideal fluids. *Arch. Rational Mech. Anal.* **27**, 329–348.

Moore, D.W. (1978) The equation of motion of a vortex layer of small thickness. *Studies in Appl. Math.* **58**, 119–140.

Moore, D.W. (1979) The spontaneous appearance of a singularity in the shape of an evolving vortex sheet. *Proc. Roy. Soc. London Ser. A* **365**, 105–119.

Obabko, A.V. & Cassel, K.W. (2002) Navier–Stokes solutions of unsteady separation induced by a vortex. *J. Fluid Mech.* **465**, 99–130.

Oleinik, O.A. (1967a) On the mathematical theory of boundary layer for an unsteady flow of incompressible fluid. *J. Appl. Math. Mech.* **30**, 951–974.

Oleinik, O.A. (1967b) Solutions of the system of equations of boundary layer theory by the method of straight lines. *Soviet Physics Dokl.* **12**, 525–528.

Oleinik, O.A. & Samokhin, V.N. (1999) *Mathematical models in boundary layer theory.* Applied Mathematics and Mathematical Computation, **15**, Chapman & Hall/CRC, Boca Raton, FL.

Paicu, M. (2005) Équation periodique de Navier–Stokes sans viscosité dans une direction. *Comm. Partial Differential Equations* **30**, 1107–1140.

Pauls, W., Matsumoto, T., Frisch, U., & Bec, J. (2006) Nature of complex singularities for the $2D$ Euler equation. *Physica D* **219**, 40–59.

Saffman, P.G. (1992) *Vortex dynamics*. Cambridge University Press, Cambridge.

Safonov, M.V. (1995) The abstract Cauchy–Kovalevskaya theorem in a weighted Banach space. *Comm. Pure Appl. Math.* **48**, 629–637.

Sammartino, M. & Caflisch, R.E. (1998a) Zero viscosity limit for analytic solutions of the Navier–Stokes equation on a half-space I. Existence for Euler and Prandtl equations, *Comm. Math. Phys.* **192**, 433–461.

Sammartino, M. & Caflisch, R.E. (1998b) Zero viscosity limit for analytic solutions of the Navier–Stokes equation on a half-space II. Construction of the Navier–Stokes solution, *Comm. Math. Phys.* **192**, 463–491.

Schochet, S. (1996) Point-vortex method for periodic weak solutions of the $2D$ Euler equations. *Comm. Pure Appl. Math.* **49**, 911–965.

Sears, W.R. & Telionis, D.P. (1975) Boundary layer separation in unsteady flow. *J. on Appl. Math.* **28**, 215–235.

Sulem C., Sulem, P.L., Bardos, C., & Frisch, U. (1981) Finite time analyticity for the two and three–dimensional Kelvin–Helmholtz instability. *Comm. Math. Phys.* **80**, 485–516.

Sulem, C., Sulem, P.L., & Frisch, U. (1983) Tracing complex singularities with spectral methods. *J. Comput. Phys.* **50**, 138–161.

Swann, H.S.G. (1971) The convergence with vanishing viscosity of nonstationary Navier–Stokes flow to ideal flow in $\mathbb{R}^3$. *Trans. Amer. Math. Soc.* **157**, 373–397.

Temam, R. & Wang, X. (1997) On the behavior of the solutions of the Navier–Stokes equations at vanishing viscosity. *Ann. Scuola Norm. Sup. Pisa Cl. Sci.* **25**, 807–828.

Van Dommelen, L.L. & Shen, S. (1980) The spontaneous generation of the singularity in a separating laminar boundary layer. *J. Comput. Phys.* **38**, 125–140.

Van Dyke, M. (1975) *Perturbation methods in fluid mechanics*. The Parabolic Press, Stanford.

Walker, J.D.A. (1978) The boundary layer due to rectilinear vortex. *Proc. R. Soc. Lond. A.* **359**, 167–188.

Wang, X. (2001) A Kato type theorem on zero viscosity limit of Navier–Stokes flows. *Indiana Univ. Math. J.* **50**, 223–241.

Xin, Z. & Zhang, L. (2004) On the global existence of solutions to the Prandtl's system. *Adv. Math.* **181**, 88–133.

6

Partial regularity results for solutions of the Navier-Stokes system

Igor Kukavica

Department of Mathematics
University of Southern California
Los Angeles, CA 90089, USA
`kukavica@usc.edu`

Abstract

In this paper we provide a self-contained proof of the partial regularity result for the Navier-Stokes system when the force satisfies $f \in L^{5/3}$. We do so by estimating the size of the space-time singularities when a force belongs to a more general Morrey-type class and then proving that the condition for regularity agrees with the classical condition due to Serrin when $f \in L^{5/3}$.

6.1 Introduction

In this paper we address the partial regularity of solutions of the three-dimensional Navier–Stokes system

$$\partial_t u - \Delta u + \partial_j (u_j u) + \nabla p = f$$
$$\nabla \cdot u = 0 \qquad\qquad (6.1)$$

in the whole of $\mathbb{R}^3$, a bounded smooth domain in $\mathbb{R}^3$ with Dirichlet boundary conditions, or a periodic domain. Here, u and p represent the unknown velocity and the pressure in the incompressible fluid, while the function f stands for the external force. Given an initial condition $u(\cdot, 0) = u_0 \in L^2$, it was proven by Leray (1934) and Hopf (1951) that there exists a global weak solution $u \in L_t^\infty L_x^2 \cap L_{t\,loc}^2 H_x^1$ if, for instance, $f \in L_t^\infty L_x^2$. Such solutions satisfy the Navier–Stokes system and an energy inequality, but they are not known to be unique and are not known to belong to a higher regularity space $L_t^\infty H_x^1 \cap L_{t\,loc}^2 H_x^2$. Also, Leray and Hopf proved existence of a local strong solution $u \in L_t^\infty H_x^1 \cap L_t^2 H_x^2$ if the initial data satisfies $u_0 \in H^1$. Strong solutions agree with weak solutions on the common interval of existence (Serrin, 1963), but it is not known whether they can be extended to global strong solutions. For a more detailed presentation on the existence theory we refer the

Published in *Partial Differential Equations and Fluid Mechanics*, edited by James C. Robinson and José L. Rodrigo. © Cambridge University Press 2009.

reader to Constantin & Foias (1988), Fabes et al. (1972), Lemarié-Rieusset (2002), Serrin (1962, 1963), Sohr (1983), and Temam (2001).

In a series of papers Scheffer (1976a,b, 1977) took a different approach towards regularity by estimating the size of a (possible) set $\mathcal{S}$ of space-time singularities for suitable weak solutions, i.e. solutions that satisfy a local version of the energy inequality. For instance, he proved that when $f = 0$, we have $\mathcal{H}^{5/3}(\mathcal{S}) = 0$, where $\mathcal{H}^a$ denotes the a-dimensional Hausdorff measure. In a classical paper Caffarelli et al. (1982) proved that $\mathcal{P}^1(\mathcal{S}) = 0$ where $\mathcal{P}^1$ is the one-dimensional parabolic measure (in particular $\mathcal{P}^1(A) \leq \mathcal{H}^1(A)$ for every set A) if the force satisfies $f \in L^{5/2+\delta}$ for some $\delta > 0$. A simpler argument leading to the same result was presented by Lin (1998) (see also Ladyzhenskaya & Seregin (1999) for f satisfying a Morrey-type condition with the same scaling as the Caffarelli–Kohn–Nirenberg condition $f \in L^{5/2+\delta}$). In Kukavica (2008a) we give a simple proof of the statement $\mathcal{P}^1(\mathcal{S}) = 0$ with the condition on the force $f \in L^{5/3+\delta}$ for $\delta > 0$. For other results on partial regularity, see Lemarié-Rieusset (2002), Robinson & Sadowski (2007), Struwe (1998), and Vasseur (2007).

The purpose of the present paper is three-fold. The first goal is to present a concise and self-contained proof of the partial regularity theorem when a force belongs to a Morrey-type space in the same scaling class as $L^{5/3}$. The second goal is to prove the partial regularity theorem for forces in the borderline class $f \in L^{5/3}$, the case left open in Kukavica (2008a). The third goal of the paper is to clarify the role of the energy inequality—namely to separate the proof into parts where the energy inequality is necessary and the parts where it is not.

The structure of the paper is as follows. In Section 6.2, we specify when (x, t) is a singular point and prove that the singular set has one-dimensional parabolic measure zero. In Sections 6.3 and 6.4, we prove that the definition of the singular set agrees with the classical one when the force belongs to $f \in L^{5/3+\delta}$ and $f \in L^{5/3}$ respectively. Thus Sections 6.2 and 6.3 constitute a self contained proof of the partial regularity result when $f \in L^{5/3+\delta}$, while Sections 6.2 and 6.4 contain the proof in the case $f \in L^{5/3}$.

6.2 Notation and the main theorem

Let D be an open, bounded, and connected subset of $\mathbb{R}^3 \times \mathbb{R}$. We always assume that (u, p) is a suitable weak solution in D by which we mean

(i) $u \in L_t^\infty L_x^2(D) \cap L_t^2 H_x^1(D)$ and $p \in L^{3/2}(D)$,

(ii) $f \in L_{\text{loc}}^{10/7}(D)$ is divergence free,

(iii) the Navier–Stokes equations (6.1) are satisfied in D in the weak sense, and

(iv) the local energy inequality holds in D, i.e.

$$\int |u|^2 \phi|_T + 2 \iint_{\mathbb{R}^3 \times (-\infty, T]} |\nabla u|^2 \phi$$
$$\leq \iint_{\mathbb{R}^3 \times (-\infty, T]} \left(|u|^2 (\phi_t + \Delta\phi) + (|u|^2 + 2p)u \cdot \nabla\phi + 2(u \cdot f)\phi \right)$$

(6.2)

for all $T \in \mathbb{R}$ and all $\phi \in C_0^\infty(D)$ such that $\phi \geq 0$ in D. In an effort to simplify the expressions in the paper, we will not include dx or dt in the integrals when it is clear from the domain of integration which variables are being integrated. Above and in the sequel, we define

$$L_t^s L_x^q(D) = \Big\{ u \text{ measurable in } D :$$
$$\|u\|_{L_t^s L_x^q(D)} = \big\| \|u(x,t)\|_{L_x^q} \big\|_{L_t^s} < \infty \Big\}.$$

By the Gagliardo–Nirenberg inequality, (i) implies that $u \in L^{10/3}(D)$, and therefore the assumption (ii) is a natural one to guarantee that the last integral in (6.2) exists. The assumption $p \in L^{3/2}(D)$ on the pressure can be weakened to $p \in L_t^q L_x^1(D)$, where $q > 1$ (cf. Remark 6.2.5 below).

Above and in the sequel, we write $\nabla = (\partial_1, \partial_2, \partial_3)$. We denote by $B_r(x_0)$ the standard Euclidean ball in $\mathbb{R}^3$ with centre x_0 and radius r and abbreviate $B_r = B_r(0)$. By $Q_r(x_0, t_0) = \overline{B}_r(x_0) \times [t_0 - r^2, t_0]$ we denote the parabolic cylinder in $\mathbb{R}^4$ labelled by the top-centre point $(x_0, t_0) \in D$. Also, let $Q_r^*(x_0, t_0) = \overline{B}_r(x_0) \times [t_0 - r^2, t_0 + r^2]$. For simplicity, we write $Q_r = Q_r(0,0)$ and $B_r = B_r(0)$.

Let μ be a fixed universal constant (in Sections 6.3 and 6.4 it will be assumed to be sufficiently small). We write $(x, t) \in \mathcal{R}$ (and we say that (x, t) is regular) if there exists $r_0 > 0$ such that

$$\frac{1}{r^2} \iint_{Q_r(x,t) \cap D} \left(|u|^3 + |p|^{3/2} \right) \leq \mu^3$$

(6.3)

for all $(x, t) \in Q_{r_0}^*(x_0, t_0)$ and $r \in (0, r_0]$. We also define $\mathcal{S} = D \backslash \mathcal{R}$.

For the force f, we assume that

$$\sup_{(x,t)\in D}\sup_{r>0}\frac{1}{r^{5/7}}\iint_{Q_r(x,t)\cap D}|f|^{10/7}<\infty. \qquad (6.4)$$

By Hölder's inequality, a sufficient condition for (6.4) is $f\in L^{5/3}(D)$.
Let $(x_0,t_0)\in D$ and $r>0$ be such that $Q_r(x_0,t_0)\subseteq D$. Define

$$\alpha_{(x_0,t_0)}(r)=\frac{1}{r^{1/2}}\|u\|_{L_t^\infty L_x^2(Q_r(x_0,t_0))},$$

$$\beta_{(x_0,t_0)}(r)=\frac{1}{r^{1/2}}\|\nabla u\|_{L_t^2 L_x^2(Q_r(x_0,t_0))},$$

$$\gamma_{(x_0,t_0)}(r)=\frac{1}{r^{2/3}}\|u\|_{L_t^3 L_x^3(Q_r(x_0,t_0))},$$

$$\delta_{(x_0,t_0)}(r)=\frac{1}{r^{2/3}}\|p\|_{L_t^{3/2} L_x^{3/2}(Q_r(x_0,t_0))}^{1/2},$$

$$\lambda_{(x_0,t_0)}(r)=\frac{1}{r^{1/2}}\|f\|_{L_t^{10/7} L_x^{10/7}(Q_r(x_0,t_0))},$$

and

$$\Gamma_{(x_0,t_0)}(r)=\frac{1}{r^{1/2}}\|u\|_{L_t^{10/3} L_x^{10/3}(Q_r(x_0,t_0))}.$$

If the label (x_0,t_0) is omitted, it is understood to be $(0,0)$, that is, $\alpha(r)=\alpha_{(0,0)}(r)$. All the quantities above, as well as the left-hand sides of (6.3) and (6.4), are dimensionless when following the usual convention that the dimension exponents of x, t, u, p, and f are 1, 2, -1, -2, and -3 respectively. (The dimension exponents are obtained from the following scaling property of the Navier–Stokes equation: if $u(x,t)$ is a solution with the pressure $p(x,t)$ and force $f(x,t)$, so is $Lu(Lx,L^2t)$ with the pressure $L^2p(Lx,L^2t)$ and the force $L^3f(Lx,L^2t)$ for any $L>0$.) Also, the exponents are chosen so that the the expressions are of order 1 as far as the dependence on u is concerned; thus it is easier to track which expressions arise from linear and which from nonlinear terms.

Let $\mathcal{P}^1$ be the one-dimensional parabolic measure (see Caffarelli et al. (1982) for the definition). The following implies the partial regularity result if the force f belongs to a Morrey space.

Theorem 6.2.1 *There exists a sufficiently small universal constant $\epsilon_*>0$ with the following property: if $(x_0,t_0)\in D$, and if there is $r_0>0$ such that*

$$\lambda_{(x,t)}(r)\le\epsilon_* \qquad (6.5)$$

for $(x,t) \in Q_r^*(x_0, t_0)$ *with* $r \in (0, r_0]$, *and if*

$$\limsup_{r \to 0+} \beta_{(x_0, t_0)}(r) \leq \epsilon_*, \tag{6.6}$$

then $(x_0, t_0) \in \mathcal{R}$. *In particular,* $\mathcal{P}^1(\mathcal{S}) = 0$.

From Caffarelli et al. (1982), recall that $\mathcal{P}^1(S) = 0$ means that for every $\epsilon > 0$ there exist points $(x_i, t_i) \in \mathbb{R}^3 \times \mathbb{R}$ and $r_i > 0$ for $i = 1, 2, \ldots$ such that $\sum_{i=1}^{\infty}(r_i) < \epsilon$ and $S \subseteq \cup_{i=1}^{\infty} Q_{r_i}(x_i, t_i)$.

The second assertion in the theorem, $\mathcal{P}^1(\mathcal{S}) = 0$, follows from the first using arguments from Caffarelli et al. (1982). Therefore, we only need to show that (6.5) and (6.6) with a sufficiently small ϵ_* imply that $(x_0, t_0) \in \mathcal{R}$.

For $0 < r < \rho/2$, define

$$\psi(x,t) = r^2 G(x, r^2 - t), \qquad (x,t) \in \mathbb{R}^3 \times \mathbb{R},$$

where G is the Gaussian kernel, i.e.

$$G(x,t) = (4\pi t)^{-3/2} \exp(-|x|^2/4t)$$

for $t > 0$ and $G(x,t) = 0$ for $t \leq 0$. Let $\eta_0 \colon \mathbb{R}^3 \times \mathbb{R} \to [0, 1]$ be a smooth function such that $\operatorname{supp} \eta_0 \subseteq B_1 \times (-1, 1)$ and $\eta_0 \equiv 1$ on $B_{1/2} \times (-1/4, 1/4)$, and let $\eta(x,t) = \eta_0(x/\rho, t/\rho^2)$. We shall use

$$\phi(x,t) = \psi(x,t)\eta(x,t)\overline{\eta}(t) \tag{6.7}$$

as a test function, where $\overline{\eta} \in C^\infty(\mathbb{R}, [0, 1])$ is such that $\overline{\eta} \equiv 1$ on $(-\infty, r^2/4)$ and $\overline{\eta} \equiv 0$ on $[r^2/2, \infty)$. In the first lemma, we collect useful upper and lower bounds from Kukavica (2008a) for this test function.

Lemma 6.2.2 *Let* $0 < r \leq \rho/2$. *Then on* $Q_r = \overline{B}_r(0) \times [-r^2, 0]$ *we have*

$$\phi(x,t) \geq \frac{1}{Cr}, \qquad (x,t) \in Q_r, \tag{6.8}$$

while on Q_ρ *we have*

$$\phi(x,t) \leq \frac{C}{r}, \qquad (x,t) \in Q_\rho, \tag{6.9}$$

and

$$|\nabla \phi(x,t)| \leq \frac{C}{r^2}, \qquad (x,t) \in Q_\rho. \tag{6.10}$$

Moreover,

$$|\phi_t(x,t) + \Delta \phi(x,t)| \leq \frac{Cr^2}{\rho^5}, \qquad (x,t) \in Q_\rho. \tag{6.11}$$

Above and in the sequel, the symbol C stands for a generic universal constant except when other dependence is stated explicitly.

The inequality (6.11) is the essential one for a simple proof the partial regularity result since in the iterative proof below it is actually the linear part that causes the main difficulty. The inequality (6.11) ensures that the first term on the right-hand side of the local energy inequality (6.2) (which is the term resulting from the linear part) has a small constant in front.

Proof of Lemma 6.2.2. The inequality (6.8) can be rewritten as

$$(4\pi(1 - t/r^2))^{-3/2} \exp(-|x|^2/4r^2(1 - t/r^2)) \geq 1/C \text{ for } (x,t) \in Q_r.$$

By rescaling, we can reduce this to the case $r = 1$, for which the inequality is clear. The inequalities (6.9) and (6.10) follow from $r\psi(x,t) \leq C$ and $r^2|\nabla\psi(x,t)| \leq C$ for $(x,t) \in \mathbb{R}^3 \times (-\infty, 0)$, which can be established by rescaling.

Regarding (6.11), note that $\phi_t + \Delta\phi = \psi(\partial_t + \Delta)(\eta\overline{\eta}) + 2\overline{\eta}\nabla\eta \cdot \nabla\psi$ due to $(\partial_t + \Delta)\psi = 0$. The functions $(\partial_t + \Delta)(\eta\overline{\eta})$ and $\nabla\eta$ vanish on $Q_{\rho/2}$, and the inequality (6.11) then follows from

$$|\psi(x,t)| \leq \frac{Cr^2}{\rho^3}, \qquad (x,t) \in Q_\rho \backslash Q_{\rho/2},$$

and

$$|\nabla\psi(x,t)| \leq \frac{Cr^2}{\rho^4}, \qquad (x,t) \in Q_\rho \backslash Q_{\rho/2},$$

which can be checked directly. □

We start with estimates on α and β. These follow from the generalized energy inequality (6.2). They are stated for quantities centred at $(0,0)$ for simplicity of notation, but they can be used around any point in D.

Lemma 6.2.3 *Assume that $(0,0) \in D$. Then for $0 < r \leq \rho/2$ such that $Q_\rho \subseteq D$, we have*

$$\alpha(r) + \beta(r) \leq C\kappa^{2/3}\big(\alpha(\rho) + \beta(\rho)\big) + \frac{C}{\kappa^6}\beta(\rho)^3 + \frac{C}{\kappa}\alpha(\rho)\beta(\rho)^{1/2}$$

$$+ \frac{C}{\kappa^{8/3}}\delta(\rho)^2 + \frac{C}{\kappa^{5/3}}\lambda(\rho) \tag{6.12}$$

and

$$\alpha(r) + \beta(r) \leq C\kappa\gamma(\rho) + \frac{C}{\kappa}\gamma(\rho)^{3/2} + \frac{C}{\kappa}\delta(\rho)\gamma(\rho)^{1/2} + \frac{C}{\kappa^{1/2}}\Gamma(\rho)^{1/2}\lambda(\rho)^{1/2},$$

$$\tag{6.13}$$

where $\kappa = r/\rho$.

Proof of Lemma 6.2.3. Let ϕ be as in (6.7). Then for all $t_0 \in [-r^2, 0]$ (6.2) yields

$$\int_{B_r} |u|^2 \phi|_{t_0} + 2 \iint_{Q_r \cap \{t \leq t_0\}} |\nabla u|^2 \phi \leq \iint_{Q_\rho \cap \{t \leq t_0\}} |u|^2 (\phi_t + \Delta\phi) + \iint_{Q_\rho \cap \{t \leq t_0\}} |u|^2 u \cdot \nabla\phi$$

$$+ 2 \iint_{Q_\rho \cap \{t \leq t_0\}} p\, u \cdot \nabla\phi + 2 \iint_{Q_\rho \cap \{t \leq t_0\}} (u \cdot f)\phi$$

$$= I_1 + I_2 + I_3 + I_4. \tag{6.14}$$

In order to obtain (6.13), note that we have

$$I_1 \leq Cr^2\rho^{-5} \iint_{Q_\rho} |u|^2 \leq Cr^2\rho^{-10/3} \left(\iint_{Q_\rho} |u|^3 \right)^{2/3} = C\kappa^2 \gamma(\rho)^2$$

by definition of γ, (6.11), and Hölder's inequality. Similarly,

$$I_2 \leq \frac{C}{r^2} \iint_{Q_\rho} |u|^3 = C\kappa^{-2}\gamma(\rho)^3,$$

and

$$I_3 \leq \frac{C}{r^2} \iint_{Q_\rho} |p|\,|u| \leq \frac{C}{r^2} \|p\|_{L^{3/2}(Q_\rho)} \|u\|_{L^3(Q_\rho)}$$

$$= \frac{C\rho^2}{r^2} \delta(\rho)^2 \gamma(\rho) = \frac{C}{\kappa^2} \delta(\rho)^2 \gamma(\rho) \tag{6.15}$$

by (6.10) and Hölder's inequality; also, by (6.9),

$$I_4 \leq \frac{C}{r} \iint_{Q_\rho} |u|\,|f| \leq \frac{C}{r} \|u\|_{L^{10/3}(Q_\rho)} \|f\|_{L^{10/7}(Q_\rho)}$$

$$\leq C\kappa^{-1}\lambda(\rho)\Gamma(\rho). \tag{6.16}$$

Using the above inequalities and (6.8) then gives (6.13). For (6.12), we estimate I_1 and I_2 differently. By (6.11), we get

$$I_1 \leq \frac{Cr^2}{\rho^5} \iint_{Q_\rho} |u|^2 \leq C\kappa^2 \alpha(\rho)^2. \tag{6.17}$$

Next, defining

$$A_\rho g = A_\rho g(\cdot, t) = |B_\rho|^{-1} \int_{B_\rho} g(\cdot, t),$$

we have (as in Caffarelli et al., 1982), using

$$\int_{Q_\rho \cap \{t \leq t_0\}} |u|^2 u \cdot \nabla\phi = \int_{Q_\rho \cap \{t \leq t_0\}} (|u|^2 - A_\rho |u|^2) u \cdot \nabla\phi$$

and (6.10), that

$$
\begin{aligned}
I_2 &\leq \frac{C}{r^2} \iint_{Q_\rho} \big||u|^2 - A_\rho |u|^2\big|\, |u| \leq \frac{C}{r^2} \|u\|_{L^3(Q_\rho)} \big\||u|^2 - A_\rho|u|^2\big\|_{L^{3/2}(Q_\rho)} \\
&\leq \frac{C}{r^2} \|u\|_{L^3(Q_\rho)} \big\|\nabla(|u|^2)\big\|_{L_t^{3/2}L_x^1(Q_\rho)} \\
&\leq \frac{C}{r^2} \|u\|_{L^3(Q_\rho)} \|u\|_{L_t^6 L_x^2(Q_\rho)} \|\nabla u\|_{L_t^2 L_x^2(Q_\rho)} \\
&\leq \frac{C\rho^{1/3}}{r^2} \|u\|_{L^3(Q_\rho)} \|u\|_{L_t^\infty L_x^2(Q_\rho)} \|\nabla u\|_{L^2(Q_\rho)},
\end{aligned}
$$

whence

$$
I_2 \leq \frac{C}{\kappa^2} \alpha(\rho)\beta(\rho)\gamma(\rho). \tag{6.18}
$$

Using (6.8) on the left side of (6.14), collecting the estimates (6.15), (6.16), (6.17), (6.18), and taking the square root of the resulting inequality, we get

$$
\begin{aligned}
\alpha(r) + \beta(r) &\leq C\kappa\alpha(\rho) + \frac{C}{\kappa}\alpha(\rho)^{1/2}\beta(\rho)^{1/2}\gamma(\rho)^{1/2} + \\
&\quad + \frac{C}{\kappa}\delta(\rho)\gamma(\rho)^{1/2} + \frac{C}{\kappa^{1/2}}\Gamma(\rho)^{1/2}\lambda(\rho)^{1/2}. \tag{6.19}
\end{aligned}
$$

The Gagliardo–Nirenberg inequality

$$
\|v\|_{L_x^q} \leq C\|v\|_{L_x^2}^{3/q-1/2}\|\nabla v\|_{L_x^2}^{3/2-3/q} + \frac{C}{r^{3/2-3/q}}\|v\|_{L_x^2}
$$

for $2 \leq q \leq 6$ implies that

$$
\frac{1}{\rho^{1/2}}\|u\|_{L_t^s L_x^q(Q_\rho)} \leq C\alpha(\rho)^{3/q-1/2}\beta(\rho)^{3/2-3/q} + C\alpha(\rho)
$$

for $2 \leq q \leq 6$ and $2/s + 3/q = 3/2$. Therefore,

$$
\gamma(\rho) \leq C\alpha(\rho)^{1/2}\beta(\rho)^{1/2} + C\alpha(\rho), \tag{6.20}
$$

whence

$$
\frac{C}{\kappa}\alpha(\rho)^{1/2}\beta(\rho)^{1/2}\gamma(\rho)^{1/2} \leq C\kappa^{2/3}\alpha(\rho) + \frac{C}{\kappa^6}\beta(\rho)^3 + \frac{C}{\kappa}\alpha(\rho)\beta(\rho)^{1/2}
$$

and

$$
\frac{C}{\kappa}\delta(\rho)\gamma(\rho)^{1/2} \leq C\kappa^{2/3}\big(\alpha(\rho) + \beta(\rho)\big) + \frac{C}{\kappa^{8/3}}\delta(\rho)^2. \tag{6.21}
$$

Similarly to above, we have

$$
\Gamma(\rho) \leq C\alpha(\rho)^{2/5}\beta(\rho)^{3/5} + C\alpha(\rho). \tag{6.22}
$$

Hence,

$$\frac{C}{\kappa^{1/2}}\Gamma(\rho)^{1/2}\lambda(\rho)^{1/2} \le C\kappa(\alpha(\rho) + \beta(\rho)) + \frac{C}{\kappa^2}\lambda(\rho)$$

and (6.12) follows. □

Next, we state an inequality for the pressure term. It is based solely on the identity

$$-\Delta p = \partial_i\partial_j(u_iu_j), \tag{6.23}$$

which is obtained by taking the divergence of the Navier–Stokes equation (6.1).

Lemma 6.2.4 *For $0 < r \le \rho/2$, we have*

$$\delta(r) \le \frac{C}{\kappa^{1/2}}\alpha(\rho)^{1/2}\beta(\rho)^{1/2} + C\kappa^{1/3}\delta(\rho) \tag{6.24}$$

where $\kappa = r/\rho$.

The pressure estimate follows Caffarelli et al. (1982), Kukavica (2008a), Lemarié-Rieusset (2002), and Lin (1998).

Proof of Lemma 6.2.4. Let $\eta_0 \colon \mathbb{R}^3 \to [0,1]$ be a smooth function such that $\eta_0 \equiv 1$ on $B_{3/5}$ and $\eta_0 \equiv 0$ on $B^c_{4/5}$, and let let $\eta(x) = \eta_0(x/\rho)$. Denoting $U_{ij} = -u_i(u_j - A_\rho u_j)$, the equation (6.23) can be written as[1] $\Delta p = \partial_{ij}U_{ij}$, from where it follows by a short calculation that

$$\Delta(\eta p) = \partial_{ij}(\eta U_{ij}) + (\partial_{ij}\eta)U_{ij} - \partial_j(U_{ij}\partial_i\eta) - \partial_i(U_{ij}\partial_j\eta) \\ - p\Delta\eta + 2\partial_j((\partial_j\eta)p). \tag{6.25}$$

(An advantage of writing the equation for the localized pressure ηp this way rather than, say, as $\Delta(\eta p) = p\Delta\eta + 2\partial_i\eta\partial_i p + \partial_{ij}(\eta U_{ij}) - U_{ij}\partial_{ij}\eta - \partial_i\eta\partial_j U_{ij} - \partial_j\eta\partial_i U_{ij}$ is that in this way no derivative falls directly on U or p.) Using $g = N * (\Delta g)$, where $N(x) = -1/4\pi|x|$ is the Newtonian potential, valid for compactly supported and smooth g, and the fact that $N * \partial_i g = (\partial_i N) * g$ we get

$$\eta p = -R_iR_j(\eta U_{ij}) + N * ((\partial_{ij}\eta)U_{ij}) - \partial_j N * (U_{ij}\partial_i\eta) \\ - \partial_i N * (U_{ij}\partial_j\eta) - N * (p\Delta\eta) + 2\partial_j N * ((\partial_j\eta)p), \tag{6.26}$$

where R_i is the standard i-th Riesz transform (cf. Stein, 1993, p. 26). Denote the terms on the right by p_1 through p_6. For p_1, we use the Calderón–Zygmund theorem and obtain

---

[1]We use the notation $\partial_{ij} = \partial_i\partial_j$.

$$\|p_1\|_{L^{3/2}(Q_r)} \leq \|p_1\|_{L^{3/2}(\mathbb{R}^3 \times (-r^2,0))} \leq C \sum_{i,j} \|\eta U_{ij}\|_{L^{3/2}(\mathbb{R}^3 \times (-r^2,0))}$$

$$\leq C\|u\|_{L_t^6 L_x^2(\mathbb{R}^3 \times (-r^2,0))}\|u - A_\rho u\|_{L_t^2 L_x^6(Q_\rho)}$$

$$\leq Cr^{1/3}\|u\|_{L_t^\infty L_x^2(\mathbb{R}^3 \times (-r^2,0))}\|u - A_\rho u\|_{L_t^2 L_x^6(Q_\rho)}$$

$$\leq Cr^{1/3}\|u\|_{L_t^6 L_x^2(Q_\rho)}\|\nabla u\|_{L_t^2 L_x^2(Q_\rho)} \leq Cr^{1/3}\rho\alpha(\rho)\beta(\rho).$$

In order to bound $p_2 = N * ((\partial_{ij}\eta)U_{ij})$, note that $\partial_{ij}\eta$ vanishes on $\overline{B}_{3\rho/5} \cup B_{4\rho/5}^c$, so the convolution does not contain a singularity. We get

$$\|p_2\|_{L^{3/2}(Q_r)} \leq Cr^{7/3}\|p_2\|_{L_t^2 L_x^\infty(B_r \times (-r^2,0))}$$

$$\leq \frac{Cr^{7/3}}{\rho}\|U_{ij}\partial_{ij}\eta\|_{L_t^2 L_x^1(B_\rho \times (-r^2,0))}$$

$$\leq \frac{Cr^{7/3}}{\rho^2}\|u\|_{L_t^\infty L_x^2(Q_\rho)}\|\nabla u\|_{L_t^2 L_x^2(Q_\rho)} = \frac{Cr^{7/3}}{\rho}\alpha(\rho)\beta(\rho).$$

The estimate for p_3 is analogous; namely,

$$\|p_3\|_{L^{3/2}(Q_r)} \leq Cr^{7/3}\|p_3\|_{L_t^2 L_x^\infty(B_r \times (-r^2,0))}$$

$$\leq \frac{Cr^{7/3}}{\rho^2} \sum_j \|U_{ij}\partial_i\eta\|_{L_t^2 L_x^1(B_\rho \times (-r^2,0))}$$

$$\leq \frac{Cr^{7/3}}{\rho^3} \sum_{i,j} \|U_{ij}\|_{L_t^2 L_x^1(B_\rho \times (-r^2,0))}$$

$$\leq \frac{Cr^{7/3}}{\rho^2}\|u\|_{L_t^\infty L_x^2(Q_\rho)}\|\nabla u\|_{L_t^2 L_x^2(Q_\rho)} \leq \frac{Cr^{7/3}}{\rho}\alpha(\rho)\beta(\rho).$$

The bound for p_4 is the same as for p_3. The estimate for p_5 is similar as that for p_2, namely,

$$\|p_5\|_{L^{3/2}(Q_r)} \leq Cr^2\|p_5\|_{L_t^{3/2} L_x^\infty(B_r \times (-r^2,0))}$$

$$\leq \frac{Cr^2}{\rho}\|p\Delta\eta\|_{L_t^{3/2} L_x^1(B_\rho \times (-r^2,0))}$$

$$\leq \frac{Cr^2}{\rho^2}\|p\|_{L_t^{3/2} L_x^{3/2}(Q_\rho)} \leq \frac{Cr^2}{\rho^{2/3}}\delta(\rho)^2.$$

Similarly,

$$\|p_6\|_{L^{3/2}(Q_r)} \le Cr^2\|p_6\|_{L_t^{3/2}L_x^\infty(B_r\times(-r^2,0))}$$

$$\le \frac{Cr^2}{\rho^2}\|p\nabla\eta\|_{L_t^{3/2}L_x^1(\mathbb{R}^3\times(-r^2,0))}$$

$$\le \frac{Cr^2}{\rho^2}\|p\|_{L_t^{3/2}L_x^{3/2}(Q_\rho)} \le \frac{Cr^2}{\rho^{2/3}}\delta(\rho)^2$$

and (6.24) follows by collecting all the estimates above. $\square$

Proof of Theorem 6.2.1. Without loss of generality, set $(x_0, t_0) = (0, 0)$. Let $0 < r \le \rho/2 \le r_0/4$, and denote $\kappa = r/\rho$. By (6.12) and (6.24), the quantity

$$\theta(r) = \alpha(r) + \beta(r) + \frac{1}{\kappa^4}\delta(r)^2 \tag{6.27}$$

satisfies

$$\theta(r) \le C_0\kappa^{2/3}\theta(\rho) + \frac{C_0}{\kappa^5}\beta(\rho)\theta(\rho) + \frac{C_0}{\kappa^6}\beta(\rho)^3 + \frac{C_0}{\kappa^{5/3}}\lambda(\rho). \tag{6.28}$$

Now, fix $\kappa = \min\{1/2, 1/(4C_0)^{3/2}\}$ in order to ensure that $\kappa \le 1/2$ and that the factor in front of $\theta(\rho)$ in the first term of (6.28) is less than or equal to $1/4$. By possibly reducing r_0, we may assume without loss of generality that $\beta(r) \le 2\epsilon_*$ for $r \in (0, r_0)$. Using also $\lambda(r) \le \epsilon_*$ for $r \in (0, r_0)$, we get

$$\theta(\kappa\rho) \le \frac{1}{4}\theta(\rho) + \frac{C_0}{\kappa^5}\epsilon_*\theta(\rho) + \frac{C_0}{\kappa^6}\epsilon_*^3 + \frac{C_0}{\kappa^{5/3}}\epsilon_*. \tag{6.29}$$

Now, assume that ϵ_* is so small that $C_0\epsilon_*/\kappa^5 \le 1/4$ and $\epsilon_* \le 1$. Then in fact

$$\theta(\kappa\rho) \le \frac{1}{2}\theta(\rho) + C\epsilon_*.$$

Let $\epsilon \in (0, 1]$. Then there exist $\epsilon_* \in (0, \epsilon]$ and $m \in \mathbb{N}$ such that $\theta(\kappa^m r_0) \le \epsilon$. Define $r_1 = \kappa^m r_0$. By (6.20), we get $\gamma(r_1) \le C_1\epsilon$ and by (6.22), $\Gamma(r_1) \le C_1\epsilon$. Also, by the definition of θ, we have $\delta(r_1) \le C_1\epsilon^{1/2}$. By the continuity of the integral, there exists $r_2 > 0$ so small that $\gamma_{(x,t)}(r_1) \le 2C_1\epsilon$, $\Gamma_{(x,t)}(r_1) \le 2C_1\epsilon$, and $\delta_{(x,t)}(r_1) \le 2C_1\epsilon^{1/2}$ for $(x, t) \in Q_{r_2}^*(0, 0)$. Additionally, we may assume that $\lambda_{(x,t)}(r) \le \epsilon_*$ for $(x, t) \in Q_{r_2}^*(0, 0)$ and $r \in (0, r_1]$. (To summarize the proof so far, we are able to make $\gamma(r_1)$, $\Gamma(r_1)$, and $\delta(r_1)$ as small as we wish for some $r_1 > 0$ and for (x, t) in a sufficiently small neighbourhood of $(0, 0)$.)

Fix $(x, t) \in Q^*_{r_2}(0, 0)$. By (6.13), $\max\{\alpha_{(x,t)}(\kappa r_1), \beta_{(x,t)}(\kappa r_1)\} \le C\epsilon$. Similarly, using the monotonicity of the integral (or by (6.24)), we obtain $\delta_{(x,t)}(\kappa r_1) \le C\epsilon^{1/2}$. Therefore, $\theta_{(x,t)}(\kappa r_1) \le C_2\epsilon$, where $\theta_{(x,t)}$ denotes the analog of (6.27) centred at (x, t) instead of $(0, 0)$. Now, by (6.28) and by the definition of κ, we have

$$\theta_{(x,t)}(\kappa\rho) \le \frac{1}{4}\theta_{(x,t)}(\rho) + C_3\theta_{(x,t)}(\rho)^3 + C_3\epsilon_*.$$

If ϵ is so small that $C_3(C_2\epsilon)^2 \le 1/4$ and $\epsilon_* > 0$ is so small that $C_3\epsilon_* \le C_2\epsilon/4$, we get $\theta_{(x,t)}(\kappa^n r_1) \le C_2\epsilon$ for every $n \in \mathbb{N}$ as long as $(x, t) \in Q^*_{r_2}(0, 0)$. By the monotonicity of the integral, it then follows that $\theta_{(x,t)}(r) \le C\epsilon$ for all $r \in (0, r_1]$. This implies that

$$\alpha_{(x,t)}(r) + \beta_{(x,t)}(r) + \delta_{(x,t)}(r)^2 \le C\epsilon$$

for all $r \in (0, r_1]$. By (6.20) and (6.22), we get

$$\max\{\gamma_{(x,t)}(r), \Gamma_{(x,t)}(r)\} \le C\epsilon$$

for all $r \in (0, r_1]$. Since $(x, t) \in Q^*_{r_2}(0, 0)$ was arbitrary, we conclude that $(x_0, t_0) \in \mathcal{R}$ as claimed. $\square$

Remark 6.2.5 Theorem 6.2.1 in this section remains valid if (i) is replaced by
(i)′ $u \in L^\infty_t L^2_x(D) \cap L^2_t H^1_x(D)$ and $p \in L^q_t L^1_x(D)$, where $q > 1$,
and the inequality (6.3) in the definition of a regular point is replaced by

$$\frac{1}{r^{2/3}}\|u\|_{L^3(Q_r)} + \frac{1}{r^{2/q-1/2}}\|p\|_{L^q_t L^2_x(Q_r)} \le \mu.$$

Observe that, by elliptic regularity, we get

$$p \in \bigcap_{1<s\le 3} L^{2s/(3s-3)}_t L^s_x(Q_\rho) + L^q_t L^\infty_x(Q_\rho).$$

The proof is similar to the proof of Theorem 6.2.1. The inequality (6.12) is replaced by

$$\alpha(r) + \beta(r) \le C\kappa^{2/3}\big(\alpha(\rho) + \beta(\rho)\big) + \frac{C}{\kappa^6}\beta(\rho)^3 + \frac{C}{\kappa}\alpha(\rho)\beta(\rho)^{1/2}$$
$$+ \frac{1}{\kappa^{8/3}}\Delta(\rho)^2 + \frac{C}{\kappa^{5/3}}\lambda(\rho),$$

where

$$\Delta(r) = \frac{1}{r^{1/q-1/4}}\|p\|_{L^q_t L^2_x(Q_r)}^{1/2}.$$

Similarly, (6.13) is replaced by

$$\alpha(r)+\beta(r) \le C\kappa\gamma(\rho)+\frac{C}{\kappa}\gamma(\rho)^{3/2}+\frac{C}{\kappa}\Delta(\rho)\widetilde{\Gamma}(\rho)^{1/2}+\frac{C}{\kappa^{1/2}}\Gamma(\rho)^{1/2}\lambda(\rho)^{1/2},$$

where $\widetilde{\Gamma}(r) = r^{-1/2}\|u\|_{L_t^{q/(q-1)}L_x^{6q/(4-q)}(Q_r)}$, while instead of (6.24) we have

$$\Delta(r) \le \frac{C}{\kappa}\alpha(\rho)^{1/2}\beta(\rho)^{3/2} + \frac{C}{\kappa}\alpha(\rho)\beta(\rho) + C\kappa^{2-2/q}\Delta(\rho)$$

as one can readily verify. The proof of Theorem 6.2.1 is then the same except that δ is replaced by Δ.

6.3 A partial regularity result for $f \in L^{5/3+q}$ with $q > 0$

In this section, we show that if $(x_0, t_0) \in \mathcal{R}$ with μ sufficiently small, then u is regular in a neighborhood of (x_0, t_0) in a classical sense, that is, it belongs to Serrin's L^5 class. Assume that

$$f \in L^{5/3+q}(D), \tag{6.30}$$

where $q > 0$. Although this case is less general than the case $f \in L^{5/3}(D)$ discussed in the next section, we address it separately since the proof is much simpler.

It is convenient to introduce the following parabolic type Morrey spaces (O'Leary, 2003). For a measurable function g on $\mathbb{R}^n$, $n \ge 3$, $\lambda \in [0, n+2]$, and $q \in [1, \infty)$, define

$$\|g\|_{\mathcal{L}_\lambda^q} = \sup_{(x,t)\in\mathbb{R}^n\times\mathbb{R}} \sup_{\rho>0} \frac{1}{\rho^{\lambda/q}}\|g\|_{L^q(Q_\rho^*(x,t))}.$$

Let $\mathcal{L}_\lambda^q$ be the set of measurable functions g for which $\|g\|_{\mathcal{L}_\lambda^q} < \infty$. Also, for any open nonempty set $\mathcal{V} \subseteq \mathbb{R}^n \times \mathbb{R}$, write $\|g\|_{\mathcal{L}_\lambda^q(\mathcal{V})} = \|g\chi_\mathcal{V}\|_{\mathcal{L}_\lambda^q}$ and denote $\mathcal{L}_\lambda^q(\mathcal{V}) = \{g \text{ measurable on } \mathcal{V} : \|g\|_{\mathcal{L}_\lambda^q(\mathcal{V})} < \infty\}$, identifying functions that are equal almost everywhere .

From Kukavica (2008b) we recall the following inequality.

Lemma 6.3.1 *Let* $g \in L^p(\mathbb{R}^n \times \mathbb{R})\cap\mathcal{L}_\lambda^q(\mathbb{R}^n \times \mathbb{R})$, *where* $1 \le q \le p < \infty$ *and* $\lambda \in [0, n+2)$ *with* $p > 1$. *Define*

$$h(x,t) = \iint_{\mathbb{R}^n\times\mathbb{R}} \frac{g(y,s)}{(|x-y|+|t-s|^{1/2})^{n+2-\alpha}} \, dy \, ds, \tag{6.31}$$

where $\alpha > 0$. Assume $q < (n + 2 - \lambda)/\alpha$. Then for $\widetilde{p} = \frac{p}{(1-q\alpha/(n+2-\lambda))}$ we have $h \in L^{\widetilde{p}}(\mathbb{R}^n \times \mathbb{R})$ and

$$\|h\|_{L^{\widetilde{p}}} \leq C\|g\|_{\mathcal{L}_\lambda^q}^{1-p/\widetilde{p}}\|g\|_{L^p}^{p/\widetilde{p}},$$

where the constant C depends on α, λ, q, and p.

Proof. Since the inequality is proved in Kukavica (2008b), we only provide a sketch of the proof. We may assume that g is not the null function. For $m \in \mathbb{Z}$, write

$$h_m(x,t) = \iint_{Q_{2^m}^*(x,t)\setminus Q_{2^{m-1}}^*(x,t)} \frac{g(y,s)}{(|x-y| + |t-s|^{1/2})^{n+2-\alpha}}\, dy\, ds.$$

Fix $m_0 \in \mathbb{Z}$, which is to be determined below. If $m \leq m_0$, we use

$$|h_m(x,y)| \leq C2^{m\alpha}(M_{\text{par}}g)(x,t),$$

where

$$(M_{\text{par}}g)(x,t) = \sup_{r>0} |Q_r^*(x,t)|^{-1} \iint_{Q_r^*(x,t)} |g(y,s)|\, dy\, ds$$

is the parabolic maximal function. For $m \geq m_0 + 1$, we estimate

$$|h_m(x,t)| \leq \frac{C}{2^{m(n+2-\alpha)}} \iint_{Q_{2^m}^*} |g(y,s)|\, dy\, ds$$

$$\leq C2^{-m(-\lambda/q-\alpha+(n+2)/q)}\|g\|_{\mathcal{L}_\lambda^q}.$$

We get

$$|h(x,t)| \leq C(M_{\text{par}}g)(x,t)\sum_{m=-\infty}^{m_0} 2^{m\alpha}$$

$$+ C\|g\|_{\mathcal{L}_\lambda^q}\sum_{m=m_0+1}^{\infty} 2^{-m(-\lambda/q-\alpha+(n+2)/q)}.$$

Summing up the geometric series, we get

$$|h(x,t)| \leq C2^{m_0\alpha}(M_{\text{par}}g)(x,t) + C2^{-m_0(-\lambda/q-\alpha+(n+2)/q)}\|g\|_{\mathcal{L}_\lambda^q},$$

where the constants depend on α, λ, and q. Choosing a suitable m_0, we get

$$|h(x,t)| \leq C(n,\alpha,\lambda,q)\|g\|_{\mathcal{L}_\lambda^q}^{\alpha q/(n+2-\lambda)}(M_{\text{par}}g)(x,t)^{1-\alpha q/(n+2-\lambda)}$$

and the rest follows by the L^r-boundedness property of M_{par} (Stein, 1993, p. 83). $\qquad\square$

Remark 6.3.2 An analogous statement is valid also for the classical Morrey spaces. First, for measurable $g \colon \mathbb{R}^n \to \mathbb{R}$, we define

$$\|g\|_{\mathcal{M}_\lambda^q} = \sup_{x \in \mathbb{R}^n} \sup_{\rho > 0} \frac{1}{\rho^{\lambda/q}} \|g\|_{L^q(B_\rho(x))}.$$

Now let $g \in L^p(\mathbb{R}^n) \cap \mathcal{M}_\lambda^q(\mathbb{R}^n)$, where $1 \leq q \leq p < \infty$ and $\lambda \in [0, n)$ with $p > 1$. Define

$$h(x) = \int_{\mathbb{R}^n} \frac{g(y)\,\mathrm{d}y}{|x - y|^{n - \alpha}},$$

where $\alpha > 0$. Assume $q < (n - \lambda)/\alpha$. Then for $\tilde{p} = p/(1 - q\alpha/(n - \lambda))$ we have $h \in L^{\tilde{p}}(\mathbb{R}^n)$ and

$$\|h\|_{L^{\tilde{p}}} \leq C \|g\|_{\mathcal{M}_\lambda^q}^{1 - p/\tilde{p}} \|g\|_{L^p}^{p/\tilde{p}},$$

where the constant C depends on α, λ, q, and p. Note that the statement reduces to the classical Hardy–Littlewood–Sobolev inequality if $p = q$ and $\lambda = 0$. For other properties of classical Morrey spaces cf. Olsen (1995) and Taylor (1992).

We shall also need below space-time convolution inequalities in the Lebesgue spaces. For the integral (6.31) and $\alpha \in (0, n + 2]$, we have

$$\|g\|_{L_t^{s_3} L_x^{q_3}} \leq C(s_2, s_3, q_2, q_3) \|f\|_{L_t^{s_2} L_x^{q_2}} \tag{6.32}$$

where $1 \leq s_1, s_2, s_3, q_1, q_2, q_3 \leq \infty$ satisfy

$$\frac{1}{q_3} + 1 = \frac{1}{q_2} + \frac{1}{q_1}$$
$$\frac{1}{s_3} + 1 = \frac{1}{s_2} + \frac{1}{s_1},$$

with the conditions $q_1 > n/(n + 2 - \alpha)$ and $1 < s_2 < s_3 < \infty$ with $s_1 = 2q_1/((n + 2 - \alpha)q_1 - n)$. The proof is a simple application of Young's inequality in space and the Hardy–Littlewood–Sobolev inequality in time. The inequality also holds in the case $q_1 = n/(n + 2 - \alpha)$ and $(s_1, s_2, s_3) = (\infty, 1, \infty)$, but we need to add the condition $1 < q_2 < q_3 < \infty$. We shall refer to (6.32) with the stated conditions as the parabolic Hardy–Littlewood–Sobolev inequality.

Theorem 6.3.3 *Assume that* (6.30) *holds for some* $q > 0$. *If* $\mu > 0$ *in* (6.3) *is a small enough universal constant, then* $(x_0, t_0) \in \mathcal{R}$ *implies that* $u \in L^5(\mathcal{V})$ *for some neighborhood* $\mathcal{V}$ *of* (x_0, t_0).

It is possible to prove the above theorem without using the local energy inequality (6.2) (cf. the next section)—however, the proof is simpler if we use it.

Proof. Without loss of generality, assume that $(x_0, t_0) = (0, 0)$. By reducing D, we may assume that $\|f\|_{L^{5/3}(D)} \leq 1$. Since f has more regularity than that required by Theorem 6.2.1, we may improve (6.13) slightly. Namely, I_4 in (6.14) can be estimated as

$$I_4 \leq \frac{C}{r}\|f\|_{L^{3/2}(Q_\rho)}\|u\|_{L^3(Q_\rho)} = \frac{C\rho^{2/3}}{r}\|f\|_{L^{3/2}(Q_\rho)}\gamma(\rho)$$

$$\leq \frac{C\rho}{r}\|f\|_{L^{5/3}(Q_\rho)}\gamma(\rho)$$

and thus (6.13) may be replaced by

$$\alpha_{(x,t)}(r) + \beta_{(x,t)}(r) \leq C\kappa\gamma_{(x,t)}(\rho) + \frac{C}{\kappa}\gamma_{(x,t)}(\rho)^{3/2} + \frac{C}{\kappa}\delta_{(x,t)}(\rho)\gamma_{(x,t)}(\rho)^{1/2}$$

$$+ \frac{C}{\kappa^{1/2}}\|f\|_{L^{5/3}(Q_\rho(x,t))}^{1/2}\gamma_{(x,t)}(\rho)^{1/2}, \tag{6.33}$$

where as before $\kappa = r/\rho$. Therefore, if $(x, t) \in Q_{r_0}^*(x_0, t_0)$ and we have $0 < 2r \leq \rho \leq r_0$ then

$$\alpha(r) + \beta(r) \leq C\kappa\mu + \frac{C}{\kappa}\mu^{3/2} + \frac{C\mu^{1/2}}{\kappa^{1/2}}\|f\|_{L^{5/3}(Q_\rho)}^{1/2}.$$

Let $\kappa > 0$ be as in the proof of Theorem 6.2.1. If $\mu > 0$ is sufficiently small, we get

$$\alpha_{(x,t)}(r) + \beta_{(x,t)}(r) + \gamma_{(x,t)}(r) + \delta_{(x,t)}(r) \leq C\mu^{1/2}$$

for all r and (x, t) as above. Now, the assumption (6.30) on f implies that

$$\lambda_{(x,t)}(r) \leq C\|f\|_{L^{5/3+q}(D)}r^{q_0}$$

for all $r > 0$ and all (x, t), where $q_0 = 9q/(5 + 3q)$. By (6.28), we have

$$\theta_{(x,t)}(r) \leq \frac{1}{4}\theta_{(x,t)}(\rho) + C\theta_{(x,t)}(\rho)^3 + C\|f\|_{L^{5/3+q}(D)}\rho^{q_0}.$$

Iterating this inequality, we obtain $u \in \mathcal{L}_{2+q_1}^3(\mathcal{V})$ and $p \in \mathcal{L}_{2+q_1}^{3/2}(\mathcal{V})$ for some $q_1 > 0$ and a neighbourhood $\mathcal{V}$ of (x_0, t_0). Without loss of generality $q_1 < 3$.

With this last information, $u \in L^{10/3}(D)$ can be bootstrapped to a higher L^r space in a neighbourhood of (x_0, t_0). Namely, let $\eta \in C_0^\infty(\mathbb{R}^n)$

be a function that is identically 1 on a neighbourhood D_0 of (x_0, t_0) and is supported in $\mathcal{V}$. Consider

$$v_k(x, t) = \int_{-\infty}^t \int \partial_j G(x - y, t - s) \eta(y, s) u_j(y, s) u_k(y, s) \, dy \, ds$$

$$+ \int_{-\infty}^t \int \partial_k G(x - y, t - s) \eta(y, s) p(y, s) \, dy \, ds$$

$$+ \int_{-\infty}^t \int G(x - y, t - s) \eta(y, s) f_k(y, s) \, dy \, ds$$

$$= v_k^{(1)} + v_k^{(2)} + v_k^{(3)}.$$

Then $u - v$ is smooth in D_0 since it solves the heat equation there. On the other hand using $G(x, t) \leq C(|x| + \sqrt{t})^{-3}$ and $|\nabla G(x, t)| \leq C(|x| + \sqrt{t})^{-4}$ for all $(x, t) \in \mathbb{R}^3 \times (0, \infty)$ and Lemma 6.3.1, we get $v_k^{(1)} \in L^{(10/3)/(1 - q_1/(3 - q_1))}(\mathcal{V}_1)$, where $\mathcal{V}_1$ is a neighborhood of (x_0, t_0). Since $u \in L_{\text{loc}}^q(\mathcal{V})$ and $p \in L^{3/2}(D)$ imply $p \in L_{\text{loc}}^{q/2}(\mathcal{V}) + (L_t^{3/2} L_x^\infty)_{\text{loc}}(\mathcal{V})$, we get

$$v_k^{(2)} \in L^{(10/3)/(1 - q_1/(3 - q_1))}(\mathcal{V}_1) + (L_t^6 L_x^\infty)_{\text{loc}}(\mathcal{V}_1),$$

by shrinking $\mathcal{V}_1$ if required. On the other hand, by Young's inequality, $v_k^{(3)} \in L^5(D)$. Repeating this argument finitely many times, we conclude that u belongs to L^5 in a neighborhood of (x_0, t_0) as claimed. $\qquad\square$

6.4 A partial regularity result for $f \in L^{5/3}$

In this section, we prove that the same theorem as in Section 6.3 holds even if $q = 0$. The theorem below is proved in Kukavica (2008b) for the case of zero forcing.

Theorem 6.4.1 *Assume that*

$$f \in L^{5/3}(D). \tag{6.34}$$

If $\mu > 0$ in (6.3) is a small enough universal constant, then $(x_0, t_0) \in \mathcal{R}$ implies that $u \in L^5(\mathcal{V})$ for some neighbourhood $\mathcal{V}$ of (x_0, t_0).

Instead of (6.3), we shall only need the assumption

$$\frac{1}{r^2} \iint_{Q_r^*(x, t)} |u|^3 \leq \mu^3 \tag{6.35}$$

for $(x, t) \in Q_{r_0}^*(x_0, t_0)$ and $r \in (0, r_0]$ where μ is sufficiently small constant which is to be determined. (We are allowed to take the centred parabolic cubes $Q_r^*(x,t)$ rather than $Q_r(x,t)$ since (6.3) is assumed to hold for (x,t) in a neighbourhood of (x_0,t_0).)

Proof. Using a standard fixed point argument, we may assume that $r_0 > 0$ is so small that the equation

$$\partial_t u_f - \Delta u_f + u_f \cdot \nabla u_f + \nabla p_f = f\chi_{Q_{r_0}^*(x_0,t_0)}$$

$$u_f(x, t_0 - r_0^2) = 0$$

is uniquely solvable on $\mathbb{R}^3 \times [t_0 - r_0^2, \infty)$. By taking r_0 sufficiently small, we may assume that $\|u_f\|_{L^5(\mathbb{R}^3 \times (t_0 - r_0^2, \infty))}$ and $\|p_f\|_{L^{5/2}(\mathbb{R}^3 \times (t_0 - r_0^2, \infty))}$ are as small as we wish. Let $\widetilde{u} = u - u_f$ and $\widetilde{p} = p - p_f$. Then

$$\partial_t \widetilde{u} - \Delta \widetilde{u} + u_f \cdot \nabla \widetilde{u} + \widetilde{u} \cdot \nabla u_f + \widetilde{u} \cdot \nabla \widetilde{u} + \nabla \widetilde{p} = \widetilde{f}$$

$$\nabla \cdot \widetilde{u} = 0,$$

where $\widetilde{f}$ vanishes in $Q_{r_0}^*(x_0,t_0)$. Since $\|u_f\|_{L^5(\mathbb{R}^3 \times (t_0 - r_0^2, \infty))}$ is as small as we wish, we may assume that $\|u_f\|_{L^5(\mathbb{R}^3 \times (t_0 - r_0^2, \infty))}$ and

$$\frac{1}{r^2} \iint\limits_{Q_r^*(x,t)} |\widetilde{u}|^3 \leq 2\mu^3$$

for all $(x,t) \in Q_{r_0}^*(x_0,t_0)$ and $r \leq r_0$. For simplicity of notation, we omit from here on the tilde on $\widetilde{u}$ and $\widetilde{p}$. First, we have

$$\partial_t u - \Delta u + u_f \cdot \nabla u + u \cdot \nabla u_f + u \cdot \nabla u + \nabla p = 0$$

$$\nabla \cdot u = 0$$

in $Q_{r_0}^*(x_0,t_0)$. Assume that $0 < 100r \leq \widetilde{\rho} \leq 100\widetilde{\rho} \leq \rho \leq 100\rho \leq r_0/4$. (Further below we shall choose $\widetilde{\rho} = \sqrt{r\rho}$.) Let $(x_1,t_1) \in Q_{r_0/2}^*(x_0,t_0)$ be arbitrary. For simplicity of notation, we translate in (x,t) so that $(x_1,t_1) = (0,0)$. Let $\eta_0(x,t) = \eta_1(x)\eta_2(t)$, where $\eta_1 \in C_0^\infty(B_{4/5}, [0,1])$ is such that $\eta_1 \equiv 1$ on $B_{3/5}$, and $\eta_2 \in C_0^\infty((-16/25, 16/25), [0,1])$ is such that $\eta_1 \equiv 1$ on $(-9/25, 9/25)$. Denote

$$\widetilde{\eta}(x,t) = \eta_0\left(\frac{x}{\widetilde{\rho}}, \frac{t}{\widetilde{\rho}^2}\right)$$

and

$$\eta(x,t) = \eta_0\left(\frac{x}{\rho}, \frac{t}{\rho^2}\right)$$

and let $v = \widetilde{\eta} u = (\widetilde{\eta} u_1, \widetilde{\eta} u_2, \widetilde{\eta} u_3)$ and $U = \chi_{Q_{\tilde{\rho}}^*} u$. Then

$$\partial_t v_k - \Delta v_k = -\partial_j(U_j v_k) - \partial_k(\widetilde{\eta} p) - \partial_j(U_{fj} v_k) - \partial_j(v_j U_{fk}) + U_j U_k \partial_j \widetilde{\eta}$$
$$+ p \partial_k \widetilde{\eta} + U_{fj} U_k \partial_j \widetilde{\eta} + U_j U_{fk} \partial_j \widetilde{\eta} + U_k(\widetilde{\eta}_t + \Delta \widetilde{\eta}) - 2\partial_j(U_k \partial_j \widetilde{\eta}),$$

where $U_f = (U_{f1}, U_{f2}, U_{f3}) = u_f \chi_{Q_{\tilde{\rho}}^*}$. Recalling that $G(x, t) = 0$ for $t \le 0$, we get

$$v_k(x, t) = - \iint \partial_j G(x - y, t - s)\big(U_j(y, s)v_k(y, s) + U_{fj}(y, s)v_k(y, s)$$
$$+ U_{fk}(y, s)v_j(y, s)\big)\, dy\, ds$$
$$+ \iint G(x - y, t - s)\big(U_j U_k \partial_j \widetilde{\eta} + U_{fj} U_k \partial_j \widetilde{\eta} + U_j U_{fk} \partial_j \widetilde{\eta} + p \partial_k \widetilde{\eta}$$
$$+ U_k(\widetilde{\eta}_t + \Delta \widetilde{\eta})\big)(y, s)\, dy\, ds$$
$$- 2 \iint \partial_j G(x - y, t - s)(U_k \partial_j \widetilde{\eta})(y, s)\, dy\, ds$$
$$- \iint \partial_k G(x - y, t - s)(\widetilde{\eta} p)(y, s)\, dy\, ds$$
$$= I_1 + I_2 + I_3 + I_4.$$

Note that

$$|I_1(x, t)| \le C \iint \frac{(|v|\,|U| + |v|\,|U^f|)(y, s)}{(|x - y| + \sqrt{t - s})^4}\, dy\, ds,$$

whence, by Lemma 6.3.1 and the parabolic Hardy–Littlewood–Sobolev inequality,

$$\|I_1\|_{L^3(Q_r^*)} \le C\|Uv\|_{L_2^{3/2}}^{1/2}\|Uv\|_{L^{3/2}}^{1/2} + C\|v\|_{L^3}\|U^f\|_{L^5} \le C\mu\|u\|_{L^3(Q_{\tilde{\rho}}^*)}.$$

Note that each term of the integrand in I_2 contains a derivative of $\widetilde{\eta}$ and thus vanishes for $(y, s) \in Q_{3\tilde{\rho}/5}^* \cup (Q_{4\tilde{\rho}/5}^*)^c$; therefore, the integrand is bounded if $(x, t) \in Q_r^*$. In fact, for $(x, t) \in Q_r^*$,

$$|I_2(x, t)| \le \frac{C}{\tilde{\rho}^4} \sum_{j,k} \|U_j U_k\|_{L^1(Q_{\tilde{\rho}}^*)} + \frac{C}{\tilde{\rho}^4}\|p\|_{L^1(Q_{\tilde{\rho}}^*)}$$
$$+ \frac{C}{\tilde{\rho}^4} \sum_{j,k} \|U_{fj} U_k\|_{L^1(Q_{\tilde{\rho}}^*)} + \frac{C}{\tilde{\rho}^5}\|U\|_{L^1(Q_{\tilde{\rho}}^*)}$$

and thus, assuming $\mu \leq 1$,

$$\|I_2\|_{L^3(Q_r^*)} \leq \frac{Cr^{5/3}}{\widetilde{\rho}^{7/3}}\left(\|U\|_{L^3(Q_{\widetilde{\rho}}^*)}^2 + \|p\|_{L^{3/2}(Q_{\widetilde{\rho}}^*)}\right.$$

$$\left. + \|U_f\|_{L^3(Q_{\widetilde{\rho}}^*)}\|U\|_{L^3(Q_{\widetilde{\rho}}^*)} + \widetilde{\rho}^{2/3}\|U\|_{L^3(Q_{\widetilde{\rho}}^*)}\right)$$

$$\leq C\widetilde{\kappa}^{5/3}\|u\|_{L^3(Q_{\widetilde{\rho}}^*)} + \frac{Cr^{5/3}\rho^{2/3}}{\widetilde{\rho}^{7/3}}\frac{1}{\rho^{2/3}}\|p\|_{L^{3/2}(Q_{\widetilde{\rho}}^*)}$$

$$\leq C\widetilde{\kappa}^{5/3}\|u\|_{L^3(Q_{\widetilde{\rho}}^*)} + \frac{C\widetilde{\kappa}^{5/3}}{\overline{\kappa}^{2/3}}\frac{1}{\rho^{2/3}}\|p\|_{L^{3/2}(Q_{\rho}^*)},$$

where $\widetilde{\kappa} = r/\widetilde{\rho}$ and $\overline{\kappa} = \widetilde{\rho}/\rho$. Similarly, for $(x,t) \in Q_r^*$,

$$|I_3(x,t)| \leq \frac{C}{\widetilde{\rho}^5}\|U\|_{L^1(Q_{\widetilde{\rho}}^*)},$$

from where

$$\|I_3\|_{L^3(Q_r^*)} \leq \frac{Cr^{5/3}}{\widetilde{\rho}^{5/3}}\|U\|_{L^3} \leq C\widetilde{\kappa}^{5/3}\|u\|_{L^3(Q_{\widetilde{\rho}}^*)}.$$

Since $\widetilde{\eta} = \eta\widetilde{\eta}$, we may rewrite

$$I_4(x,t) = -\iint \partial_k G(x-y, t-s)(\widetilde{\eta}\eta p)(y,s)\,dy\,ds.$$

Then we use a representation of the type (6.26), which takes the form

$$\eta p = -R_i R_j(\eta U_{ij}) + N * ((\partial_{ij}\eta)U_{ij}) - \partial_j N * (U_{ij}\partial_i\eta)$$

$$- \partial_i N * (U_{ij}\partial_j\eta) - N * (p\Delta\eta) + 2\partial_j N * ((\partial_j\eta)p)$$

$$= \pi_1 + \pi_2 + \pi_3 + \pi_4 + \pi_5 + \pi_6,$$

where

$$U_{ij} = u_i u_j + u_{fi} u_j + u_i u_{fj}. \tag{6.36}$$

Define

$$J_m(x,t) = -\iint \partial_k G(x-y, t-s)\widetilde{\eta}(y,s)\pi_m(y,s)\,dy\,ds, \qquad m = 1,\ldots,6.$$

For the first term J_1, we have

$$J_1(x,t) = \iint \partial_k G(x-y, t-s)\widetilde{\eta}(y,s)R_i R_j(\eta U_{ij})(y,s)\,dy\,ds$$

$$= \iint \partial_k R_i R_j G(x-y, t-s)(\eta U_{ij})(y,s)\,dy\,ds$$

$$+ \iint \partial_k G(x-y, t-s)(\widetilde{\eta}(y,s)-1)R_i R_j(\eta U_{ij})(y,s)\,dy\,ds$$

$$= J_{11} + J_{12}. \tag{6.37}$$

For the term J_{11}, we have by Fabes et al. (1972),

$$|\nabla R_i R_j G(x,t)| \leq \frac{C}{(|x| + \sqrt{t})^4}, \qquad x \in \mathbb{R}^3, \quad t > 0, \quad i,j = 1,2,3.$$

Using Lemma 6.3.1 and the parabolic Hardy–Littlewood–Sobolev inequality, we obtain

$$\|J_{11}\|_{L^3(Q_r^*)} \leq C \|\eta u_i u_j\|_{L_2^{3/2}}^{1/2} \|\eta u_i u_j\|_{L^{3/2}}^{1/2} + C \sum_{i,j} \|\eta u_i u_{fj}\|_{L^{15/8}}$$

$$\leq C\mu \|u\|_{L^3(Q_\rho^*)}.$$

For the term J_{12}, note that $\widetilde{\eta} - 1$ vanishes on $Q_{\widetilde{\rho}/2}^*$. Therefore,

$$|J_{12}(x,t)| \leq C \iint\limits_{Q_{\widetilde{\rho}/2}^*(x,t)^c} \frac{|R_i R_j (\eta u_i u_j)(y,s)| + |R_i R_j (\eta u_{fi} u_j)(y,s)|}{(|x-y| + |t-s|^{1/2})^4} \, dy \, ds$$

for $(x,t) \in Q_r^*$. Then write

$$Q_{\widetilde{\rho}/2}^*(x,t)^c = \bigcup_{m=0}^{\infty} \left(Q_{2^m \widetilde{\rho}}^*(x,t) \cap Q_{2^{m-1}\widetilde{\rho}}^*(x,t)^c \right)$$

and obtain

$$|J_{12}(x,t)| \leq \sum_{m=0}^{\infty} \frac{C}{2^{4m}\widetilde{\rho}^4} \left(\|R_i R_j(\eta u_i u_j)\|_{L^1(Q_{2^m\widetilde{\rho}}^*(x,t) \cap Q_{2^{m-1}\widetilde{\rho}}^*(x,t)^c)} \right.$$

$$\left. + \|R_i R_j(\eta u_{fi} u_j)\|_{L^1(Q_{2^m\widetilde{\rho}}^*(x,t) \cap Q_{2^{m-1}\widetilde{\rho}}^*(x,t)^c)} \right)$$

$$\leq \sum_{m=0}^{\infty} \frac{C}{2^{7m/3}\widetilde{\rho}^{7/3}} \left(\|R_i R_j(\eta u_i u_j)\|_{L^{3/2}(Q_{2^m\widetilde{\rho}}^*(x,t) \cap Q_{2^{m-1}\widetilde{\rho}}^*(x,t)^c)} \right.$$

$$\left. + \|R_i R_j(\eta u_{fi} u_j)\|_{L^{3/2}(Q_{2^m\widetilde{\rho}}^*(x,t) \cap Q_{2^{m-1}\widetilde{\rho}}^*(x,t)^c)} \right)$$

$$\leq \sum_{m=0}^{\infty} \frac{C}{2^{7m/3}\widetilde{\rho}^{7/3}} \left(\sum_{i,j} \|\eta u_i u_j\|_{L^{3/2}(\mathbb{R}^3 \times (-\rho^2, \rho^2))} \right.$$

$$\left. + \sum_{i,j} \|\eta u_{fi} u_j\|_{L^{3/2}(\mathbb{R}^3 \times (-\rho^2, \rho^2))} \right)$$

$$\leq \sum_{m=0}^{\infty} \frac{C}{2^{7m/3}\widetilde{\rho}^{7/3}} \left(\|u\|_{L^3(Q_\rho^*)}^2 + \|u\|_{L^3(Q_\rho^*)} \|u_f\|_{L^3(Q_\rho^*)} \right)$$

$$\leq \frac{C\mu\rho^{2/3}}{\widetilde{\rho}^{7/3}} \|u\|_{L^3(Q_\rho^*)}$$

for $(x,t) \in Q_r^*$. Therefore,

$$\|J_{12}\|_{L^3(Q_r^*)} \le Cr^{5/3}\|J_{12}\|_{L^\infty(Q_r^*)} \le \frac{C\mu\widetilde{\kappa}^{5/3}}{\overline{\kappa}^{2/3}}\|u\|_{L^3(Q_r^*)}.$$

Next, using the parabolic Hardy–Littlewood–Sobolev inequality, we get

$$\|J_2\|_{L^3(Q_r^*)} \le Cr^{7/10}\|J_2\|_{L_t^3 L_x^{10}(Q_r^*)} \le Cr^{7/10}\|\widetilde{\eta}\pi_2\|_{L_t^{3/2} L_x^{90/19}}$$
$$\le Cr^{7/10}\widetilde{\rho}^{19/30}\|N*((\partial_{ij}\eta)U_{ij})\|_{L_t^{3/2} L_x^\infty(Q_{\widetilde{\rho}}^*)}.$$

Now, note that, by the construction of η, we have $\partial_{ij}\eta(y,s) = 0$ on the set $\{(y,s) : |y| \le 3/5\}$. Therefore, the convolution above is not singular, and we have $\|N*((\partial_{ij}\eta)U_{ij})\|_{L_t^{3/2} L_x^\infty(Q_{\widetilde{\rho}}^*)} \le C\rho^{-3}\|U\|_{L_t^{3/2} L_x^1(Q_\rho^*)}$, and thus

$$\|J_2\|_{L^3(Q_r^*)} \le \frac{Cr^{7/10}\widetilde{\rho}^{19/30}}{\rho^3}\|U\|_{L_t^{3/2} L_x^1(Q_\rho^*)} \le \frac{Cr^{7/10}\widetilde{\rho}^{19/30}}{\rho^2}\|U\|_{L^{3/2}(Q_\rho^*)}$$
$$\le \frac{C\mu r^{7/10}\widetilde{\rho}^{19/30}}{\rho^{4/3}}\|u\|_{L^3(Q_\rho^*)} \le C\mu\|u\|_{L^3(Q_\rho^*)}.$$

The same upper bounds can be derived for J_3 and J_4. As for J_5, we have by the parabolic Hardy–Littlewood–Sobolev inequality

$$\|J_5\|_{L^3(Q_r^*)} \le Cr^{7/10}\|J_5\|_{L_t^3 L_x^{10}(Q_r^*)} \le Cr^{7/10}\|\widetilde{\eta}\pi_5\|_{L_t^{3/2} L_x^{90/19}}$$
$$\le Cr^{7/10}\widetilde{\rho}^{19/30}\|N*(p\Delta\eta)\|_{L_t^{3/2} L_x^\infty(Q_{\widetilde{\rho}}^*)}$$
$$\le \frac{Cr^{7/10}\widetilde{\rho}^{19/30}}{\rho^3}\|p\|_{L_t^{3/2} L_x^1(Q_\rho^*)}$$
$$\le \frac{Cr^{7/10}\widetilde{\rho}^{19/30}}{\rho^2}\|p\|_{L^{3/2}(Q_\rho^*)} = C\widetilde{\kappa}^{7/10}\overline{\kappa}^{4/3}\frac{1}{\rho^{2/3}}\|p\|_{L^{3/2}(Q_\rho^*)}.$$

The same estimate holds also for J_6. In summary, we get

$$\|u\|_{L^3(Q_r^*)} \le C\left(\mu + \widetilde{\kappa}^{5/3} + \frac{\mu\widetilde{\kappa}^{5/3}}{\overline{\kappa}^{2/3}}\right)\|u\|_{L^3(Q_\rho^*)}$$
$$+ C\left(\frac{\widetilde{\kappa}^{5/3}}{\overline{\kappa}^{2/3}} + \widetilde{\kappa}^{7/10}\overline{\kappa}^{4/3}\right)\frac{1}{\rho^{2/3}}\|p\|_{L^{3/2}(Q_\rho^*)}.$$

Setting $\widetilde{\rho} = \sqrt{\rho r}$, we have $\widetilde{\kappa} = \overline{\kappa} = \kappa^{1/2}$, where $\kappa = r/\rho$, and thus the inequality reduces to

$$\|u\|_{L^3(Q_r^*)} \le C(\mu + \kappa^{5/6})\|u\|_{L^3(Q_\rho^*)} + \frac{C\kappa^{1/2}}{\rho^{2/3}}\|p\|_{L^{3/2}(Q_\rho^*)}$$

whence

$$\frac{1}{r^{2/3}}\|u\|_{L^3(Q_r^*)} \le C\left(\frac{\mu}{\kappa^{2/3}} + \kappa^{1/6}\right)\frac{1}{\rho^{2/3}}\|u\|_{L^3(Q_\rho^*)} + \frac{C}{\kappa^{1/6}\rho^{4/3}}\|p\|_{L^{3/2}(Q_\rho^*)}.$$

By a similar estimate for the pressure, we get

$$\|p\|_{L^{3/2}(Q_r^*)} \le C\|u\|_{L^3(Q_\rho^*)}^2 + C\|u\|_{L^3(Q_\rho^*)}\|u_f\|_{L^3(Q_\rho^*)} + C\kappa^2\|p\|_{L^3(Q_\rho^*)}$$

$$\le C\mu\rho^{2/3}\|u\|_{L^3(Q_\rho^*)} + C\kappa^2\|p\|_{L^3(Q_\rho^*)},$$

from where

$$\frac{1}{r^{4/3}}\|p\|_{L^{3/2}(Q_\rho^*)} \le \frac{C\mu}{\kappa^{4/3}\rho^{2/3}}\|u\|_{L^3(Q_\rho^*)} + \frac{C\kappa^{2/3}}{\rho^{4/3}}\|p\|_{L^{3/2}(Q_\rho^*)}. \tag{6.38}$$

The function $\theta(r) = r^{-2/3}\|u\|_{L^3(Q_r^*)} + \kappa^{-1/3}r^{-4/3}\|p\|_{L^{3/2}(Q_r^*)}$ satisfies

$$\theta(r) \le C\left(\frac{\mu}{\kappa^{5/3}} + \kappa^{1/6}\right)\theta(\rho). \tag{6.39}$$

Next, choosing κ and μ as appropriate constants (first we choose κ sufficiently small and then μ sufficiently small compared to $\kappa^{5/3}$), the equation (6.39) becomes

$$\theta(\kappa\rho) \le \frac{1}{2}\theta(\rho) \tag{6.40}$$

provided that $0 < \rho < r_0/C$ with C a sufficiently large universal constant. Iterating this inequality, we get existence of $\epsilon > 0$ such that

$$\frac{1}{r^{2+\epsilon}}\iint\limits_{Q_r^*(x_1,t_1)}\left(|u|^3 + |p|^{3/2}\right) \le \mu^3 \tag{6.41}$$

for all (x_1, t_1) is a neighbourhood of (x_0, t_0) and all sufficiently small $r > 0$. Now, this condition is no longer critical, so we may bootstrap as in the previous section and obtain $u \in L^5(D_0)$, where D_0 is a neighborhood of (x_0, t_0). $\qquad\square$

Acknowledgements

I would like to thank James Robinson, José Rodrigo, and Vlad Vicol for useful remarks and suggestions. The work was supported in part by the NSF grant DMS-0604886.

References

Caffarelli, L., Kohn, R. & Nirenberg, L. (1982) Partial regularity of suitable weak solutions of the Navier–Stokes equations. *Comm. Pure Appl. Math.* **35**, no. 6, 771–831.

Constantin, P. & Foias, C. (1988) *Navier–Stokes equations*. Chicago Lectures in Mathematics, University of Chicago Press, Chicago, IL.

Fabes, E.B., Jones, B.F. & Riviere, N.M. (1972) The initial value problem for the Navier–Stokes equations with data in L^p. *Arch. Rational Mech. Anal.* **45**, 222–240.

Hopf, E. (1951) Über die Anfangswertaufgabe für die hydrodynamischen Grundgleichungen. *Math. Nachr.* **4**, 213–231.

Kukavica, I. (2008a) On partial regularity for the Navier–Stokes equations. *Discrete Contin. Dynam. Systems* **21**, no. 3, 717–728.

Kukavica, I. (2008b) Regularity for the Navier–Stokes equations with a solution in a Morrey space. *Indiana Univ. Math. J.* **57**, no. 6, 2843–2860.

Ladyzhenskaya O.A. & Seregin, G.A. (1999) On partial regularity of suitable weak solutions to the three-dimensional Navier–Stokes equations. *J. Math. Fluid Mech.* **1**, no. 4, 356–387.

Lemarié-Rieusset, P.G. (2002) *Recent developments in the Navier–Stokes problem*. Chapman & Hall/CRC Research Notes in Mathematics, **431**, Chapman & Hall/CRC, Boca Raton, FL.

Leray, J. (1934) Sur le mouvement d'un liquide visqueux emplissant l'espace. *Acta Math.* **63**, no. 1, 193–248.

Lin, F. (1998) A new proof of the Caffarelli–Kohn–Nirenberg theorem. *Comm. Pure Appl. Math.* **51**, no. 3, 241–257.

O'Leary, M. (2003) Conditions for the local boundedness of solutions of the Navier–Stokes system in three dimensions. *Comm. Partial Differential Equations* **28**, no. 3-4, 617–636.

Olsen, P.A. (1995) Fractional integration, Morrey spaces and a Schrödinger equation. *Comm. Partial Differential Equations* **20**, no. 11–12, 2005–2055.

Robinson J.C. & Sadowski, W. (2007) Decay of weak solutions and the singular set of the three-dimensional Navier–Stokes equations. *Nonlinearity* **20**, no. 5, 1185–1191.

Scheffer, V. (1976a) Partial regularity of solutions to the Navier–Stokes equations. *Pacific J. Math.* **66**, no. 2, 535–552.

Scheffer, V. (1976b) Turbulence and Hausdorff dimension. *Turbulence and Navier–Stokes equations* (Proc. Conf., Univ. Paris-Sud, Orsay, 1975). Lecture Notes in Math. **565**, 174–183. Springer, Berlin.

Scheffer, V. (1977) Hausdorff measure and the Navier–Stokes equations. *Comm. Math. Phys.* **55**, no. 2, 97–112.

Serrin, J. (1962) On the interior regularity of weak solutions of the Navier–Stokes equations. *Arch. Rational Mech. Anal.* **9**, 187–195.

Serrin, J. (1963) The initial value problem for the Navier–Stokes equations. *Nonlinear Problems* (Proc. Sympos., Madison, Wis.) 69–98. Univ. of Wisconsin Press, Madison, Wis.

Sohr, H. (1983) Zur Regularitätstheorie der instationären Gleichungen von Navier–Stokes. *Math. Z.* **184**, no. 3, 359-375.

Struwe, M. (1998) On partial regularity results for the Navier–Stokes equations. *Comm. Pure Appl. Math.* **41**, no. 4, 437–458.

Stein, E.M. (1993) *Harmonic analysis: real-variable methods, orthogonality, and oscillatory integrals.* Princeton Mathematical Series, **43**, Princeton University Press, Princeton, NJ.

Taylor, M.E. (1992) Analysis on Morrey spaces and applications to Navier–Stokes and other evolution equations *Comm. Partial Differential Equations* **17** (1992), no. 9-10, 1407–1456.

Temam, R. (2001) *Navier–Stokes equations: Theory and numerical analysis.* AMS Chelsea Publishing, Providence, RI. Reprint of the 1984 edition.

Vasseur, A.F.(2007) A new proof of partial regularity of solutions to Navier-Stokes equations. *NoDEA Nonlinear Differential Equations Appl.* **14**, no. 5-6, 753–785.

7

Anisotropic Navier–Stokes equations in a bounded cylindrical domain

Marius Paicu

Univ Paris-Sud and CNRS, Laboratoire de Mathématiques d'Orsay,
Orsay Cedex, F-91405. France.
Marius.Paicu@math.u-psud.fr

Geneviève Raugel

CNRS and Univ Paris-Sud, Laboratoire de Mathématiques d'Orsay,
Orsay Cedex, F-91405. France.
Genevieve.Raugel@math.u-psud.fr

Abstract

In this paper, we study the global and local existence and uniqueness of solutions to the Navier–Stokes equations with anisotropic viscosity in a bounded cylindrical domain $Q = \Omega \times (0,1)$, where Ω is a star-shaped domain in $\mathbb{R}^2$. Here we consider the case of homogeneous Dirichlet boundary conditions on the lateral boundary and vanishing normal trace on the top and the bottom of the cylinder.

7.1 Introduction

The Navier–Stokes equations with anisotropic viscosity are classical in geophysical fluid dynamics. Instead of choosing the more standard viscosity $-\nu(\partial_1^2 + \partial_2^2 + \partial_3^2)$ in the case of three-dimensional fluids, meteorologists often model turbulent flows using a viscosity of the form $-\nu_h(\partial_1^2 + \partial_2^2) - \nu_v\partial_3^2$, where ν_v is usually much smaller than ν_h and thus can be neglected (see Chapter 4 in Pedlosky (1979), for a detailed discussion).

More precisely, in geophysical fluids, the rotation of the earth plays a crucial role. The Coriolis force introduces a penalized skew-symmetric term $\varepsilon^{-1}u \times e_3$ into the equations, where $\varepsilon > 0$ is the Rossby number and $e_3 = (0,0,1)$ is the unit vertical vector. This leads to an asymmetry between the horizontal and vertical motions. By the Taylor–Proudman Theorem (see Pedlosky (1979) and Taylor (1923)), the fluid tends to have a two-dimensional behaviour far from the boundary of the domain. When the fluid evolves between two parallel plates with homogeneous Dirichlet boundary conditions, Ekman boundary layers of the

Published in *Partial Differential Equations and Fluid Mechanics*, edited by James C. Robinson and José L. Rodrigo. © Cambridge University Press 2009.

form $U_{BL}(x_1, x_2, \varepsilon^{-1}x_3)$ appear near the boundary. In order to compensate the term $\varepsilon^{-1}U_{BL} \times e_3$ by the term $-\nu_v \partial_3^2 U_{BL}$, we need to impose the requirement that $\nu_v = \beta\varepsilon$, for $\beta > 0$ (Grenier & Masmoudi, 1997; Chemin et al., 2006).

When the fluid occupies the whole space, the Navier–Stokes equations with vanishing or small vertical viscosity are as follows:

$$\partial_t u + u\nabla u - \nu_h(\partial_1^2 + \partial_2^2)u - \nu_v \partial_3^2 u = -\nabla p \ \ \text{in } \mathbb{R}^3, \quad t > 0$$
$$\text{div } u = 0 \ \ \text{in } \mathbb{R}^3, \quad t > 0 \tag{7.1}$$
$$u|_{t=0} = u_0,$$

where $\nu_h > 0$ and $\nu_v \geq 0$ represent the horizontal and vertical viscosities and where $u = (u_1, u_2, u_3)$ and p are the vector field of the velocities and the pressure respectively. In the case of vanishing vertical viscosity, the classical theory of the Navier–Stokes equations does not apply and new difficulties arise. Some partial L^2-energy estimates still hold for system (7.1), but they do not allow one to pass to the limit and to obtain a weak solution as in the well-known construction of weak Leray solutions nor to use directly the results of Fujita & Kato (1964) on the existence of strong solutions. Of course, neglecting the horizontal viscosity and requiring a lot of regularity of the initial data, one can prove the local existence of strong solutions within the framework of hyperbolic symmetric systems. Thus, new methods have to be developed.

In the case $\nu_v = 0$, this system was first studied by Chemin et al. (2000), who showed local and global existence of solutions in anisotropic Sobolev spaces which take into account this anisotropy. More precisely, for $s \geq 0$, let us introduce the anisotropic Sobolev spaces,

$$H^{0,s} = \{u \in L^2(\mathbb{R}^3)^3 \mid \|u\|_{H^{0,s}}^2 = \int_{\mathbb{R}^3}(1 + |\xi_3|^2)^s |\hat{u}(\xi)|^2 \, d\xi < +\infty\}$$

and $H^{1,s} = \{u \in H^{0,s} \mid \partial_i u \in H^{0,s}, \ i = 1, 2\}$ (such anisotropic spaces were introduced in Iftimie (1999) for the study of the Navier–Stokes system in thin domains).

Chemin et al. (2000) showed that, for any $s_0 > 1/2$, and any $u_0 \in H^{0,s_0}$, there exist $T > 0$ and a local solution

$$u \in L^\infty((0, T), H^{0,s_0}) \cap L^2((0, T), H^{1,s_0})$$

of the anisotropic Navier–Stokes equations (7.1). If $\|u_0\|_{H^{0,s_0}} \leq c\nu_h$, where $c > 0$ is a small constant, then the solution is global in time. In the same paper, the authors proved that there exists at most one solution $u(t)$ of the equations (7.1) in the space $L^\infty((0, T), H^{0,s}) \cap L^2((0, T), H^{1,s})$

for $s > 3/2$. In these results, there was a gap between the regularity required for the existence of solutions and that required for uniqueness. This gap was closed by Iftimie (2002) who showed the uniqueness of the solution in the space $L^\infty((0,T), H^{0,s}) \cap L^2((0,T), H^{1,s})$ for $s > 1/2$.

Like the classical Navier–Stokes equations, the system (7.1) on the whole space $\mathbb{R}^3$ has a scaling. Indeed, if u is a solution of the equations (7.1) on a time interval $[0,T]$ with initial data u_0, then $u_\mu(t,x) = \mu u(\mu^2 t, \mu x)$ is also a solution of (7.1) on the time interval $[0, \mu^{-2}T]$, with initial data $\mu u_0(\mu x)$. This was one of the motivations of Paicu for considering initial data in the scaling invariant Besov space $\mathcal{B}^{0,1/2}$. Paicu (2005b) proved the local existence and uniqueness of the solutions of (7.1), for initial data $u_0 \in \mathcal{B}^{0,1/2}$. He showed global existence of the solution when the initial data in $\mathcal{B}^{0,1/2}$ are small, compared to the horizontal viscosity ν_h (for further details see Paicu, 2004, 2005a,b). Very recently Chemin & Zhang (2007) introduced the scaling invariant Besov–Sobolev spaces $\mathcal{B}_4^{-1/2,1/2}$ and showed existence of global solutions when the initial data u_0 in $\mathcal{B}_4^{-1/2,1/2}$ are small compared to the horizontal viscosity ν_h. This result implies the global well-posedness of (7.1) with highly oscillatory initial data.

Notice that in all the above results, as well as in this paper, one of the key observations is that, in the various essential energy estimates, the partial derivative ∂_3 appears only when applied to the component u_3 in terms like $u_3 \partial_3 u_3$. Even if there is no vertical viscosity and thus no smoothing in the vertical variable, the divergence-free condition implies that $\partial_3 u_3$ is regular enough to get good estimates of the nonlinear term.

Considering the anisotropic Navier–Stokes equations on the whole space $\mathbb{R}^3$ (or on the torus $\mathbb{T}^3$) instead of on a bounded domain with boundary leads to some simplifications. For example, the Stokes operator coincides with the operator $-\Delta$ on the space of smooth divergence-free vector fields. Also one can use Fourier transforms and Littlewood–Paley theory.

One of the few papers considering the anisotropic Navier–Stokes equations on a domain with a boundary is the article by Iftimie & Planas (2006), who studied the anisotropic equations (7.1) on a half-space $\mathcal{H}$, supplemented with the boundary condition

$$u_3 = 0 \text{ in } \partial\mathcal{H}, \quad t > 0. \tag{7.2}$$

This system of equations can be reduced to the case of the whole space $\mathbb{R}^3$. Indeed, let w be a solution of the anisotropic Navier–Stokes equations on the half-space $\mathcal{H}$. Extending the components w_1 and w_2 to

$\mathbb{R}^3$ by an even reflection and the third component w_3 by an odd reflection with respect to the plane $x_3 = 0$, we obtain a vector field $\widetilde{w}$, which is a solution of the equations (7.1) on the whole space $\mathbb{R}^3$. Conversely, if $u(t)$ is a solution of the anisotropic Navier–Stokes equations on the whole space $\mathbb{R}^3$ with initial data $u_0 \in H^{0,1}$ satisfying the condition $u_{0,3} = 0$ on $\partial\mathcal{H}$, then the restriction of $u(t)$ to $\mathcal{H}$ is a solution of the anisotropic Navier–Stokes equations on $\mathcal{H}$, satisfying the condition (7.2). For initial data $u_0 \in L^2(\mathcal{H})^3$, satisfying the condition $\partial_3 u_0 \in L^2(\mathcal{H})^3$, Iftimie & Planas (2006) showed that the solutions of the anisotropic Navier–Stokes equations on $\mathcal{H}$ are limits of solutions u_ε of the Navier–Stokes equations on $\mathcal{H}$ with Navier boundary conditions on the boundary $\partial\mathcal{H}$ and viscosity term $-\nu_h(\partial_1^2 + \partial_2^2)u_\varepsilon - \varepsilon\partial_3^2 u_\varepsilon$ on a time interval $(0, T_0)$, where $T_0 > 0$ is independent of ε. If the initial data u_0 are small with respect to the horizontal viscosity ν_h, they showed the convergence on the infinite time interval $(0, +\infty)$. In our study of the anisotropic Navier–Stokes equations on a bounded domain, we are also going to introduce an auxiliary Navier–Stokes system with viscosity $-\nu_h(\partial_1^2 + \partial_2^2)u_\varepsilon - \varepsilon\partial_3^2 u_\varepsilon$ (see system (NS$_\varepsilon$) below), but instead of considering Navier boundary conditions in the vertical variable, we will choose periodic conditions.

In this paper, we study the global and local existence and uniqueness of solutions to the anisotropic Navier–Stokes equations on a bounded product domain of the type $Q = \Omega \times (0,1)$, where Ω is a smooth two-dimensional domain, with homogeneous Dirichlet boundary conditions on the lateral boundary $\partial\Omega \times (0,1)$. For the sake of simplicity, we assume that Ω is a star-shaped domain. We denote by $\Gamma_0 = \Omega \times \{0\}$ and $\Gamma_1 = \Omega \times \{1\}$ the top and the bottom of Q. More precisely, we consider the system of equations

$$(\text{NS}_h) \begin{cases} \partial_t u + u\nabla u - \nu_h \Delta_h u = -\nabla p \text{ in } Q , \quad t > 0 \\ \operatorname{div} u = 0 \text{ in } Q , \quad t > 0 \\ u|_{\partial\Omega \times (0,1)} = 0 , \quad t > 0 \\ u_3|_{\Gamma_0 \cup \Gamma_1} = 0 , \quad t > 0 \\ u|_{t=0} = u_0 \in H^{0,1}(Q), \end{cases}$$

where the operator $\Delta_h = \partial_{x_1}^2 + \partial_{x_2}^2$ denotes the horizontal Laplacian and $\nu_h > 0$ is the horizontal viscosity. Here $u \equiv (u_1, u_2, u_3) \equiv (u_h, u_3)$ is the vector field of velocities and p denotes the pressure term. To simplify the discussion, we suppose that the forcing term f vanishes (the case of a non-vanishing forcing term as well as the asymptotic behaviour in time of the solutions of (NS$_h$) will be studied in a subsequent paper).

Since the viscosity is anisotropic, we want to solve the above system in the anisotropic function space

$$\widetilde{H}^{0,1}(Q) = \{u \in L^2(Q)^3 \,|\, \operatorname{div} u = 0 \,;\, \gamma_n u = 0 \text{ on } \partial Q \,;\, \partial_3 u \in L^2(Q)^3\},$$

where $\gamma_n u$ is the extension of the normal trace $u \cdot n$ to $H^{-1/2}(\partial Q)$. Since u belongs to $L^2(Q)^3$ and $\operatorname{div} u = 0$, $\gamma_n u$ is well defined and belongs to $H^{-1/2}(\partial Q)$. For later use, we also introduce the space

$$H^{0,1}(\Omega \times (a,b)) = \{u \in L^2(\Omega \times (a,b))^3 \,|\, \operatorname{div} u = 0 \,;\, \partial_3 u \in L^2(\Omega \times (a,b))^3\},$$

where $-\infty < a < b < +\infty$. Clearly, $\widetilde{H}^{0,1}(Q)$ is a closed subspace of $H^{0,1}(Q)$.

Considering a bounded domain Q with lateral Dirichlet boundary conditions instead of working with periodic boundary conditions introduces a new difficulty. In particular, we have to justify that these Dirichlet boundary conditions make sense. Before stating existence results for solutions $u(t)$ to the system (NS$_h$), we describe our strategy to solve this problem. One way of solving the system (NS$_h$) consists of adding an artificial viscosity term $-\varepsilon \partial_3^2 u$ (where $\varepsilon > 0$) to the first equation in (NS$_h$) and in replacing the initial data u_0 by more regular data u_0^ε, that is, in solving the system

$$(\text{NS}_{h,\varepsilon}) \begin{cases} \partial_t u + u \nabla u - \nu_h \Delta_h u - \varepsilon \partial_3^2 u = -\nabla p \;\text{ in } Q \,, \quad t > 0 \\ \operatorname{div} u = 0 \;\text{ in } Q \,, \quad t > 0 \\ u|_{\partial\Omega \times (0,1)} = 0 \,, \quad t > 0 \\ u_3|_{\Gamma_0 \cup \Gamma_1} = 0 \,, \quad t > 0 \\ u|_{t=0} = u_0^\varepsilon \,, \end{cases}$$

where the initial data u_0^ε are chosen in the function space

$$\widetilde{H}_0^1(Q) = \{u \in H^1(Q)^3 \,|\, \operatorname{div} u = 0 \,;\, u|_{\partial\Omega \times (0,1)} = 0 \,;\, u_3 = 0 \text{ on } \Gamma_0 \cup \Gamma_1\},$$

and are close to u_0. In what follows, we will actually consider a sequence of positive numbers ε_n converging to 0, as n goes to infinity. Thus, we will also choose a sequence of initial data $u_0^{\varepsilon_n} \in \widetilde{H}_0^1(Q)$ converging to u_0 in the space $\widetilde{H}^{0,1}(Q)$ when n goes to infinity. A choice of such a sequence is possible since the space $\widetilde{H}_0^1(Q)$ is dense in $\widetilde{H}^{0,1}(Q)$ (see Lemma 7.2.1 in the next section). Notice that, for $\varepsilon > 0$, we need to replace u_0 by more regular initial data u_0^ε in order to be able to apply the Fujita–Kato Theorem.

However, the system of equations (NS$_{h,\varepsilon}$) is still not a classical system. Indeed, we need to impose boundary conditions on the horizontal part

$u_h(t, x_h, x_3)$ on the top $x_3 = 1$ and the bottom $x_3 = 0$ (that is on $\Gamma_0 \cup \Gamma_1$). As in Iftimie & Planas (2006), we could impose Navier-type boundary conditions on the horizontal part of u on Γ_0 and Γ_1, but here we will take another path. We will extend the velocity field u by symmetry to the domain $\widetilde{Q} = \Omega \times (-1, 1)$ and then solve the equations $(\mathrm{NS}_{h,\varepsilon})$ on the symmetrical domain $\widetilde{Q}$ by imposing homogeneous Dirichlet boundary conditions on the lateral boundary and periodic conditions in the vertical variable x_3. More precisely, let u be a vector in $\widetilde{H}_0^1(Q)$. We extend u to a vector $\widetilde{u} \equiv \Sigma u$ on $\widetilde{Q} = \Omega \times (-1, 1)$ by setting

$$(\Sigma u)_i(x_h, -x_3) \equiv \widetilde{u}_i(x_h, -x_3) = u_i(x_h, x_3), \quad i = 1, 2, \quad 0 \le x_3 \le 1$$
$$(\Sigma u)_3(x_h, -x_3) = \widetilde{u}_3(x_h, -x_3) = -u_3(x_h, x_3), \quad 0 \le x_3 \le 1,$$

and $(\Sigma u)(x_h, x_3) \equiv \widetilde{u}(x_h, x_3) = u(x_h, x_3)$ for $0 \le x_3 \le 1$; notice that the vector $\widetilde{u} = \Sigma u$ belongs to the space $H^1(\widetilde{Q})^3$. We introduce the function space

$$\widetilde{V} \equiv H_{0,\mathrm{per}}^1(\widetilde{Q}) = \{u \in H^1(\widetilde{Q})^3 \mid \mathrm{div}\, u = 0; u|_{\partial\Omega \times (0,1)} = 0;$$
$$u(x_h, x_3) = u(x_h, x_3 + 2)\}.$$

The vector $\widetilde{u} = \Sigma u$ clearly belongs to the space $H_{0,\mathrm{per}}^1(\widetilde{Q})$.

We finally consider the problem

$$(\mathrm{NS}_\varepsilon) \begin{cases} \partial_t u_\varepsilon + u_\varepsilon \nabla u_\varepsilon - \nu_h \Delta_h u_\varepsilon - \varepsilon \partial_{x_3}^2 u_\varepsilon = -\nabla p_\varepsilon & \text{in } \widetilde{Q}, \quad t > 0 \\ \mathrm{div}\, u_\varepsilon = 0 \ \text{ in } \widetilde{Q}, \quad t > 0 \\ u_\varepsilon|_{\partial\Omega \times (-1,1)} = 0, \quad t > 0 \\ u_\varepsilon(x_h, x_3) = u_\varepsilon(x_h, x_3 + 2), \quad t > 0 \\ u_\varepsilon|_{t=0} = u_{\varepsilon,0} \in H_{0,\mathrm{per}}^1(\widetilde{Q}). \end{cases}$$

According to the classical theorem of Fujita & Kato (1964), for any initial condition $u_{\varepsilon,0} \in \widetilde{V}$ there exists a unique local strong solution $u_\varepsilon(t) \in C^0([0, T_\varepsilon), \widetilde{V})$ of the Navier–Stokes equations $(\mathrm{NS}_\varepsilon)$. Moreover, this solution is classical and belongs to $C^0((0, T_\varepsilon), H^2(\widetilde{Q})^3) \cap C^1((0, T_\varepsilon), L^2(\widetilde{Q})^3)$. If the time interval of existence is bounded, that is, if $T_\varepsilon < +\infty$, then

$$\|u_\varepsilon(t)\|_{\widetilde{V}} \longrightarrow +\infty, \qquad \text{as } t \to T_\varepsilon^-.$$

We next introduce the "symmetry map" $S : u \in \widetilde{V} \mapsto Su \in \widetilde{V}$ defined as follows

$$(Su)_i(x_h, -x_3) = u_i(x_h, x_3), \quad i = 1, 2,$$
$$(Su)_3(x_h, -x_3) = -u_3(x_h, x_3).$$

We remark that if $u_\varepsilon(t) \in C^0([0, T_\varepsilon), \widetilde{V})$ is a solution of the Navier–Stokes equations (NS$_\varepsilon$), then $Su_\varepsilon \in C^0([0, T_\varepsilon), \widetilde{V})$ is a solution of the equations (NS$_\varepsilon$) with $Su_\varepsilon(0) = Su_{\varepsilon,0}$. If $u_{\varepsilon,0} = \Sigma u_0$, where u_0 belongs to $\widetilde{H}_0^1(Q)$, then, $Su_{\varepsilon,0} = u_{\varepsilon,0}$ and by the above uniqueness property, the solutions $Su_\varepsilon(t)$ and $u_\varepsilon(t)$ coincide. This implies in particular that $u_{\varepsilon,3}(t)$ vanishes on $\Gamma_0 \cup \Gamma_1$ for any $t \in [0, T_\varepsilon)$. Since $u_\varepsilon(t) = Su_\varepsilon(t)$ belongs to $C^0((0, T_\varepsilon), H^2(\widetilde{Q}))$, this also implies that $\partial_3 u_{\varepsilon,h}$ vanishes on $\Gamma_0 \cup \Gamma_1$ for any $t \in (0, T_\varepsilon)$ and for any $\varepsilon > 0$.

In what follows, we denote by ∇_h the gradient operator in the horizontal direction, that is, the gradient with respect to the variables x_1 and x_2. To summarize, for any $u_0 \in \widetilde{H}^{0,1}(Q)$, we will construct a (unique) local (respectively global) solution $\bar{u} \in L^\infty((0, T_0), \widetilde{H}^{0,1}(Q))$, with $\nabla_h \bar{u} \in L^2((0, T_0), H^{0,1}(Q))$ (respectively $\bar{u} \in L^\infty((0, +\infty), \widetilde{H}^{0,1}(Q))$, with $\nabla_h \bar{u}$ in the space $L_{loc}^2((0, +\infty), H^{0,1}(Q)))$ by proceeding as follows. We consider a (decreasing) sequence of positive numbers ε_m converging to zero and, using Lemma 7.2.1, a sequence of initial data $u_0^m \in \widetilde{H}_0^1(Q)$ converging to u_0 in $\widetilde{H}^{0,1}(Q)$, when m goes to infinity. Then, for each m, we solve the problem (NS$_{\varepsilon_m}$) with initial data $u_{\varepsilon_m,0} = \Sigma u_0^m$. We thus obtain a unique local (respectively global) solution $u_{\varepsilon_m}(t) \in C^0((0, T), \widetilde{V})$ of the problem (NS$_{\varepsilon_m}$) where $0 < T < +\infty$ (respectively $T = +\infty$). We show that this sequence $u_{\varepsilon_m}(t)$ is uniformly bounded in $L^\infty((0, T_0), H^{0,1}(\widetilde{Q}))$ and that the sequence of derivatives $\nabla_h u_{\varepsilon_m}(t)$ is uniformly bounded in $L^2((0, T_0), H^{0,1}(\widetilde{Q}))$, where $T_0 > 0$ is independent of m (depending only on u_0). If the initial data are small enough, we show the global existence of these solutions, again with bounds in $L^\infty((0, +\infty), H^{0,1}(\widetilde{Q}))$ and $L^2((0, +\infty), H^{0,1}(\widetilde{Q}))$, independent of m. Using these uniform bounds, we show that the sequences $u_{\varepsilon_m}(t)$ and $\nabla_h u_{\varepsilon_m}(t)$ are Cauchy sequences in $L^\infty((0, T_0), L^2(\widetilde{Q}))$ and in $L^2((0, T_0), L^2(\widetilde{Q}))$ respectively, which implies the existence of a solution $\bar{u} \in L^\infty((0, T_0), \widetilde{H}^{0,1}(Q))$ of (NS$_h$), with $\nabla_h \bar{u} \in L^2((0, T_0), H^{0,1}(Q))$. The uniqueness of the solution $\bar{u}$ is straightforward and is proved in the same way as the Cauchy property of the sequence $u_{\varepsilon_m}(t)$.

This paper is organized as follows. In the second section, we introduce several notations and spaces. We also prove auxiliary results, which will be used in the following sections. In the third section, we show global existence results under various smallness assumptions. The fourth section is devoted to the proof of the local existence of solutions for general initial data.

7.2 Preliminaries and auxiliary results

In what follows, for any u in $H^{0,1}(Q)$, we will often use the notation $\operatorname{div}_h u \equiv \operatorname{div}_h u_h = \partial_1 u_1 + \partial_2 u_2$. This quantity is well defined since, by the divergence-free condition, $\operatorname{div}_h u_h = -\partial_3 u_3$.

We begin this section by proving the density of $\widetilde{H}_0^1(Q)$ in the space $\widetilde{H}^{0,1}(Q)$. For the sake of simplicity, we assume below that Ω is star-shaped. This hypothesis allows us to give a constructive proof of the density result. This density result should be true even without this additional assumption.

Lemma 7.2.1 *Let Ω be a smooth bounded star-shaped domain. Then the space $\widetilde{H}_0^1(Q)$ is dense in the space $\widetilde{H}^{0,1}(Q)$.*

Proof Without loss of generality, we will assume that Ω is star-shaped with respect to the origin 0.

Let u be an element of $\widetilde{H}^{0,1}(Q)$. We denote by $u^*(x_h, x_3)$ the extension of $u(x_h, x_3)$ by zero on $\mathbb{R}^2 \times (0,1)$, that is,

$$u^*(x_h, x_3) = u(x_h, x_3), \qquad \forall (x_h, x_3) \in \Omega \times (0,1)$$
$$u^*(x_h, x_3) = 0, \qquad \forall (x_h, x_3) \in (\mathbb{R}^2 \backslash \Omega) \times (0,1).$$

We notice that the property $\gamma_n u = 0$ on $\partial\Omega \times (0,1)$ implies that, for any $\Phi \in \mathcal{D}(\mathbb{R}^2 \times (0,1))$ with $\Phi_{|\overline{\Omega}} = \varphi$, we have

$$\langle \operatorname{div} u^*, \Phi \rangle_{\mathcal{D}',\mathcal{D}} = -\langle u^*, \nabla\Phi \rangle_{\mathcal{D}',\mathcal{D}} = -\int_{\Omega \times (0,1)} u \cdot \nabla\varphi \, dx_h \, dx_3$$

$$= \int_{\Omega \times (0,1)} \operatorname{div} u \, \varphi \, dx_h \, dx_3 - \langle \gamma_n u, \varphi_{|\partial\Omega \times (0,1)} \rangle_{H^{-1/2}, H^{1/2}}$$

$$= \int_{\Omega \times (0,1)} \operatorname{div} u \, \varphi \, dx_h \, dx_3,$$

which implies that $\operatorname{div} u^* = 0$. We also notice that u^* and $\partial_3 u^*$ belong to $L^2(\mathbb{R}^2 \times (0,1))^3$ and that thus $\operatorname{div}_h u_h^* = -\partial_3 u_3^*$ is in $L^2(\mathbb{R}^2 \times (0,1))$. We next want to approximate the vector u^* by a vector with compact support in $\Omega \times [0,1]$. To this end, inspired by Remark 1.7 in Chapter I of Temam (1979), we introduce the vector u_λ^*, for $\lambda > 1$, defined by

$$u_{\lambda,i}^*(x_h, x_3) = u_i^*(\lambda x_h, x_3), \quad i = 1, 2,$$
$$u_{\lambda,3}^*(x_h, x_3) = \lambda u_3^*(\lambda x_h, x_3).$$

We remark that both u_λ^* and $\partial_3 u_\lambda^*$ belong to $L^2(\mathbb{R}^2 \times (0,1))^3$ and that $u_{\lambda,3}^*(x_h, 0) = u_{\lambda,3}^*(x_h, 1) = 0$. Moreover,

$$\text{div}_h u_\lambda^*(x_h, x_3) = \lambda(\text{div}_h u^*)(\lambda x_h, x_3),$$
$$\partial_3 u_{\lambda,3}^*(x_h, x_3) = \lambda(\partial_3 u_3^*)(\lambda x_h, x_3), \qquad \text{and}$$
$$\text{div}\, u_\lambda^*(x_h, x_3) = \lambda((\text{div}_h u^*)(\lambda x_h, x_3) + (\partial_3 u_3^*)(\lambda x_h, x_3)) = 0.$$

For any $\lambda > 1$, the support of u_λ^* is contained in $(\frac{1}{\lambda}\bar{\Omega}) \times [0,1]$ and therefore is a compact strict subset of $\Omega \times [0,1]$. Furthermore, using the Lebesgue Dominated Convergence Theorem, one easily shows that u_λ^* converges to u^* and thus to u in $\tilde{H}^{0,1}(Q)$, when λ converges to 1.

We next introduce a smooth bump function with compact support $\rho \in \mathcal{D}(\mathbb{R}^2)$ such that

$$\rho(x_h) \geq 0 \quad \text{and} \quad \int_{\mathbb{R}^2} \rho(x_h)\,dx_h = 1.$$

For any small positive number η, we set

$$\rho_\eta(x_h) = \frac{1}{\eta^2}\rho\left(\frac{x_h}{\eta}\right).$$

It is well-known that $\rho_\eta(x_h)$ converges in the sense of distributions to the Dirac distribution $\delta_{\mathbb{R}^2}$. For any $\lambda > 1$ and any $\eta > 0$, where η is small with respect to $1 - \lambda$, we consider the vector $u_{\lambda,\eta}^*$, which is the "horizontal convolution" (denoted by $\star_h$) of u_λ^* with ρ_η, that is,

$$u_{\lambda,\eta}^*(x_h, x_3) = (\rho_\eta \star_h u_\lambda^*)(x_h, x_3)$$
$$= \int_{\mathbb{R}^2} \rho_\eta(y_h) u_\lambda^*(x_h - y_h, x_3)dy_h.$$

Since, for any $i = 1, 2, 3$,

$$\partial_i u_{\lambda,\eta}^*(x) \equiv \partial_i(\rho_\eta \star_h u_\lambda^*)(x) = (\rho_\eta \star_h \partial_i(u_\lambda^*))(x), \qquad (7.3)$$

it follows directly that, for any $(x_h, x_3) \in \mathbb{R}^2 \times (0,1)$,

$$\text{div}\, u_{\lambda,\eta}^* = \rho_\eta \star_h (\text{div}\, u_\lambda^*) = 0.$$

Using Young's inequality,

$$\|\rho_\eta \star_h f(x_h)\|_{L^2(\mathbb{R}^2)} \leq c\|\rho_\eta\|_{L^1(\mathbb{R}^2)}\|f(x_h)\|_{L^2(\mathbb{R}^2)} \leq c\|f(x_h)\|_{L^2(\mathbb{R}^2)},$$

we can write

$$\|u_{\lambda,\eta}^*\|_{L^2}^2 = \int_0^1 \left(\int_{\mathbb{R}^2} |u_{\lambda,\eta}^*(x_h, x_3)|^2 dx_h\right) dx_3$$
$$\leq c^2 \int_0^1 \left(\int_{\mathbb{R}^2} |u_\lambda^*(x_h, x_3)|^2 dx_h\right) dx_3 = c^2\|u_\lambda^*\|_{L^2}^2. \qquad (7.4)$$

Properties (7.3) and (7.4) also imply that

$$\|\partial_3 u^*_{\lambda,\eta}\|_{L^2} \le c \|\partial_3 u^*_\lambda\|_{L^2}.$$

We remark that $u^*_{\lambda,\eta}$ is a C^∞-function in the horizontal variable. Indeed, for any integers k_1, k_2, we have

$$\frac{\partial^{k_1+k_2} u^*_{\lambda,\eta}}{\partial x_1^{k_1} \partial x_2^{k_2}}(x_h, x_3) = \int_{\mathbb{R}^2} \frac{\partial^{k_1+k_2} \rho_\eta}{\partial x_1^{k_1} \partial x_2^{k_2}}(x_h - y_h) u^*_\lambda(y_h, x_3) \, dy_h.$$

If $\eta > 0$ is small with respect to $\lambda - 1$, the support of $u^*_{\lambda,\eta}(x_h, x_3) = (\rho_\eta \star_h u^*_\lambda)(x_h, x_3)$ is a compact set strictly contained in $\Omega \times [0, 1]$. All these properties imply in particular that $u^*_{\lambda,\eta}$ belongs to the Sobolev space $H^1(Q)$. We also check that $u^*_{\lambda,\eta}$ vanishes on $\Gamma_0 \cup \Gamma_1$. Thus, $u^*_{\lambda,\eta}$ belongs to the space $\widetilde{H}^1_0(Q)$, for $\eta > 0$ small enough with respect to $\lambda - 1$.

For any fixed $\lambda > 0$, one shows, as in the classical case of convolutions in all the variables, that as $\eta \to 0$

$$\rho_\eta \star_h u^*_\lambda \to u^*_\lambda \text{ in } L^2(Q)$$
$$\partial_3(\rho_\eta \star_h u^*_\lambda) \to \partial_3 u^*_\lambda \text{ in } L^2(Q). \tag{7.5}$$

A quick proof of the first property of (7.5) is as follows. Let $w_n \in \mathcal{D}(\bar{Q})^3$ be a sequence of smooth vectors converging to u^*_λ in $L^2(Q)^3$. Arguing as in (7.4), by using Young's inequality, one proves that, for any positive number δ, there exists an integer n_δ such that, for $n \ge n_\delta$, for any $\eta > 0$,

$$\|w_n - u^*_\lambda\|_{L^2} + \|\rho_\eta \star_h w_n - \rho_\eta \star_h u^*_\lambda\|_{L^2} \le \frac{\delta}{2}.$$

It thus remains to show, for instance, that $\|w_{n_\delta} - \rho_\eta \star_h w_{n_\delta}\|_{L^2}$ converges to zero as η tends to zero. Using the C^1-regularity of the vector w_{n_δ} as well as the properties of ρ_η, one easily shows that

$$\lim_{\eta \to 0}(\rho_\eta \star_h w_{n_\delta})(x_h, x_3) = w_{n_\delta}(x_h, x_3) \text{ for a.e. } (x_h, x_3).$$

Moreover, by (7.4),

$$\|w_{n_\delta} - \rho_\eta \star_h w_{n_\delta}\|_{L^2} \le (c + 1)\|w_{n_\delta}\|_{L^2}.$$

These two properties imply, due to the Lebesgue Dominated Convergence Theorem, that $\|w_{n_\delta} - \rho_\eta \star_h w_{n_\delta}\|_{L^2}$ converges to zero as η tends to zero, that is, there exists $\eta_0 > 0$ such that, for any $0 < \eta \le \eta_0$,

$$\|w_{n_\delta} - \rho_\eta \star_h w_{n_\delta}\|_{L^2} \le \frac{\delta}{2}.$$

The first property in (7.5) is thus proved. The second property in (7.5) is shown in the same way.

Finally, let $\lambda_n > 1$ and $\eta_n > 0$ be two sequences converging to 1 and zero respectively as n tends to infinity. To complete the proof of the lemma, it suffices to notice that, by a diagonal procedure, one can extract two subsequences λ_{n_k} and η_{n_k} such that $u^*_{\lambda_{n_k}, \eta_{n_k}}$ converges to u^* in $\widetilde{H}^{0,1}(Q)$ as n_k tends to infinity. The lemma is thus proved. $\qquad\square$

Remark 7.2.2 *We can also define spaces with higher regularity in the vertical variable. For instance, let*

$$\widetilde{H}^{0,2}(Q) = \{u \in L^2(Q)^3 \,|\, \operatorname{div} u = 0 \,; \gamma_n u = 0 \ \text{on } \partial Q \,;$$
$$\partial_3^i u \in L^2(Q)^3 \,, \, i = 1, 2\}$$

and

$$\widetilde{H}_0^{0,2}(Q) = \{u \in L^2(Q)^3 \,|\, \operatorname{div} u = 0 \,; \gamma_n u = 0 \ \text{on } \partial Q \,;$$
$$\partial_3 u_h = 0 \ \text{on } \Gamma_0 \cup \Gamma_1 \,; \partial_3^i u \in L^2(Q)^3 \,, \, i = 1, 2\}.$$

For later use, we also introduce the space

$$H^{0,2}(\Omega \times (a,b)) = \{u \in L^2(\Omega \times (a,b))^3 \,|\, \operatorname{div} u = 0 \,;$$
$$\partial_3^i u \in L^2(\Omega \times (a,b))^3 \,, \, i = 1, 2\},$$

where $-\infty < a < b < +\infty$.

Arguing as in the proof of Lemma 7.2.1, one shows that, under the same hypothesis, $\widetilde{H}_0^1(Q) \cap H^2(Q)$ is dense in $\widetilde{H}^{0,2}(Q)$ and that $\widetilde{H}_0^1(Q) \cap H^2(Q) \cap \widetilde{H}_0^{0,2}(Q)$ is dense in $\widetilde{H}_0^{0,2}(Q)$.

Let $1 \le p \le +\infty$ and $1 \le q \le +\infty$. We use the notation $L_v^q L_h^p(\widetilde{Q}) = L^q((-1,+1); L^p(\Omega))$ or simply $L_v^q L_h^p$ for the space of (equivalence classes of) functions g such that

$$\|g\|_{L_v^q L_h^p} = \left(\int_{-1}^{+1} \left(\int_\Omega |g(x_h, x_3)|^p \, dx_h \right)^{q/p} dx_3 \right)^{1/q} < \infty.$$

We point out that the order of integration is important. Of course, $L_v^q L_h^q$ is the usual space $L^q(\widetilde{Q})$ and the norm $\|g\|_{L_v^q L_h^q}$ is denoted by $\|g\|_{L^q}$. Likewise we define the spaces $L_v^q L_h^p(Q) = L^q((0,+1); L^p(\Omega))$.

Lemma 7.2.3 *The following anisotropic estimates hold.*
1) For any function g in $L^2(\widetilde{Q})$ (with $\nabla_h g \in L^2(\widetilde{Q})$) satisfying homogeneous Dirichlet boundary conditions on the boundary $\partial\Omega \times (-1,+1)$,

$$\|g\|_{L_v^2(L_h^4)} \le C_0 \|g\|_{L^2}^{\frac{1}{2}} \|\nabla_h g\|_{L^2}^{\frac{1}{2}}, \tag{7.6}$$

2) *For any function g in $L^2(\widetilde{Q})$, with $\partial_3 g \in L^2(\widetilde{Q})$,*

$$\|g\|_{L_v^\infty(L_h^2)} \le C_0\Big(\|g\|_{L^2}^{\frac{1}{2}}\|\partial_3 g\|_{L^2}^{\frac{1}{2}} + \|g\|_{L^2}\Big), \tag{7.7}$$

where $C_0 > 1$ is a constant independent of g.

Proof We first prove inequality (7.6). Since g vanishes on the lateral boundary, using the Gagliardo–Nirenberg and the Poincaré inequalities in the horizontal variable and also the Cauchy–Schwarz inequality in the vertical variable, we obtain,

$$\|g\|_{L_v^2(L_h^4)}^2 \le C \int_{-1}^{+1}\Big(\int_\Omega |g(x_h,x_3)|^2 dx_h\Big)^{1/2}\Big(\int_\Omega |\nabla_h g(x_h,x_3)|^2 dx_h\Big)^{1/2} dx_3$$

$$\le C_0^2 \Big(\int_{-1}^{+1}\!\!\int_\Omega |g(x_h,x_3)|^2 dx_h dx_3\Big)^{1/2}\Big(\int_{-1}^{+1}\!\!\int_\Omega |\nabla_h g(x_h,x_3)|^2 dx_h dx_3\Big)^{1/2}.$$

To prove inequality (7.7), we first apply the Agmon inequality in the vertical variable and then the Cauchy–Schwarz inequality in the horizontal variable to obtain,

$$\sup_{x_3\in(-1,+1)}\Big(\int_\Omega |g(x_h,x_3)|^2 dx_h\Big)^{1/2} \le \Big(\int_\Omega \sup_{x_3\in(-1,+1)} |g(x_h,x_3)|^2 dx_h\Big)^{1/2}$$

$$\le C\Big(\int_\Omega \Big[\Big(\int_{-1}^{+1} |g(x_h,x_3)|^2 dx_3\Big)^{1/2}\Big(\int_{-1}^{+1} |\partial_3 g(x_h,x_3)|^2 dx_3\Big)^{1/2}$$

$$+ \int_{-1}^{+1} |g(x_h,x_3)|^2 dx_3\Big] dx_h\Big)^{1/2}$$

$$\le C_0\Big(\|g\|_{L^2}^{1/2}\|\partial_3 g\|_{L^2}^{1/2} + \|g\|_{L^2}\Big). \qquad \square$$

The previous lemma allows us to estimate the term $(u\nabla u, u)_{H^{0,1}}$, which will often appear in the estimates given below. More precisely, we can prove the following lemma.

Lemma 7.2.4 *There exists a positive constant C_1 such that, for any smooth enough divergence-free vector field u and any smooth enough vector field v, both satisfying homogeneous Dirichlet boundary conditions on the boundary $\partial\Omega \times (-1,+1)$, we have*

$$|(u\nabla v, v)_{H^{0,1}}| \le C_1\big(\|u\|_{H^{0,1}}^{1/2}\|\nabla_h u\|_{H^{0,1}}^{1/2}\|v\|_{H^{0,1}}^{1/2}\|\nabla_h v\|_{H^{0,1}}^{3/2}$$

$$+ \|\nabla_h u\|_{H^{0,1}}\|v\|_{H^{0,1}}\|\nabla_h v\|_{H^{0,1}}\big).$$

Proof The proof of this lemma is very simple. Integrating by parts and using the divergence-free condition on u, we can write

$$
\begin{aligned}
(u\nabla v, v)_{H^{0,1}} &= (\partial_3(u\nabla v), \partial_3 v) = (\partial_3 u \nabla v, \partial_3 v) \\
&= (\partial_3 u_h \nabla_h v, \partial_3 v) + (\partial_3 u_3 \partial_3 v, \partial_3 v) \qquad (7.8) \\
&= (\partial_3 u_h \nabla_h v, \partial_3 v) - (\operatorname{div}_h u_h \partial_3 v, \partial_3 v).
\end{aligned}
$$

Applying Lemma 7.2.3, we obtain the estimate,

$$
\begin{aligned}
|(\operatorname{div}_h u_h \partial_3 v, \partial_3 v)_{L^2}| &\le C\|\nabla_h u\|_{L^\infty_v(L^2_h)} \|\partial_3 v\|^2_{L^2_v(L^4_h)} \\
&\le C\Big(\|\nabla_h u\|^{1/2}_{L^2}\|\nabla_h \partial_3 u\|^{1/2}_{L^2} + \|\nabla_h u\|_{L^2}\Big)\|\partial_3 v\|_{L^2}\|\nabla_h \partial_3 v\|_{L^2} \\
&\le C\|\nabla_h u\|_{H^{0,1}}\|\partial_3 v\|_{L^2}\|\nabla_h \partial_3 v\|_{L^2}
\end{aligned}
$$
$$(7.9)$$

Furthermore, using Lemma 7.2.3 once more, we get

$$
\begin{aligned}
|(\partial_3 u_h \nabla_h v, \partial_3 v)| &\le C\|\partial_3 u\|_{L^2_v(L^4_h)}\|\nabla_h v\|_{L^\infty_v(L^2_h)}\|\partial_3 v\|_{L^2_v(L^4_h)} \\
&\le C\|\partial_3 u\|^{1/2}_{L^2}\|\nabla_h \partial_3 u\|^{1/2}_{L^2}\|\partial_3 v\|^{1/2}_{L^2}\|\nabla_h \partial_3 v\|^{1/2}_{L^2} \\
&\qquad \times \Big(\|\nabla_h v\|^{1/2}_{L^2}\|\nabla_h \partial_3 v\|^{1/2}_{L^2} + \|\nabla_h v\|_{L^2}\Big) \\
&\le C\|\partial_3 u\|^{1/2}_{L^2}\|\nabla_h \partial_3 u\|^{1/2}_{L^2}\|\partial_3 v\|^{1/2}_{L^2}\|\nabla_h \partial_3 v\|^{1/2}_{L^2}\|\nabla_h v\|_{H^{0,1}}.
\end{aligned}
$$
$$(7.10)$$

Together, estimates (7.8), (7.9) and (7.10) imply the lemma. $\qquad\square$

The next proposition shows that sequences of uniformly bounded (with respect to ε_m) classical solutions of the equations (NS$_{\varepsilon_m}$) converge to solutions of the system (NS$_h$) when ε_m goes to zero. The same type of proof implies the uniqueness of the solutions of system (NS$_h$). In order to state the result, we introduce the space

$$H^{1,0}(\Omega\times(a,b)) = \{u \in L^2(\Omega\times(a,b))^3 \mid \operatorname{div} u = 0\,;\, \nabla_h u \in L^2(\Omega\times(a,b))^3\},$$

where $-\infty < a < b < +\infty$.

Proposition 7.2.5 *1) Let $u_0 \in \widetilde{H}^{0,1}(Q)$ be given. Let $\varepsilon_m > 0$ be a (decreasing) sequence converging to zero and $u_0^m \in \widetilde{H}_0^1(Q)$ be a sequence of initial data converging to u_0 in $\widetilde{H}^{0,1}(Q)$, as $m \to \infty$. Assume that the system (NS$_{\varepsilon_m}$), with initial data Σu_0^m, admits a strong solution $u_{\varepsilon_m}(t) \in C^0((0,T_0), \widetilde{V})$ where T_0 does not depend on ε_m and that the sequences $u_{\varepsilon_m}(t)$ and $\nabla_h u_{\varepsilon_m}(t)$ are uniformly bounded in $L^\infty((0,T_0), H^{0,1}(\widetilde{Q}))$ and in $L^2((0,T_0), H^{0,1}(\widetilde{Q}))$ respectively.*

Then $u_{\varepsilon_m}(t)$ *converges in* $L^\infty((0,T_0), L^2(\widetilde{Q})^3) \cap L^2((0,T_0), H^{1,0}(\widetilde{Q}))$ *to a solution* $u^* \in L^\infty((0,T_0), H^{0,1}(\widetilde{Q}))$ *of the problem* (NS$_h$), *such that* $\nabla_h u^*$ *belongs to* $L^2((0,T_0), H^{0,1}(\widetilde{Q}))$. *In particular, the vector field* u^* *belongs to* $L^\infty((0,T_0), \widetilde{H}^{0,1}(Q))$.

2) *The problem* (NS$_h$) *has at most one solution* u^* *in* $L^\infty((0,T_0), H^{0,1}(\widetilde{Q}))$ *with* $\nabla_h u^*$ *in* $L^2((0,T_0), H^{0,1}(\widetilde{Q}))$.

Proof We recall that $u_{\varepsilon_m}(t) \in C^0((0,T_0), \widetilde{V})$ is a classical solution of the equations

$$\partial_t u_{\varepsilon_m} + u_{\varepsilon_m} \nabla u_{\varepsilon_m} - \nu_h \Delta_h u_{\varepsilon_m} - \varepsilon_m \partial_{x_3}^2 u_{\varepsilon_m} = -\nabla p_{\varepsilon_m}$$

$$\operatorname{div} u_{\varepsilon_m} = 0$$

$$u_{\varepsilon_m}|_{t=0} = \Sigma u_0^m.$$

We first want to show that, under the hypotheses of the proposition, $u_{\varepsilon_m}(t)$ and $\nabla_h u_{\varepsilon_m}(t)$ are Cauchy sequences in $L^\infty((0,T_0), L^2(\widetilde{Q})^3)$ and in $L^2((0,T_0), L^2(\widetilde{Q})^3)$ respectively.

In order to simplify the notation in the estimates below, we will simply denote the vector u_{ε_m} by u_m. Let $m > k$. Since the sequence ε_n is decreasing, $\varepsilon_m < \varepsilon_k$. The vector $w_{m,k} = u_m - u_k$ satisfies the equation

$$\partial_t w_{m,k} - \nu_h \Delta_h w_{m,k} - \varepsilon_k \partial_3^2 w_{m,k} = (\varepsilon_m - \varepsilon_k)\partial_3^2 u_m - w_{m,k}\nabla u_m$$
$$- u_k \nabla w_{m,k} - \nabla(p_{\varepsilon_m} - p_{\varepsilon_k}).$$

Taking the inner product in $L^2(\widetilde{Q})^3$ of the previous equality with $w_{m,k}$, we obtain the equality

$$\frac{1}{2}\partial_t \|w_{m,k}\|_{L^2}^2 + \nu_h \|\nabla_h w_{m,k}\|_{L^2}^2 + \varepsilon_k \|\partial_3 w_{m,k}\|_{L^2}^2 =$$
$$= (\varepsilon_k - \varepsilon_m)(\partial_3 u_m, \partial_3 w_{m,k})_{L^2} + B_1 + B_2, \tag{7.11}$$

where

$$B_1 = -(w_{m,k,h}\nabla_h u_m, w_{m,k})_{L^2}$$
$$B_2 = -(w_{m,k,3}\partial_3 u_m, w_{m,k})_{L^2}.$$

Applying the Hölder and Young inequalities and Lemma 7.2.3, we estimate B_1 as follows,

$$|B_1| \leq \|w_{m,k,h}\|_{L_v^2(L_h^4)}\|w_{m,k}\|_{L_v^2(L_h^4)}\|\nabla_h u_m\|_{L_v^\infty(L_h^2)}$$

$$\leq \|w_{m,k}\|_{L^2}\|\nabla_h w_{m,k}\|_{L^2}\left(\|\nabla_h u_m\|_{L^2} + \|\nabla_h u_m\|_{L^2}^{1/2}\|\nabla_h \partial_3 u_m\|_{L^2}^{1/2}\right)$$

$$\leq \frac{\nu_h}{4}\|\nabla_h w_{m,k}\|_{L^2}^2 + \frac{4}{\nu_h}\|w_{m,k}\|_{L^2}^2\left(\|\nabla_h u_m\|_{L^2}^2 + \|\nabla_h \partial_3 u_m\|_{L^2}^2\right). \tag{7.12}$$

Using the same arguments as above and also the fact that $\partial_3 w_{m,k,3} = -\mathrm{div}\,_h w_{m,k,h}$, we can bound B_2 as follows,

$$
\begin{aligned}
|B_2| &\leq \|\partial_3 u_m\|_{L_v^2(L_h^4)} \|w_{m,k}\|_{L_v^2(L_h^4)} \|w_{m,k,3}\|_{L_v^\infty(L_h^2)} \\
&\leq \|\partial_3 u_m\|_{L^2}^{1/2} \|\nabla_h \partial_3 u_m\|_{L^2}^{1/2} \|w_{m,k}\|_{L^2}^{1/2} \|\nabla_h w_{m,k}\|_{L^2}^{1/2} \\
&\quad \times \left(\|w_{m,k,3}\|_{L^2} + \|w_{m,k,3}\|_{L^2}^{1/2} \|\partial_3 w_{m,k,3}\|_{L^2}^{1/2} \right) \\
&\leq \|\partial_3 u_m\|_{L^2}^{1/2} \|\nabla_h \partial_3 u_m\|_{L^2}^{1/2} \|w_{m,k}\|_{L^2}^{3/2} \|\nabla_h w_{m,k}\|_{L^2}^{1/2} \\
&\quad + \|\partial_3 u_m\|_{L^2}^{1/2} \|\nabla_h \partial_3 u_m\|_{L^2}^{1/2} \|w_{m,k}\|_{L^2} \|\nabla_h w_{m,k}\|_{L^2} \\
&\leq \frac{\nu_h}{4} \|\nabla_h w_{m,k}\|_{L^2}^2 + \frac{3}{2\nu_h^{1/3}} \|\partial_3 u_m\|_{L^2}^{2/3} \|\nabla_h \partial_3 u_m\|_{L^2}^{2/3} \|w_{m,k}\|_{L^2}^2 \\
&\quad + \frac{2}{\nu_h} \|\partial_3 u_m\|_{L^2} \|\nabla_h \partial_3 u_m\|_{L^2} \|w_{m,k}\|_{L^2}^2 . \tag{7.13}
\end{aligned}
$$

The estimates (7.11), (7.12), and (7.13) together with the Cauchy–Schwarz inequality imply that, for $t \in [0, T_0]$,

$$
\begin{aligned}
\partial_t \|w_{m,k}\|_{L^2}^2 &+ \nu_h \|\nabla_h w_{m,k}\|_{L^2}^2 + (\varepsilon_k + \varepsilon_m) \|\partial_3 w_{m,k}\|_{L^2}^2 \\
&\leq (\varepsilon_k - \varepsilon_m) \|\partial_3 u_m\|_{L^2}^2 + \frac{8}{\nu_h} \|w_{m,k}\|_{L^2}^2 \left(\|\nabla_h u_m\|_{L^2}^2 + \|\nabla_h \partial_3 u_m\|_{L^2}^2 \right) \\
&\quad + \frac{3}{\nu_h^{1/3}} \|\partial_3 u_m\|_{L^2}^{2/3} \|\nabla_h \partial_3 u_m\|_{L^2}^{2/3} \|w_{m,k}\|_{L^2}^2 \\
&\quad + \frac{4}{\nu_h} \|\partial_3 u_m\|_{L^2} \|\nabla_h \partial_3 u_m\|_{L^2} \|w_{m,k}\|_{L^2}^2 . \tag{7.14}
\end{aligned}
$$

Integrating the inequality (7.14) from 0 to t, and applying the Gronwall Lemma, we obtain, for $0 < t \leq T_0$,

$$
\begin{aligned}
\|w_{m,k}(t)\|_{L^2}^2 &+ \nu_h \int_0^t \|\nabla_h w_{m,k}(s)\|_{L^2}^2 \mathrm{d}s + (\varepsilon_k + \varepsilon_m) \int_0^t \|\partial_3 w_{m,k}(s)\|_{L^2}^2 \mathrm{d}s \\
&\leq \left[(\varepsilon_k - \varepsilon_m) \int_0^{T_0} \|\partial_3 u_m(s)\|_{L^2}^2 \mathrm{d}s + \|u_0^m - u_0^k\|_{L^2}^2 \right] \\
&\quad \times \exp\left(\frac{c_0}{\nu_h} \int_0^{T_0} \left(\|\nabla_h u_m(s)\|_{L^2}^2 + \|\nabla_h \partial_3 u_m(s)\|_{L^2}^2 + \|\partial_3 u_m(s)\|_{L^2}^2 \right) \mathrm{d}s \right) \\
&\quad \times \exp\left[\frac{c_1}{\nu_h^{1/3}} \left(\int_0^{T_0} \|\nabla_h \partial_3 u_m(s)\|_{L^2}^2 \mathrm{d}s \right)^{1/3} \left(\int_0^{T_0} \|\partial_3 u_m(s)\|_{L^2} \mathrm{d}s \right)^{2/3} \right] , \\
&\tag{7.15}
\end{aligned}
$$

where c_0 and c_1 are two positive constants independent of m and k.

Since the sequences $u_{\varepsilon_m}(t)$ and $\nabla_h u_{\varepsilon_m}(t)$ are uniformly bounded in $L^\infty((0, T_0), H^{0,1}(\widetilde{Q}))$ and in $L^2((0, T_0), H^{0,1}(\widetilde{Q}))$ respectively, the estimate (7.15) implies that u_{ε_m} and $\nabla_h u_{\varepsilon_m}$ are Cauchy sequences in $L^\infty((0, T_0), L^2(\widetilde{Q})^3)$ and $L^2((0, T_0), L^2(\widetilde{Q})^3)$ respectively. Thus u_{ε_m} converges in $L^\infty((0, T_0), L^2(\widetilde{Q})^3) \cap L^2((0, T_0), H^{1,0}(\widetilde{Q}))$ to an element u^* in this same space. Moreover, u^* and $\nabla_h u^*$ are bounded in $L^\infty((0, T_0), H^{0,1}(\widetilde{Q}))$ and in $L^2((0, T_0), H^{0,1}(\widetilde{Q}))$ respectively. The convergence in the sense of distributions of u_{ε_m} to u^* and the divergence-free property of the sequence u_{ε_m} imply that u^* is also divergence-free. Furthermore, one easily shows that the restriction of u^* to Q is a weak solution of the system (NS_h). From the equality $\Sigma u_{\varepsilon_m}(t) = u_{\varepsilon_m}(t)$, it follows that $\Sigma u^*(t) = u^*(t)$. In particular, $u_3^*(t)$ vanishes on $\Gamma_0 \cup \Gamma_1$. Finally, we notice that, since $u_{\varepsilon_m}(t) \in C^0((0, T_0), \widetilde{V})$ converges in $L^2((0, T_0), H^{1,0}(\widetilde{Q}))$, $u^*(t)$ satisfies the homogeneous Dirichlet boundary condition on the lateral boundary $\partial\Omega \times (-1, 1)$ for almost all $t \in (0, T_0)$. One proves the uniqueness of the solution u^* in $L^\infty((0, T_0), H^{0,1}(\widetilde{Q}))$ with $\nabla_h u^*$ in $L^2((0, T_0), H^{0,1}(\widetilde{Q}))$ in the same way as the above Cauchy property.　□

We now state the classical energy estimate which will be widely used in the subsequent sections.

Lemma 7.2.6 *Let $u_\varepsilon(t) \in C^0([0, T_0], \widetilde{V})$ be the classical solution of the equations (NS_ε) with initial data $u_{\varepsilon,0} \in \widetilde{V}$. Then the following estimates are satisfied, for any $t \in [0, T_0]$, for any $0 \leq t_0 \leq t$,*

$$\|u_\varepsilon(t)\|_{L^2}^2 \leq \|u_\varepsilon(0)\|_{L^2}^2 \exp(-2\nu_h \lambda_0^{-1} t), \qquad and$$

$$\nu_h \int_{t_0}^t \|\nabla_h u_\varepsilon(s)\|_{L^2}^2 ds + \varepsilon \int_{t_0}^t \|\partial_3 u_\varepsilon(s)\|_{L^2}^2 ds \qquad (7.16)$$

$$\leq \frac{1}{2} \|u_\varepsilon(0)\|_{L^2}^2 \exp(-2\nu_h \lambda_0^{-1} t_0),$$

where λ_0 is the constant from the Poincaré inequality.

Proof Since $u_\varepsilon \in C^0([0, T_0], \widetilde{V})$ is the classical solution of (NS_ε), we can take the inner product in $L^2(\widetilde{Q})^3$ of the first equation in (NS_ε) with u_ε and integrate by parts. We thus obtain, for $0 \leq t \leq T_0$,

$$\partial_t \|u_\varepsilon(t)\|_{L^2}^2 + 2\nu_h \|\nabla_h u_\varepsilon(t)\|_{L^2}^2 + 2\varepsilon \|\partial_3 u_\varepsilon(t)\|_{L^2}^2 \leq 0. \qquad (7.17)$$

Since u_ε satisfies homogeneous Dirichlet boundary conditions on the lateral boundary, there exists a positive constant λ_0 depending only on

Ω such that

$$\|u_\varepsilon\|_{L^2}^2 \le \lambda_0 \|\nabla_h u_\varepsilon\|_{L^2}^2. \tag{7.18}$$

The inequalities (7.17) and (7.18) imply that, for $0 \le t \le T_0$,

$$\partial_t \|u_\varepsilon(t)\|_{L^2}^2 + 2\nu_h \lambda_0^{-1} \|u_\varepsilon(t)\|_{L^2}^2 \le 0.$$

Integrating the previous inequality and applying the Gronwall Lemma, we obtain the first inequality in (7.16). Integrating now inequality (7.17) from t_0 to t and taking into account the first estimate in (7.16), we obtain the second estimate of (7.16). □

We continue this section with an auxiliary proposition, which will be used several times in the proof of global existence of solutions of the system (NS$_\varepsilon$).

Proposition 7.2.7 *Let $u_\varepsilon \in C^0([0, T_0), \widetilde{V})$ be a classical solution of* (NS$_\varepsilon$). *Let $T_n < T_0$ be a sequence converging to T_0 as $n \to \infty$. If $u_\varepsilon(t)$ is uniformly bounded in $L^\infty((0, T_n), H^{0,1}(\widetilde{Q})) \cap L^2((0, T_n), H^{0,1}(\widetilde{Q}))$ and if $\nabla_h u_\varepsilon$ and $\varepsilon \partial_3 u_\varepsilon$ are uniformly bounded in $L^2((0, T_n), H^{0,1}(\widetilde{Q}))$ as $n \to \infty$, then u_ε is uniformly bounded in $C^0([0, T_n], \widetilde{V})$ and the classical solution u_ε exists on a time interval $[0, T_\varepsilon)$ where $T_\varepsilon > T_0$. In particular, if $T_n \to \infty$ as $n \to \infty$, then the classical solution u_ε exists globally.*

Proof Let $u_\varepsilon \in C^0([0, T_0), \widetilde{V})$ be a (local) classical solution of (NS$_\varepsilon$). In order to prove the proposition, we have to show that $\nabla_h u_\varepsilon$ is uniformly bounded in $L^\infty((0, T_n), L^2(\widetilde{Q})^3)$ as $n \to \infty$. Since u_ε is a classical solution, all the *a priori* estimates made below can be justified. Let P be the classical Leray projection of $L^2(\widetilde{Q})^3$ onto $\widetilde{H}$, where

$$\widetilde{H} = \{u \in L^2(\widetilde{Q})^3 \mid \text{div } u = 0 \, ; \gamma_n u|_{\partial\Omega \times (0,1)} = 0 \, ;$$

$$u(x_h, x_3) = u(x_h, x_3 + 2)\}.$$

Taking the inner product in $L^2(\widetilde{Q})$ of the first equation of (NS$_\varepsilon$) with $-P\Delta_h u_\varepsilon$, we obtain the equality

$$-(\partial_t u_\varepsilon, \Delta_h u_\varepsilon) + \nu_h \|P\Delta_h u_\varepsilon\|_{L^2}^2 + \varepsilon(\partial_3^2 u_\varepsilon, P\Delta_h u_\varepsilon) = -(u_\varepsilon \cdot \nabla u_\varepsilon, P\Delta_h u_\varepsilon).$$

We remark that, for $0 < t < T_0$, $\partial_t u_\varepsilon$, $\partial_3 u_\varepsilon$ and $\partial_3^2 u_\varepsilon$ vanish on the lateral boundary and are periodic in x_3. Moreover, the divergence of $\partial_3^2 u_\varepsilon$ vanishes. These properties imply on the one hand that

$$-\int_{\widetilde{Q}} \partial_t u_\varepsilon \cdot \Delta_h u_\varepsilon \, dx = \int_{\widetilde{Q}} \partial_t \nabla_h u_\varepsilon \cdot \nabla_h u_\varepsilon \, dx = \frac{1}{2} \partial_t \|\nabla_h u_\varepsilon\|_{L^2}^2.$$

On the other hand, we can write, for $0 < t < T_0$,

$$\int_{\tilde{Q}} \partial_3^2 u_\varepsilon \cdot P\Delta_h u_\varepsilon \mathrm{d}x = \int_{\tilde{Q}} \partial_3^2 u_\varepsilon \cdot \Delta_h u_\varepsilon \mathrm{d}x$$

$$= -\int_{\tilde{Q}} \partial_3 u_\varepsilon \cdot \Delta_h \partial_3 u_\varepsilon \mathrm{d}x = \|\nabla_h \partial_3 u_\varepsilon\|_{L^2}^2.$$

The previous three equalities imply that, for $0 < t < T_0$,

$$\partial_t \|\nabla_h u_\varepsilon\|_{L^2}^2 + \nu_h \|P\Delta_h u_\varepsilon\|_{L^2}^2 + 2\varepsilon \|\nabla_h \partial_3 u_\varepsilon\|_{L^2}^2 \leq \frac{1}{\nu_h} \|u_\varepsilon \cdot \nabla u_\varepsilon\|_{L^2}^2. \quad (7.19)$$

To estimate the term $\|u_\varepsilon \cdot \nabla u_\varepsilon\|_{L^2}^2$, we write

$$\|u_\varepsilon \cdot \nabla u_\varepsilon\|_{L^2}^2 = \int_{\tilde{Q}} (u_{\varepsilon,h} \cdot \nabla_h u_\varepsilon + u_{\varepsilon,3} \cdot \partial_3 u_\varepsilon)^2 \mathrm{d}x \qquad (7.20)$$

$$\leq 2\|u_{\varepsilon,h} \cdot \nabla_h u_\varepsilon\|_{L^2}^2 + 2\|u_{\varepsilon,3} \cdot \partial_3 u_\varepsilon\|_{L^2}^2.$$

It remains to bound both terms on the right-hand side of the inequality (7.20). Using the Gagliardo–Nirenberg and the Poincaré inequalities, we can write

$$\|u_{\varepsilon,h} \cdot \nabla_h u_\varepsilon\|_{L^2}^2 \leq \int_{-1}^{1} \|u_\varepsilon(\cdot, x_3)\|_{L_h^4}^2 \|\nabla_h u_\varepsilon(\cdot, x_3)\|_{L_h^4}^2 \, \mathrm{d}x_3$$

$$\leq c_0 \Big(\int_{-1}^{1} \|u_\varepsilon(\cdot, x_3)\|_{L_h^2} \|\nabla_h u_\varepsilon(\cdot, x_3)\|_{L_h^2}^2 \|D_h \nabla_h u_\varepsilon(\cdot, x_3)\|_{L_h^2} \, \mathrm{d}x_3$$

$$+ \int_{-1}^{1} \|u_\varepsilon(\cdot, x_3)\|_{L_h^2} \|\nabla_h u_\varepsilon(\cdot, x_3)\|_{L_h^2}^3 \, \mathrm{d}x_3 \Big)$$

$$\leq c_1 \Big(\|u_\varepsilon\|_{L_v^\infty(L_h^2)} \|\nabla_h u_\varepsilon\|_{L_v^\infty(L_h^2)} \|\nabla_h u_\varepsilon\|_{L^2} \|D_h \nabla_h u_\varepsilon\|_{L^2}$$

$$+ \|u_\varepsilon\|_{L_v^\infty(L_h^2)} \|\nabla_h u_\varepsilon\|_{L_v^\infty(L_h^2)} \|\nabla_h u_\varepsilon\|_{L^2}^2 \Big).$$

Applying Lemma 7.2.3 to the previous inequality, we obtain,

$$\|u_{\varepsilon,h} \cdot \nabla_h u_\varepsilon\|_{L^2}^2 \leq c_2 \Big(\|u_\varepsilon\|_{L^2} + \|\partial_3 u_\varepsilon\|_{L^2}^{1/2} \|u_\varepsilon\|_{L^2}^{1/2} \Big)$$

$$\times \Big(\|\nabla_h u_\varepsilon\|_{L^2} + \|\partial_3 \nabla_h u_\varepsilon\|_{L^2}^{1/2} \|\nabla_h u_\varepsilon\|_{L^2}^{1/2} \Big) \quad (7.21)$$

$$\times \Big(\|\nabla_h u_\varepsilon\|_{L^2} + \|D_h \nabla_h u_\varepsilon\|_{L^2} \Big) \|\nabla_h u_\varepsilon\|_{L^2}.$$

The classical regularity theorem for the stationary Stokes problem (see for example Constantin & Foias, 1988; Solonnikov & Ščadilov, 1973; or Temam, 1979) implies that there exists a positive constant $K_0(\varepsilon)$, which could depend on ε, such that,

$$\|D_h \nabla_h u_\varepsilon\|_{L^2} \leq K_0(\varepsilon) \big(\|P\Delta_h u_\varepsilon\|_{L^2} + \frac{\varepsilon}{\nu_h} \|\partial_3^2 u_\varepsilon\|_{L^2} \big). \quad (7.22)$$

Using the inequality $2ab \leq a^2 + b^2$, we deduce from (7.21) and (7.22) that

$$\|u_{\varepsilon,h} \cdot \nabla_h u_\varepsilon\|_{L^2}^2 \leq 4c_2 \Big(\|u_\varepsilon\|_{L^2} + \|\partial_3 u_\varepsilon\|_{L^2} \Big) \Big(\|\nabla_h u_\varepsilon\|_{L^2} + \|\partial_3 \nabla_h u_\varepsilon\|_{L^2} \Big)$$
$$\times \Big(\|\nabla_h u_\varepsilon\|_{L^2} + K_0(\varepsilon)(\|P\Delta_h u_\varepsilon\|_{L^2} + \frac{\varepsilon}{\nu_h}\|\partial_3^2 u_\varepsilon\|_{L^2}) \Big) \|\nabla_h u_\varepsilon\|_{L^2},$$

and also

$$\frac{2}{\nu_h}\|u_{\varepsilon,h} \cdot \nabla_h u_\varepsilon\|_{L^2}^2 \leq \frac{\nu_h}{2}\|P\Delta_h u_\varepsilon\|_{L^2}^2 + \frac{\varepsilon^2}{\nu_h}\|\partial_3^2 u_\varepsilon\|_{L^2}^2 + \frac{\nu_h}{2}\|\nabla_h u_\varepsilon\|_{L^2}^2$$
$$+ c_5 \frac{K_0(\varepsilon)^2 + 1}{\nu_h^3}(\|u_\varepsilon\|_{L^2}^2 + \|\partial_3 u_\varepsilon\|_{L^2}^2)$$
$$\times (\|\nabla_h u_\varepsilon\|_{L^2}^2 + \|\partial_3 \nabla_h u_\varepsilon\|_{L^2}^2)\|\nabla_h u_\varepsilon\|_{L^2}^2. \tag{7.23}$$

Likewise, using the Gagliardo–Nirenberg and the Poincaré inequalities, we can write

$$\|u_{\varepsilon,3} \cdot \partial_3 u_\varepsilon\|_{L^2}^2 \leq c_1 \|u_{\varepsilon,3}\|_{L_v^\infty(L_h^2)} \|\nabla_h u_{\varepsilon,3}\|_{L_v^\infty(L_h^2)} \|\partial_3 u_\varepsilon\|_{L^2} \|\nabla_h \partial_3 u_\varepsilon\|_{L^2},$$

which implies, due to Lemma 7.2.3, that

$$\|u_{\varepsilon,3} \cdot \partial_3 u_\varepsilon\|_{L^2}^2 \leq c_2 \Big(\|u_\varepsilon\|_{L^2} + \|\partial_3 u_{\varepsilon,3}\|_{L^2}^{1/2}\|u_{\varepsilon,3}\|_{L^2}^{1/2} \Big)$$
$$\times \Big(\|\nabla_h u_{\varepsilon,3}\|_{L^2} + \|\partial_3 \nabla_h u_{\varepsilon,3}\|_{L^2}^{1/2}\|\nabla_h u_{\varepsilon,3}\|_{L^2}^{1/2} \Big)$$
$$\times \|\partial_3 u_\varepsilon\|_{L^2}\|\nabla_h \partial_3 u_\varepsilon\|_{L^2}.$$

Using the Young inequalities $ab \leq \frac{1}{2}a^2 + \frac{1}{2}b^2$ and $ab \leq \frac{1}{4}a^4 + \frac{3}{4}b^{4/3}$, we deduce from the previous inequality that

$$\frac{2}{\nu_h}\|u_{\varepsilon,3} \cdot \partial_3 u_\varepsilon\|_{L^2}^2 \leq \frac{c_6}{\nu_h}\Big(\|u_\varepsilon\|_{L^2} + \|\partial_3 u_{\varepsilon,3}\|_{L^2} \Big)\|\partial_3 u_\varepsilon\|_{L^2}$$
$$\times (\|\nabla_h u_{\varepsilon,3}\|_{L^2}^2 + \|\partial_3 \nabla_h u_{\varepsilon,3}\|_{L^2}^2). \tag{7.24}$$

Finally, we deduce from the estimates (7.19), (7.23) and, (7.24) that, for $0 \leq t < T_0$,

$$\partial_t \|\nabla_h u_\varepsilon(t)\|_{L^2}^2 + \frac{\nu_h}{2}\|P\Delta_h u_\varepsilon\|_{L^2}^2$$
$$\leq \frac{\varepsilon^2}{\nu_h}\|\partial_3^2 u_\varepsilon(t)\|_{L^2}^2 + \frac{\nu_h}{2}\|\nabla_h u_\varepsilon(t)\|_{L^2}^2 \tag{7.25}$$
$$+ L_\varepsilon(u_\varepsilon(t))\Big(\frac{2c_6}{\nu_h} + c_5 \frac{(K_0(\varepsilon)^2 + 1)}{\nu_h^3}\|\nabla_h u_\varepsilon(t)\|_{L^2}^2 \Big),$$

where

$$L_\varepsilon(u_\varepsilon(t)) = (\|u_\varepsilon(t)\|_{L^2}^2 + \|\partial_3 u_\varepsilon(t)\|_{L^2}^2)(\|\nabla_h u_\varepsilon(t)\|_{L^2}^2 + \|\partial_3 \nabla_h u_\varepsilon(t)\|_{L^2}^2).$$

Integrating the inequality (7.25) from 0 to T_n, where $0 < T_n < T_0$, we infer from (7.25) that, for any T_n, with $0 < T_n < T_0$,

$$\|\nabla_h u_\varepsilon(T_n)\|_{L^2}^2 + \frac{\nu_h}{2} \int_0^{T_n} \|P\Delta_h u_\varepsilon(s)\|_{L^2}^2 ds$$

$$\leq \int_0^{T_n} \left(\frac{\varepsilon^2}{\nu_h}\|\partial_3^2 u_\varepsilon(s)\|_{L^2}^2 + \frac{\nu_h}{2}\|\nabla_h u_\varepsilon(s)\|_{L^2}^2 + \frac{2c_6}{\nu_h}L_\varepsilon(u_\varepsilon(s))\right) ds$$

$$+\|\nabla_h u_\varepsilon(0)\|_{L^2}^2 + \frac{c_5(K_0(\varepsilon)^2+1)}{\nu_h^3}\int_0^{T_n} L_\varepsilon(u_\varepsilon(s))\|\nabla_h u_\varepsilon(s)\|_{L^2}^2 ds.$$

$$(7.26)$$

Using the Gronwall Lemma and taking into account the hypotheses made on u_ε, we deduce from (7.26) that, for any T_n, with $0 < T_n < T_0$,

$$\|\nabla_h u_\varepsilon(T_n)\|_{L^2}^2 \leq$$

$$\left[\int_0^{T_n}\left(\frac{\varepsilon^2}{\nu_h}\|\partial_3^2 u_\varepsilon(s)\|_{L^2}^2 + \frac{\nu_h}{2}\|\nabla_h u_\varepsilon(s)\|_{L^2}^2 + \frac{2c_6}{\nu_h}L_\varepsilon(u_\varepsilon(s))\right)ds\right.$$

$$\left. + \|\nabla_h u_\varepsilon(0)\|_{L^2}^2\right]\exp\left(\frac{c_5(K_0(\varepsilon)^2+1)}{\nu_h^3}\int_0^{T_n} L_\varepsilon(u_\varepsilon(s))ds\right)$$

$$\leq \left[\|\nabla_h u_\varepsilon(0)\|_{L^2}^2 + k_1 + \frac{2c_6}{\nu_h}k_2\right]\exp\left(\frac{c_5(K_0(\varepsilon)^2+1)}{\nu_h^3}k_2\right),$$

where k_1 and k_2 are positive constants independent of T_n (k_1 and k_2 can depend on ε). Thus the proposition is proved. $\qquad\square$

We end this section by giving an upper bound of the $H^{0,2}$-norm of the solution $u_\varepsilon(t)$ of the system (NS_ε) on any subinterval of the maximal interval of existence, when the initial data $u_{\varepsilon,0}$ belong to $\widetilde{V} \cap H^2(\widetilde{Q})^3$.

Proposition 7.2.8 *Let $u_\varepsilon \in C^0([0,T_0], \widetilde{V})$ be a classical solution of (NS_ε) with initial data $u_{\varepsilon,0}$ in $H^2(\widetilde{Q})^3 \cap \widetilde{V}$. We assume that $u_\varepsilon(t)$ and $\nabla_h u_\varepsilon$ are uniformly bounded with respect to ε in $L^\infty((0,T_0], H^{0,1}(\widetilde{Q})) \cap L^2((0,T_0], H^{0,1}(\widetilde{Q}))$ and $L^2((0,T_0), H^{0,1}(\widetilde{Q}))$ respectively. Then $\partial_3^2 u_\varepsilon$ and $\partial_3^2 \nabla_h u_\varepsilon$ are bounded in $L^\infty((0,T_0), L^2(\widetilde{Q}^3))$ and $L^2((0,T_0), L^2(\widetilde{Q}^3))$, respectively, uniformly with respect to ε and the following estimate holds* *for any* $0 \leq t \leq T_0$,

$$\|\partial_3^2 u_\varepsilon(t)\|_{L^2}^2 + \nu_h \int_0^t \|\nabla_h \partial_3^2 u_\varepsilon(s)\|_{L^2}^2 \mathrm{d}s$$

$$\leq \left[\exp(\frac{C}{\nu_h} \int_0^{T_0} \|\nabla_h u_\varepsilon(s)\|_{H^{0,1}}^2 \mathrm{d}s)\right] \left(\|\partial_3^2 u_{\varepsilon,0}\|_{L^2}^2\right.$$

$$\left. + \frac{C}{\nu_h} \sup_{0 \leq s \leq T_0} \left(\|\partial_3 u_\varepsilon(s)\|_{L^2}^2 + \nu_h \|\partial_3 u_\varepsilon(s)\|_{L^2}\right) \int_0^{T_0} \|\nabla_h u_\varepsilon(s)\|_{H^{0,1}}^2 \mathrm{d}s\right).$$

Proof Since $u_\varepsilon(t)$ is a very regular solution for $t > 0$, all the *a priori* estimates made below are justified. Differentiating the first equation in (NS$_\varepsilon$) twice with respect to x_3 and taking the inner product in $L^2(\tilde{Q})$ of the resulting equation with $\partial_3^2 u_\varepsilon$, we obtain the following equality, for $0 \leq t \leq T_0$,

$$\frac{1}{2}\partial_t \|\partial_3^2 u_\varepsilon\|_{L^2}^2 - \nu_h(\Delta_h \partial_3^2 u_\varepsilon, \partial_3^2 u_\varepsilon) - \varepsilon(\partial_3^4 u_\varepsilon, \partial_3^2 u_\varepsilon)$$

$$= -(\partial_3^2 \nabla p_\varepsilon, \partial_3^2 u_\varepsilon) - (\partial_3^2(u_\varepsilon \cdot \nabla u_\varepsilon), \partial_3^2 u_\varepsilon).$$

Since $\partial_3^2 u_\varepsilon$ vanishes on $\partial\Omega \times (-1,1)$ and is periodic in the variable x_3, the following equalities hold:

$$-\int_{\tilde{Q}} \Delta_h \partial_3^2 u_\varepsilon \cdot \partial_3^2 u_\varepsilon \mathrm{d}x_h \mathrm{d}x_3 = \int_{\tilde{Q}} |\nabla_h \partial_3^2 u_\varepsilon|^2 \mathrm{d}x_h \mathrm{d}x_3,$$

$$-\int_{\tilde{Q}} \partial_3^4 u_\varepsilon \cdot \partial_3^2 u_\varepsilon \mathrm{d}x_h \mathrm{d}x_3 = \int_{\tilde{Q}} (\partial_3^3 u_\varepsilon)^2 \mathrm{d}x_h \mathrm{d}x_3,$$

and

$$-\int_{\tilde{Q}} \nabla \partial_3^2 p_\varepsilon \partial_3^2 u_\varepsilon \mathrm{d}x_h \mathrm{d}x_3 = \int_{\tilde{Q}} \partial_3^2 p_\varepsilon \mathrm{div}\, \partial_3^2 u_\varepsilon \mathrm{d}x_h \mathrm{d}x_3,$$

$$-\int_{\partial\tilde{Q}} \partial_3^2 p_\varepsilon(\partial_3^2 u_\varepsilon \cdot n)\mathrm{d}\sigma = 0.$$

We deduce from the above equalities that, for $0 \leq t \leq T_0$,

$$\frac{1}{2}\partial_t \|\partial_3^2 u_\varepsilon\|_{L^2}^2 + \nu_h \|\nabla_h \partial_3^2 u_\varepsilon\|_{L^2}^2 + \varepsilon\|\partial_3^3 u_\varepsilon\|_{L^2}^2 \qquad (7.27)$$

$$= -(\partial_3^2 u_\varepsilon \cdot \nabla u_\varepsilon, \partial_3^2 u_\varepsilon) - 2(\partial_3 u_\varepsilon \cdot \nabla \partial_3 u_\varepsilon, \partial_3^2 u_\varepsilon).$$

As in the proof of Lemma 7.2.4, using the divergence-free condition (see (7.8)), we decompose the terms in the right hand side of (7.27) as follows:

$$(\partial_3^2 u_\varepsilon \cdot \nabla u_\varepsilon, \partial_3^2 u_\varepsilon) = (\partial_3^2 u_{\varepsilon,h} \cdot \nabla_h u_\varepsilon, \partial_3^2 u_\varepsilon) - (\partial_3 \mathrm{div}\,_h u_{\varepsilon,h} \partial_3 u_\varepsilon, \partial_3^2 u_\varepsilon)$$

$$(\partial_3 u_\varepsilon \cdot \nabla \partial_3 u_\varepsilon, \partial_3^2 u_\varepsilon) = (\partial_3 u_{\varepsilon,h} \cdot \nabla_h \partial_3 u_\varepsilon, \partial_3^2 u_\varepsilon) - (\mathrm{div}\,_h u_\varepsilon \partial_3^2 u_\varepsilon, \partial_3^2 u_\varepsilon).$$

$$(7.28)$$

Arguing as in the inequality (7.10) and applying Lemma 7.2.3, we obtain the estimate

$$|(\partial_3^2 u_{\varepsilon,h} \cdot \nabla_h u_\varepsilon, \partial_3^2 u_\varepsilon)| \leq C \|\partial_3^2 u_\varepsilon\|_{L_v^2(L_h^4)}^2 \|\nabla_h u_\varepsilon\|_{L_v^\infty(L_h^2)}$$

$$\leq C \|\partial_3^2 u_\varepsilon\|_{L^2} \|\nabla_h \partial_3^2 u_\varepsilon\|_{L^2} \|\nabla_h u_\varepsilon\|_{H^{0,1}}$$

$$\leq \frac{C}{\nu_h} \|\partial_3^2 u_\varepsilon\|_{L^2}^2 \|\nabla_h u_\varepsilon\|_{H^{0,1}}^2 + \frac{\nu_h}{8} \|\nabla_h \partial_3^2 u_\varepsilon\|_{L^2}^2. \tag{7.29}$$

In the same way

$$2|(\mathrm{div}\,_h u_\varepsilon \partial_3^2 u_\varepsilon, \partial_3^2 u_\varepsilon)| \leq \frac{C}{\nu_h} \|\partial_3^2 u_\varepsilon\|_{L^2}^2 \|\nabla_h u_\varepsilon\|_{H^{0,1}}^2 + \frac{\nu_h}{8} \|\nabla_h \partial_3^2 u_\varepsilon\|_{L^2}^2. \tag{7.30}$$

In order to estimate the term $|(\partial_3 \mathrm{div}\,_h u_{\varepsilon,h} \partial_3 u_\varepsilon, \partial_3^2 u_\varepsilon)|$, we proceed as in (7.9), by applying Lemma 7.2.3. We thus get

$$|(\partial_3 \mathrm{div}\,_h u_{\varepsilon,h} \partial_3 u_\varepsilon, \partial_3^2 u_\varepsilon)| \leq C \|\partial_3 \nabla_h u_\varepsilon\|_{L_v^\infty(L_h^2)} \|\partial_3 u_\varepsilon\|_{L_v^2(L_h^4)} \|\partial_3^2 u_\varepsilon\|_{L_v^2(L_h^4)}$$

$$\leq C \|\partial_3 u_\varepsilon\|_{L^2}^{1/2} \|\nabla_h \partial_3 u_\varepsilon\|_{L^2}^{1/2} \|\partial_3^2 u_\varepsilon\|_{L^2}^{1/2} \|\nabla_h \partial_3^2 u_\varepsilon\|_{L^2}^{1/2}$$

$$\times \left(\|\partial_3 \nabla_h u_\varepsilon\|_{L^2} + \|\partial_3 \nabla_h u_\varepsilon\|_{L^2}^{1/2} \|\partial_3^2 \nabla_h u_\varepsilon\|_{L^2}^{1/2} \right)$$

$$\leq C \|\partial_3 u_\varepsilon\|_{L^2}^{1/2} \|\nabla_h \partial_3 u_\varepsilon\|_{L^2}^{3/2} \|\partial_3^2 u_\varepsilon\|_{L^2}^{1/2} \|\nabla_h \partial_3^2 u_\varepsilon\|_{L^2}^{1/2}$$

$$+ \|\partial_3 u_\varepsilon\|_{L^2}^{1/2} \|\nabla_h \partial_3 u_\varepsilon\|_{L^2} \|\partial_3^2 u_\varepsilon\|_{L^2}^{1/2} \|\nabla_h \partial_3^2 u_\varepsilon\|_{L^2}.$$

Applying the Young inequalities $2ab \leq a^2 + b^2$ and $ab \leq (1/4)a^4 + (3/4)b^{4/3}$ to the previous estimates we obtain

$$|(\partial_3 \mathrm{div}\,_h u_{\varepsilon,h} \partial_3 u_\varepsilon, \partial_3^2 u_\varepsilon)|$$

$$\leq \frac{\nu_h}{8} \|\nabla_h \partial_3^2 u_\varepsilon\|_{L^2}^2 + \frac{C}{\nu_h} \|\nabla_h \partial_3 u_\varepsilon\|_{L^2}^2 \|\partial_3 u_\varepsilon\|_{L^2} \|\partial_3^2 u_\varepsilon\|_{L^2}$$

$$+ \frac{C}{\nu_h^{1/3}} \|\nabla_h \partial_3 u_\varepsilon\|_{L^2}^2 \|\partial_3 u_\varepsilon\|_{L^2}^{2/3} \|\partial_3^2 u_\varepsilon\|_{L^2}^{2/3}. \tag{7.31}$$

In the same way, we prove that

$$2|(\partial_3 u_{\varepsilon,h} \cdot \nabla_h \partial_3 u_\varepsilon, \partial_3^2 u_\varepsilon)|$$

$$\leq \frac{\nu_h}{8} \|\nabla_h \partial_3^2 u_\varepsilon\|_{L^2}^2 + \frac{C}{\nu_h} \|\nabla_h \partial_3 u_\varepsilon\|_{L^2}^2 \|\partial_3 u_\varepsilon\|_{L^2} \|\partial_3^2 u_\varepsilon\|_{L^2}$$

$$+ \frac{C}{\nu_h^{1/3}} \|\nabla_h \partial_3 u_\varepsilon\|_{L^2}^2 \|\partial_3 u_\varepsilon\|_{L^2}^{2/3} \|\partial_3^2 u_\varepsilon\|_{L^2}^{2/3}. \tag{7.32}$$

The equalities (7.27) and (7.28) as well as the inequalities (7.29) to (7.32) imply that, for $0 \leq t \leq T_0$,

$$\partial_t \|\partial_3^2 u_\varepsilon\|_{L^2}^2 + \nu_h \|\nabla_h \partial_3^2 u_\varepsilon\|_{L^2}^2 + 2\varepsilon \|\partial_3^3 u_\varepsilon\|_{L^2}^2$$
$$\leq \frac{C}{\nu_h} \|\nabla_h u_\varepsilon\|_{H^{0,1}}^2 \left(\|\partial_3^2 u_\varepsilon\|_{L^2}^2 + \|\partial_3 u_\varepsilon\|_{L^2}^2 + \nu_h \|\partial_3 u_\varepsilon\|_{L^2} \right).$$
(7.33)

Integrating the inequality (7.33) from 0 to t, we obtain, for $0 \leq t \leq T_0$,

$$\|\partial_3^2 u_\varepsilon(t)\|_{L^2}^2 + \nu_h \int_0^t \|\nabla_h \partial_3^2 u_\varepsilon(s)\|_{L^2}^2 ds + 2\varepsilon \int_0^t \|\partial_3^3 u_\varepsilon(s)\|_{L^2}^2 ds$$
$$\leq \|\partial_3^2 u_{\varepsilon,0}\|_{L^2}^2 + \frac{C}{\nu_h} \int_0^t \|\nabla_h u_\varepsilon(s)\|_{H^{0,1}}^2 \|\partial_3^2 u_\varepsilon(s)\|_{L^2}^2 ds$$
$$+ \frac{C}{\nu_h} \sup_{0 \leq s \leq T_0} \left(\|\partial_3 u_\varepsilon(s)\|_{L^2}^2 + \nu_h \|\partial_3 u_\varepsilon(s)\|_{L^2} \right) \int_0^{T_0} \|\nabla_h u_\varepsilon(s)\|_{H^{0,1}}^2 ds.$$

Applying the Gronwall Lemma, we deduce from the previous inequality that, for $0 \leq t \leq T_0$,

$$\|\partial_3^2 u_\varepsilon(t)\|_{L^2}^2 + \nu_h \int_0^t \|\nabla_h \partial_3^2 u_\varepsilon(s)\|_{L^2}^2 ds + 2\varepsilon \int_0^t \|\partial_3^3 u_\varepsilon(s)\|_{L^2}^2 ds$$
$$\leq \left[\exp(\frac{C}{\nu_h} \int_0^{T_0} \|\nabla_h u_\varepsilon(s)\|_{H^{0,1}}^2 ds) \right] \left(\|\partial_3^2 u_{\varepsilon,0}\|_{L^2}^2 \right.$$
$$\left. + \frac{C}{\nu_h} \sup_{0 \leq s \leq T_0} \left(\|\partial_3 u_\varepsilon(s)\|_{L^2}^2 + \nu_h \|\partial_3 u_\varepsilon(s)\|_{L^2} \right) \int_0^{T_0} \|\nabla_h u_\varepsilon(s)\|_{H^{0,1}}^2 ds \right).$$

The proposition is thus proved. $\qquad\square$

Propositions 7.2.5 and 7.2.8 together with Remark 7.2.2 imply the following $H^{0,2}$-propagation result.

Corollary 7.2.9 *Let $u_0 \in \widetilde{H}^{0,1}(Q) \cap \widetilde{H}_0^{0,2}(Q)$ be given. Let $\varepsilon_m > 0$ be a (decreasing) sequence converging to zero and $u_0^m \in \widetilde{H}_0^1(Q) \cap H^2(Q) \cap \widetilde{H}_0^{0,2}(Q)$ be a sequence of initial data converging to u_0 in $\widetilde{H}^{0,2}(Q)$, as $m \to \infty$. Assume that the system (NS_{ε_m}), with initial data Σu_0^m, has a strong solution $u_{\varepsilon_m}(t) \in C^0((0,T_0), \widetilde{V})$ where T_0 does not depend on ε_m and that the sequences $u_{\varepsilon_m}(t)$ and $\nabla_h u_{\varepsilon_m}(t)$ are uniformly bounded in $L^\infty((0,T_0), H^{0,1}(\widetilde{Q}))$ and $L^2((0,T_0), H^{0,1}(\widetilde{Q}))$ respectively. Then, the sequence $u_{\varepsilon_m}(t)$ converges in $L^\infty((0,T_0), L^2(\widetilde{Q})^3) \cap L^2((0,T_0), H^{1,0}(\widetilde{Q}))$ to a solution $u^* \in L^\infty((0,T_0), H^{0,2}(\widetilde{Q}))$ of the problem (NS_h), such that $\nabla_h \partial_3^i u^*$ belongs to $L^2((0,T_0), L^2(\widetilde{Q})^3)$, for $i = 0,1,2$. Moreover, the solution u^* belongs to $L^\infty((0,T_0), \widetilde{H}_0^{0,2}(Q))$.*

Proof Let $u_0 \in \widetilde{H}^{0,1}(Q) \cap \widetilde{H}_0^{0,2}(Q)$ be given. We notice that, by Remark 7.2.2, there exists a sequence $u_0^m \in \widetilde{H}_0^1(Q) \cap H^2(Q) \cap \widetilde{H}_0^{0,2}(Q)$ of initial data converging to u_0 in $\widetilde{H}^{0,2}(Q)$, as $m \to \infty$. Let u_0^m be such a sequence. As we have remarked in the introduction, Σu_0^m belongs to $H^2(\widetilde{Q})^3$ and $\partial_3 u_{0,h}^m$ vanishes on $\Gamma_0 \cup \Gamma_1$. By Proposition 7.2.8, the classical solution $u_{\varepsilon m}$ of $(\mathrm{NS}_{\varepsilon m})$ is more regular in the sense that $\partial_3^2 u_{\varepsilon m}$ (respectively $\nabla_h \partial_3^2 u_{\varepsilon m}$) is uniformly bounded in $L^\infty((0,T_0), L^2(\widetilde{Q})^3)$ (respectively in $L^2((0,T_0), L^2(\widetilde{Q})^3)$. Thus the limit $\partial_3^2 u^*$ belongs to $L^\infty((0,T_0), L^2(\widetilde{Q})^3)$ and $\nabla_h \partial_3^2 u^*$ belongs to $L^2((0,T_0), L^2(\widetilde{Q})^3)$.   □

7.3 Global existence results for small initial data

We begin with the simplest result.

Theorem 7.3.1 *There exists a positive constant c_0 such that, if u_0 belongs to $\widetilde{H}^{0,1}(Q)$ and $\|u_0\|_{H^{0,1}} \leq c_0 \nu_h$, then the system (NS_h) admits a (unique) global solution $u(t)$, with $u(0) = u_0$, such that*

$$u \in L^\infty(\mathbb{R}_+, \widetilde{H}^{0,1}(Q)) \quad \text{and} \quad \partial_3 \nabla_h u \in L^2(\mathbb{R}_+, L^2(Q)^3).$$

Proof According to the strategy explained in the introduction and according to Proposition 7.2.5, it is sufficient to prove that there exists a positive constant c_1 such that if $u_\varepsilon(0) = w_0$ belongs to $H_{0,\mathrm{per}}^1(\widetilde{Q})$ and satisfies

$$\|w_0\|_{H^{0,1}} \leq c_1 \nu_h,$$

then, for any $\varepsilon > 0$, the equations $(\mathrm{NS}_\varepsilon)$ admit a unique global solution $u_\varepsilon(t) \in C^0([0,+\infty), \widetilde{V})$ with $u_\varepsilon(0) = w_0$ and moreover, u_ε and $\partial_3 \nabla_h u_\varepsilon$ are uniformly bounded (with respect to ε) in $L^\infty((0,+\infty), H^{0,1}(\widetilde{Q}))$ and $L^2((0,+\infty), L^2(\widetilde{Q})^3)$.

Let u_ε be the local solution of the equations $(\mathrm{NS}_\varepsilon)$ with $u_\varepsilon(0) = w_0$. Since u_ε is a classical solution on the maximal interval of existence, all the *a priori* estimates made below can be justified rigorously. Differentiating the first equation in $(\mathrm{NS}_\varepsilon)$ with respect to x_3 and taking the inner product in $L^2(\widetilde{Q})^3$ with $\partial_3 u_\varepsilon$, we obtain, for $0 \leq t \leq T_\varepsilon$, where $T_\varepsilon > 0$ is the maximal time of existence,

$$\frac{1}{2} \partial_t \|\partial_3 u_\varepsilon\|_{L^2}^2 - \nu_h(\Delta_h \partial_3 u_\varepsilon, \partial_3 u_\varepsilon) - \varepsilon(\partial_3^3 u_\varepsilon, \partial_3 u_\varepsilon)$$
$$= -(\nabla \partial_3 p_\varepsilon, \partial_3 u_\varepsilon) - (\partial_3(u_\varepsilon \nabla u_\varepsilon), \partial_3 u_\varepsilon). \tag{7.34}$$

Since u_ε and hence $\partial_3 u_\varepsilon$ vanish on the lateral boundary $\partial\Omega \times (-1,1)$ and u_ε and p_ε are periodic in the vertical variable,

$$
-\int_{\tilde{Q}} \Delta_h \partial_3 u_\varepsilon \cdot \partial_3 u_\varepsilon \mathrm{d}x_h \mathrm{d}x_3 = \int_{\tilde{Q}} |\nabla_h \partial_3 u_\varepsilon|^2 \mathrm{d}x_h \mathrm{d}x_3
$$
$$
-\int_{\tilde{Q}} \partial_3^3 u_\varepsilon \cdot \partial_3 u_\varepsilon \mathrm{d}x_h \mathrm{d}x_3 = \int_{\tilde{Q}} (\partial_3^2 u_\varepsilon)^2 \mathrm{d}x_h \mathrm{d}x_3,
$$

(7.35)

and

$$
-\int_{\tilde{Q}} \nabla \partial_3 p_\varepsilon \partial_3 u_\varepsilon \mathrm{d}x_h \mathrm{d}x_3 = \int_{\tilde{Q}} \partial_3 p_\varepsilon \mathrm{div}\, \partial_3 u_\varepsilon \mathrm{d}x_h \mathrm{d}x_3
$$
$$
-\int_{\partial\tilde{Q}} \partial_3 p_\varepsilon (\partial_3 u_\varepsilon \cdot n) \mathrm{d}\sigma = 0 .
$$

(7.36)

The equalities (7.34), (7.35), and (7.36) together with Lemma 7.2.4 imply that, for $0 \leq t \leq T_\varepsilon$,

$$
\partial_t \|\partial_3 u_\varepsilon\|_{L^2}^2 + 2\nu_h \|\nabla_h \partial_3 u_\varepsilon\|_{L^2}^2 + 2\varepsilon \|\partial_3^2 u_\varepsilon\|_{L^2}^2
$$
$$
\leq 4C_1 \|u_\varepsilon\|_{H^{0,1}} \|\nabla_h u_\varepsilon\|_{H^{0,1}}^2.
$$

(7.37)

We deduce from the estimates (7.17) and (7.37) that, for $0 \leq t \leq T_\varepsilon$,

$$
\partial_t \big(\|u_\varepsilon\|_{L^2}^2 + \|\partial_3 u_\varepsilon\|_{L^2}^2 \big) + 2\nu_h \big(\|\nabla_h u_\varepsilon\|_{L^2}^2 + \|\nabla_h \partial_3 u_\varepsilon\|_{L^2}^2 \big)
$$
$$
+ 2\varepsilon \big(\|\partial_3 u_\varepsilon\|_{L^2}^2 + \|\partial_3^2 u_\varepsilon\|_{L^2}^2 \big)
$$
$$
\leq 8C_1 \|u_\varepsilon\|_{H^{0,1}} \big(\|\nabla_h u_\varepsilon\|_{L^2}^2 + \|\nabla_h \partial_3 u_\varepsilon\|_{L^2}^2 \big).
$$

(7.38)

Suppose now that the initial condition $u_\varepsilon(0) = w_0$ is small enough in the sense that

$$
\|w_0\|_{H^{0,1}} \leq \frac{\nu_h}{32 C_1}.
$$

Then, by continuity, there exists a time interval $[0, \tau_\varepsilon)$ such that, for $t \in [0, \tau_\varepsilon)$, $\|u_\varepsilon(t)\|_{H^{0,1}} < \nu_h/(8C_1)$. If $\tau_\varepsilon < T_\varepsilon$, then $\|u_\varepsilon(\tau_\varepsilon)\|_{H^{0,1}} = \nu_h/(8C_1)$. Assume now that $\tau_\varepsilon < T_\varepsilon$. If t belongs to the time interval $[0, \tau_\varepsilon]$, we deduce from the inequality (7.38) that

$$
\partial_t \big(\|u_\varepsilon\|_{L^2}^2 + \|\partial_3 u_\varepsilon\|_{L^2}^2 \big) + \nu_h \big(\|\nabla_h u_\varepsilon\|_{L^2}^2 + \|\nabla_h \partial_3 u_\varepsilon\|_{L^2}^2 \big)
$$
$$
+ 2\varepsilon \big(\|\partial_3 u_\varepsilon\|_{L^2}^2 + \|\partial_3^2 u_\varepsilon\|_{L^2}^2 \big) \leq 0.
$$

(7.39)

Integrating the inequality (7.39) from 0 to t, we obtain that, for $t \leq \tau_\varepsilon$,

$$\|u_\varepsilon(t)\|_{L^2}^2 + \|\partial_3 u_\varepsilon(t)\|_{L^2}^2 + \nu_h \int_0^t \left(\|\nabla_h u_\varepsilon(s)\|_{L^2}^2 + \|\nabla_h \partial_3 u_\varepsilon(s)\|_{L^2}^2 \right) ds$$

$$+ 2\varepsilon \int_0^t \left(\|\partial_3 u_\varepsilon(s)\|_{L^2}^2 + \|\partial_3^2 u_\varepsilon(s)\|_{L^2}^2 \right) ds \leq \|w_0\|_{L^2}^2 + \|\partial_3 w_0\|_{L^2}^2.$$

$$(7.40)$$

The estimate (7.40) implies that, for $t \leq \tau_\varepsilon$,

$$\|u_\varepsilon(t)\|_{H^{0,1}} \leq \frac{\nu_h}{16 C_1}.$$

In particular, $\|u_\varepsilon(\tau_\varepsilon)\|_{H^{0,1}} \leq \nu_h/(16 C_1)$, which contradicts the definition of τ_ε. Thus $\tau_\varepsilon = T_\varepsilon$ and one deduces from (7.40) that, for $0 \leq t < T_\varepsilon$,

$$\|u_\varepsilon(t)\|_{L^2}^2 + \|\partial_3 u_\varepsilon(t)\|_{L^2}^2 + \nu_h \int_0^{T_\varepsilon} \left(\|\nabla_h u_\varepsilon(s)\|_{L^2}^2 + \|\nabla_h \partial_3 u_\varepsilon(s)\|_{L^2}^2 \right) ds$$

$$+ 2\varepsilon \int_0^{T_\varepsilon} \left(\|\partial_3 u_\varepsilon(s)\|_{L^2}^2 + \|\partial_3^2 u_\varepsilon(s)\|_{L^2}^2 \right) ds \leq \|w_0\|_{L^2}^2 + \|\partial_3 w_0\|_{L^2}^2.$$

To prove that $u_\varepsilon(t)$ exists globally, that is, that $T_\varepsilon = +\infty$, it remains to show that $\|\nabla_h u_\varepsilon(t)\|_{L^2}$ is uniformly bounded with respect to $t \in [0, T_\varepsilon)$. But this property is a direct consequence of Proposition 7.2.7, and so Theorem 7.3.1 is proved. $\qquad\square$

A more careful analysis allows us to prove the following more refined global existence result.

Theorem 7.3.2 *There exist positive constants c_0 and c_0^* such that, if u_0 belongs to $\tilde{H}^{0,1}(Q)$ and satisfies the following smallness condition*

$$\|\partial_3 u_0\|_{L^2(\Omega)}^{\frac{1}{2}} \|u_0\|_{L^2(\Omega)}^{\frac{1}{2}} \exp\left(\frac{c_0 \|u_0\|_{L^2}^2}{\nu_h^2} \right) \leq c_0^* \nu_h,$$

then the system (NS_h) with $u(0) = u_0$ admits a (unique) global solution $u(t)$, such that

$$u \in L^\infty(\mathbb{R}_+, \tilde{H}^{0,1}(Q)) \quad and \quad \partial_3 \nabla_h u \in L^2(\mathbb{R}_+; L^2(Q)^3).$$

Proof As in the proof of Theorem 7.3.1, it is sufficient to prove that there exist positive constants c_1 and c_1^* such that if $u_\varepsilon(0) = w_0$ belongs to

$H_{0,\text{per}}^1(\widetilde{Q})$ and satisfies

$$\|\partial_3 w_0\|_{L^2(\Omega)}^{\frac{1}{2}} \|w_0\|_{L^2(\Omega)}^{\frac{1}{2}} \exp(c_1 \frac{\|w_0\|_{L^2}^2}{\nu_h^2}) \leq c_1^* \nu_h,$$

then, for any $\varepsilon > 0$, the equations (NS$_\varepsilon$) with $u_\varepsilon(0) = w_0$ admit a unique global solution $u_\varepsilon(t) \in C^0([0,+\infty), \widehat{V})$ and moreover, u_ε and $\nabla_h u_\varepsilon$ are uniformly bounded (with respect to ε) in $L^\infty((0,+\infty), H^{0,1}(\widetilde{Q}))$ and $L^2((0,+\infty), H^{0,1}(\widetilde{Q}))$.

Now let u_ε be the local solution of the equations (NS$_\varepsilon$) with $u_\varepsilon(0) = w_0$ and let $T_\varepsilon > 0$ be the maximal time of existence. As in the proof of Theorem 7.3.1, u_ε satisfies the equality (7.34). But here, in order to estimate the term $(\partial_3(u_\varepsilon \nabla u_\varepsilon), \partial_3 u_\varepsilon)$, we take into account the estimates (7.9) and (7.10), instead of applying Lemma 7.2.4 directly. The equalities (7.34), (7.35), (7.36), (7.8) and, the estimates (7.9) and (7.10) imply that, for $0 \leq t < T_\varepsilon$,

$$\partial_t \|\partial_3 u_\varepsilon\|_{L^2}^2 + 2\nu_h \|\nabla_h \partial_3 u_\varepsilon\|_{L^2}^2 + 2\varepsilon \|\partial_3^2 u_\varepsilon\|_{L^2}^2$$
$$\leq C_2 \big(\|\nabla_h u_\varepsilon\|_{L^2}^{1/2} \|\partial_3 u_\varepsilon\|_{L^2} \|\nabla_h \partial_3 u_\varepsilon\|_{L^2}^{3/2} + \qquad (7.41)$$
$$+ \|\nabla_h u_\varepsilon\|_{L^2} \|\partial_3 u_\varepsilon\|_{L^2} \|\nabla_h \partial_3 u_\varepsilon\|_{L^2} \big).$$

Using the Young inequalities $ab \leq \frac{3}{4}a^{\frac{4}{3}} + \frac{1}{4}b^4$ and $ab \leq \frac{1}{2}a^2 + \frac{1}{2}b^2$, we get the following estimates,

$$C_2 \|\nabla_h u_\varepsilon\|_{L^2}^{1/2} \|\partial_3 u_\varepsilon\|_{L^2} \|\nabla_h \partial_3 u_\varepsilon\|_{L^2}^{3/2}$$
$$\leq \frac{27 C_2^4}{32 \nu_h^3} \|\nabla_h u_\varepsilon\|_{L^2}^2 \|\partial_3 u_\varepsilon\|_{L^2}^4 + \frac{\nu_h}{2} \|\nabla_h \partial_3 u_\varepsilon\|_{L^2}^2 \qquad (7.42)$$

and

$$C_2 \|\nabla_h u_\varepsilon\|_{L^2} \|\partial_3 u_\varepsilon\|_{L^2} \|\nabla_h \partial_3 u_\varepsilon\|_{L^2}$$
$$\leq \frac{C_2^2}{2\nu_h} \|\nabla_h u_\varepsilon\|_{L^2}^2 \|\partial_3 u_\varepsilon\|_{L^2}^2 + \frac{\nu_h}{2} \|\nabla_h \partial_3 u_\varepsilon\|_{L^2}^2. \qquad (7.43)$$

From the estimates (7.41), (7.42), and (7.43), we deduce that, for $0 \leq t < T_\varepsilon$,

$$\frac{d}{dt} \|\partial_3 u_\varepsilon(t)\|_{L^2}^2 + \nu_h \|\nabla_h \partial_3 u_\varepsilon(t)\|_{L^2}^2 + 2\varepsilon \|\partial_3^2 u_\varepsilon\|_{L^2}^2$$
$$\leq C_3 \big(\frac{1}{\nu_h} \|\nabla_h u_\varepsilon\|_{L^2}^2 \|\partial_3 u_\varepsilon\|_{L^2}^2 + \frac{1}{\nu_h^3} \|\nabla_h u_\varepsilon\|_{L^2}^2 \|\partial_3 u_\varepsilon\|_{L^2}^4 \big), \qquad (7.44)$$

where $C_3 = \max\left(C_2^2/2, 27C_2^4/32\right)$. The inequality (7.44) shows that, if there exists $\tau_\varepsilon < T_\varepsilon$ such that $\|\partial_3 u_\varepsilon(\tau_\varepsilon)\|_{L^2}^2$ vanishes, then $\|\partial_3 u_\varepsilon(t)\|_{L^2}^2$ is identically equal to zero for $\tau_\varepsilon \leq t < T_\varepsilon$.

On the time interval $[0, \tau_\varepsilon)$, the inequality

$$\frac{\mathrm{d}}{\mathrm{d}t}\|\partial_3 u_\varepsilon(t)\|_{L^2}^2 \leq C_3\Big(\frac{1}{\nu_h}\|\nabla_h u_\varepsilon\|_{L^2}^2\|\partial_3 u_\varepsilon\|_{L^2}^2$$
$$+ \frac{1}{\nu_h^3}\|\nabla_h u_\varepsilon\|_{L^2}^2\|\partial_3 u_\varepsilon\|_{L^2}^4\Big) \tag{7.45}$$

can be written as

$$-\frac{\mathrm{d}}{\mathrm{d}t}\|\partial_3 u_\varepsilon(t)\|_{L^2}^{-2} \leq \frac{C_3}{\nu_h}\|\nabla_h u_\varepsilon\|_{L^2}^2\|\partial_3 u_\varepsilon\|_{L^2}^{-2} + \frac{C_3}{\nu_h^3}\|\nabla_h u_\varepsilon\|_{L^2}^2$$

or also

$$-\frac{\mathrm{d}}{\mathrm{d}t}\Big(\|\partial_3 u_\varepsilon\|_{L^2}^{-2}\exp(\frac{C_3}{\nu_h}\int_0^t\|\nabla_h u_\varepsilon(s)\|_{L^2}^2\mathrm{d}s)\Big)$$
$$\leq \frac{C_3}{\nu_h^3}\|\nabla_h u_\varepsilon(t)\|_{L^2}^2\exp(\frac{C_3}{\nu_h}\int_0^t\|\nabla_h u_\varepsilon(s)\|_{L^2}^2\mathrm{d}s).$$

Integrating this inequality from 0 to t, we obtain, for $0 \leq t < \tau_\varepsilon$,

$$\|\partial_3 w_0\|_{L^2}^{-2} - \|\partial_3 u_\varepsilon(t)\|_{L^2}^{-2}\exp(\frac{C_3}{\nu_h}\int_0^t\|\nabla_h u_\varepsilon(s)\|_{L^2}^2\mathrm{d}s)$$
$$\leq \frac{C_3}{\nu_h^3}\int_0^t\|\nabla_h u_\varepsilon(s)\|_{L^2}^2\mathrm{d}s \times \exp(\frac{C_3}{\nu_h}\int_0^t\|\nabla_h u_\varepsilon(s)\|_{L^2}^2\mathrm{d}s). \tag{7.46}$$

The second energy estimate in Lemma 7.2.6 and the inequality (7.46) imply that, for $0 \leq t < \tau_\varepsilon$,

$$\|\partial_3 w_0\|_{L^2}^{-2} - \frac{C_3}{\nu_h^4}\|w_0\|_{L^2}^2\exp(\frac{C_3\|w_0\|_{L^2}^2}{\nu_h^2}) \leq \|\partial_3 u_\varepsilon(t)\|_{L^2}^{-2}\exp(\frac{C_3\|w_0\|_{L^2}^2}{\nu_h^2}).$$

Thus, if we assume that,

$$\|\partial_3 w_0\|_{L^2}^{-2} - \frac{C_3\|w_0\|_{L^2}^2}{\nu_h^4}\exp(\frac{C_3\|w_0\|_{L^2}^2}{\nu_h^2}) > 0,$$

that is,

$$\|\partial_3 w_0\|_{L^2}^{1/2}\|w_0\|_{L^2}^{1/2}\exp(\frac{C_3\|w_0\|_{L^2}^2}{4\nu_h^2}) < C_3^{-1/4}\nu_h,$$

then we get the following uniform bound, for $0 \le t < \tau_\varepsilon$,

$$\|\partial_3 u_\varepsilon(t)\|_{L^2}^2$$
$$\le \exp(\frac{C_3\|w_0\|_{L^2}^2}{\nu_h^2})\Big(\|\partial_3 w_0\|_{L^2}^{-2} - \frac{C_3}{\nu_h^4}\|w_0\|_{L^2}^2 \exp(\frac{C_3\|w_0\|_{L^2}^2}{\nu_h^2})\Big)^{-1}.$$
$$(7.47)$$

Let us denote by B_0 the right-hand side of this inequality. Integrating the estimate (7.44) from 0 to t and taking into account the second energy estimate in Lemma 7.2.6 as well as the estimate (7.47) and the definition of τ_ε, we at once obtain the following inequality, for any $0 \le t < T_\varepsilon$,

$$\nu_h \int_0^t \|\nabla_h \partial_3 u_\varepsilon(s)\|_{L^2}^2 \mathrm{d}s + 2\varepsilon \int_0^t \|\partial_3^2 u_\varepsilon(s)\|_{L^2}^2 \mathrm{d}s$$
$$\le \frac{C_3}{\nu_h} B_0^2 (1 + \frac{1}{\nu_h^2} B_0^2) \int_0^t \|\nabla_h u_\varepsilon(s)\|_{L^2}^2 \mathrm{d}s$$
$$\le \frac{C_3}{2\nu_h^2} B_0^2 (1 + \frac{1}{\nu_h^2} B_0^2) \|w_0\|_{L^2}^2.$$

To prove that $u_\varepsilon(t)$ exists globally, that is, that $T_\varepsilon = +\infty$, it remains to show that $\|\nabla_h u_\varepsilon(t)\|_{L^2}$ is uniformly bounded with respect to $t \in [0, T_\varepsilon)$. As in the proof of Theorem 7.3.1, this property is a direct consequence of Proposition 7.2.7. Theorem 7.3.2 is thus proved.   □

Remark 7.3.3 *The previous theorem allows us to take large initial data in the following sense. For example, we can take* $u_0 \in \tilde{H}^{0,1}(Q)$ *such that*

$$\|u_0\|_{L^2(Q)} \le C\eta^\alpha$$

and

$$\|\partial_3 u_0\|_{L^2(Q)} \le C\eta^{-\alpha},$$

where η is a small positive constant going to 0 and $C > 0$ is an appropriate positive constant.

Remark 7.3.4 *Let us come back to the inequality (7.45). If we set*

$$y(t) = \frac{\nu_h}{2} + \frac{1}{\nu_h}\|\partial_3 u_\varepsilon(t)\|_{L^2}^2 \quad \text{and} \quad g(t) = \frac{C_3}{\nu_h}\|\nabla_h u_\varepsilon(t)\|_{L^2}^2 ,$$

then the inequality (7.45) becomes, for $0 \le t < \tau_\varepsilon$,

$$\frac{\mathrm{d}y}{\mathrm{d}t}(t) \le g(t)\, y^2(t).$$

Integrating this inequality from 0 to t, for $0 \leq t < \tau_\varepsilon$, we get

$$-\frac{1}{y(t)} + \frac{1}{y(0)} \leq \int_0^t g(s)\,ds,$$

hence

$$y(t) \leq \frac{y(0)}{1 - y(0)\int_0^t g(s)ds},$$

as long as $1 - y(0)\int_0^t g(s)\,ds > 0$. This implies that

$$\frac{\nu_h}{2} + \frac{1}{\nu_h}\|\partial_3 u_\varepsilon(t)\|_{L^2}^2 \leq \left(\frac{\nu_h}{2} + \frac{1}{\nu_h}\|\partial_3 u_\varepsilon(0)\|_{L^2}^2\right)$$
$$\times \left(1 - \frac{C_3}{\nu_h}\int_0^t \|\nabla_h u_\varepsilon(s)\|_{L^2}^2\,ds\left(\frac{\nu_h}{2} + \frac{1}{\nu_h}\|\partial_3 u_\varepsilon(0)\|_{L^2}^2\right)\right)^{-1}.$$
$$(7.48)$$

Inequality (7.48) and Lemma 7.2.6 imply that

$$\frac{\nu_h}{2} + \frac{1}{\nu_h}\|\partial_3 u_\varepsilon(t)\|_{L^2}^2 \leq \left(\frac{\nu_h}{2} + \frac{1}{\nu_h}\|\partial_3 u_\varepsilon(0)\|_{L^2}^2\right)$$
$$\times \left(1 - \frac{C_3}{2\nu_h^2}\|u_\varepsilon(0)\|_{L^2}^2\left(\frac{\nu_h}{2} + \frac{1}{\nu_h}\|\partial_3 u_\varepsilon(0)\|_{L^2}^2\right)\right)^{-1}.$$

Thus, if

$$C_3\|u_\varepsilon(0)\|_{L^2}^2\left(\frac{\nu_h}{2} + \frac{1}{\nu_h}\|\partial_3 u_\varepsilon(0)\|_{L^2}^2\right) \leq \nu_h^2, \qquad (7.49)$$

we obtain the following uniform bound, for $0 \leq t < \tau_\varepsilon$,

$$\frac{\nu_h}{2} + \frac{1}{\nu_h}\|\partial_3 u_\varepsilon(t)\|_{L^2}^2 \leq 2\left(\frac{\nu_h}{2} + \frac{1}{\nu_h}\|\partial_3 u_\varepsilon(0)\|_{L^2}^2\right).$$

As in the proof of Theorem 7.3.2, we deduce that, under the condition (7.49), the solution $u_\varepsilon(t)$ exists globally.

Theorems 7.3.1 and 7.3.2 together with Corollary 7.2.9 imply, at once, the following result of propagation of regularity.

Corollary 7.3.5 *Under the hypotheses of Theorem 7.3.1 or 7.3.2, if in addition the initial condition u_0 belongs to $\widetilde{H}_0^{0,2}(Q)$, then the solution u of system (NS$_h$) with $u(0) = u_0$ belongs to $L^\infty(\mathbb{R}_+, \widetilde{H}_0^{0,2}(Q))$ and $\partial_3^2 \nabla_h u$ belongs to $L^2(\mathbb{R}_+, L^2(Q)^3)$.*

7.4 The case of general initial data

In this section, we want to prove the local existence of the solution $u(t)$ of the equations (NS$_h$), when the initial data are not necessarily small.

Theorem 7.4.1 *Let U_0 be given in $\widetilde{H}^{0,1}(Q)$. There exist a positive time T_0 and a positive constant η such that, if u_0 belongs to $\widetilde{H}^{0,1}(Q)$ and $\|U_0 - u_0\|_{H^{0,1}} \leq \eta$, then the system (NS$_h$) admits a (unique) strong solution $u(t)$, with $u(0) = u_0$, such that*

$$u \in L^\infty((0, T_0), \widetilde{H}^{0,1}(Q)) \quad \text{and} \quad \partial_3 \nabla_h u \in L^2((0, T_0), L^2(Q)^3).$$

Proof According to the strategy explained in the introduction and according to Proposition 7.2.5, it is sufficient to prove that there exist positive constants η and T_0 such that, if $u_\varepsilon(0) = v_0$ belongs to $H^1_{0,\text{per}}(\widetilde{Q})$ and satisfies

$$\|v_0 - \Sigma U_0\|_{H^{0,1}(\widetilde{Q})} \leq \eta, \tag{7.50}$$

then, for any $\varepsilon > 0$ small enough, the equations (NS$_\varepsilon$) admit a unique (local) solution $u_\varepsilon(t) \in C^0([0, T_0], \widetilde{V})$ with $u_\varepsilon(0) = v_0$ and moreover, u_ε and $\partial_3 \nabla_h u_\varepsilon$ are uniformly bounded with respect to ε in $L^\infty((0, T_0), H^{0,1}(\widetilde{Q}))$ and $L^2((0, T_0), L^2(\widetilde{Q})^3)$.

Let $u_\varepsilon(t)$ be the strong solution of the equations (NS$_\varepsilon$) with initial data $u_\varepsilon(0) = v_0 \in H^1_{0,\text{per}}(\widetilde{Q})$ satisfying the condition (7.50). Let $T_\varepsilon > 0$ be the maximal time of existence of this solution. The proof of Theorem 7.3.2 and Remark 7.3.4 show that, if

$$\frac{C_3}{\nu_h} \int_0^\tau \|\nabla_h u_\varepsilon(s)\|_{L^2}^2 \mathrm{d}s \left(\frac{\nu_h}{2} + \frac{1}{\nu_h}\|\partial_3 u_\varepsilon(0)\|_{L^2}^2\right) < 1,$$

then $T_\varepsilon > \tau$.

It is thus sufficient to show that, for $\eta > 0$ small enough, there exists a positive constant T_0 such that, for any $\varepsilon > 0$, the strong solution u_ε of (NS$_\varepsilon$) satisfies the inequality

$$\frac{C_3}{\nu_h} \int_0^{T_0} \|\nabla_h u_\varepsilon(s)\|_{L^2}^2 \, \mathrm{d}s \left(\frac{\nu_h}{2} + \frac{1}{\nu_h}\|\partial_3 v_0\|_{L^2}^2\right) < \frac{1}{2}. \tag{7.51}$$

Actually, property (7.51) will be proved if we show that, for any positive number δ, there exist two positive numbers $T_0 = T_0(\delta)$ and $\eta_0 = \eta_0(\delta)$ such that

$$\int_0^{T_0} \|\nabla_h u_\varepsilon(s)\|_{L^2}^2 \mathrm{d}s \leq 4\delta. \tag{7.52}$$

The remaining part of the proof consists in showing (7.52). Notice that Lemma 7.2.6 gives us an estimate of the quantity $\int_0^t \|\nabla_h u_\varepsilon(s)\|_{L^2}^2 ds$, which we have used in the proofs of Theorems 7.3.1 and 7.3.2. Unfortunately, here the initial data $u_\varepsilon(0) = v_0$ are not necessarily small. In order to prove (7.52), we write the solution u_ε as

$$u_\varepsilon = v_\varepsilon + z_\varepsilon,$$

where v_ε is the solution of the linear Stokes problem

$$(\text{LS}_\varepsilon) \begin{cases} \partial_t v_\varepsilon - \nu_h \Delta_h v_\varepsilon - \varepsilon \partial_{x_3}^2 v_\varepsilon = -\nabla q_\varepsilon \text{ in } \widetilde{Q}, \ t > 0, \\ \text{div } v_\varepsilon = 0 \text{ in } \widetilde{Q}, \ t > 0, \\ v_\varepsilon|_{\partial\Omega\times(-1,1)} = 0, \ t > 0, \\ v_\varepsilon(x_h, x_3) = v_\varepsilon(x_h, x_3 + 2), \ t > 0, \\ v_\varepsilon|_{t=0} = v_0, \end{cases}$$

and where z_ε is the solution of the following auxiliary nonlinear system

$$(\text{Z}_\varepsilon) \begin{cases} \partial_t z_\varepsilon + (z_\varepsilon + v_\varepsilon)\nabla(z_\varepsilon + v_\varepsilon) - \nu_h \Delta_h z_\varepsilon - \varepsilon \partial_{x_3}^2 z_\varepsilon = -\nabla q_\varepsilon^* \text{ in } \widetilde{Q}, \ t > 0, \\ \text{div } z_\varepsilon = 0 \text{ in } \widetilde{Q}, \ t > 0, \\ z_\varepsilon|_{\partial\Omega\times(-1,1)} = 0, \ t > 0, \\ z_\varepsilon(x_h, x_3) = z_\varepsilon(x_h, x_3 + 2), \ t > 0, \\ z_\varepsilon|_{t=0} = 0. \end{cases}$$

The Stokes problem (LS_ε) admits a unique (global) classical solution v_ε in $C^0([0, +\infty), \widetilde{V})$. Lemma 7.2.6 implies that, for any $t \geq 0$,

$$\|v_\varepsilon(t)\|_{L^2}^2 + \nu_h \int_0^t \|\nabla_h v_\varepsilon(s)\|_{L^2}^2 ds + \varepsilon \int_0^t \|\partial_3 v_\varepsilon(s)\|_{L^2}^2 ds \leq \|v_0\|_{L^2}^2.$$
(7.53)

Arguing as in the proofs of Theorems 7.3.1 and 7.3.2, one at once shows that, for $t \geq 0$,

$$\|\partial_3 v_\varepsilon(t)\|_{L^2}^2 + \nu_h \int_0^t \|\nabla_h \partial_3 v_\varepsilon(s)\|_{L^2}^2 ds + \varepsilon \int_0^t \|\partial_3^2 v_\varepsilon(s)\|_{L^2}^2 ds \leq \|\partial_3 v_0\|_{L^2}^2.$$
(7.54)

Notice that (Z_ε) also admits a unique classical solution $z_\varepsilon \in C^0([0, T_\varepsilon), \widetilde{V})$, where T_ε is the maximal time of existence of u_ε.
We will prove that, for $\eta > 0$ small enough, there exists a $T_0 > 0$, independent of ε, but depending on U_0, such that

$$\int_0^{T_0} \|\nabla_h v_\varepsilon(s)\|_{L^2}^2 ds \leq \delta \quad \text{and} \quad \int_0^{T_0} \|\nabla_h z_\varepsilon(s)\|_{L^2}^2 ds \leq \delta. \qquad (7.55)$$

We introduce a positive number $\delta_0 \leq \delta$, which will be made more precise later. In order to prove the first inequality of (7.55), we proceed as follows by decomposing the linear Stokes problem (LS$_\varepsilon$) into two auxiliary linear systems, the first one with very regular initial data and the second one with small initial data. We recall that P is the classical Leray projector. Let A_0 be the Stokes operator $A_0 = -P\Delta$ with homogeneous Dirichlet boundary conditions on $\partial\Omega \times (-1, 1)$ and periodic boundary conditions in the vertical variable. The spectrum of A_0 consists of a non-decreasing sequence of eigenvalues

$$0 < \lambda_0 < \lambda_1 \leq \lambda_2 \leq \cdots \leq \lambda_m \leq \cdots,$$

which tends to infinity as m tends to infinity. We denote by $\mathbb{P}_k$ the projection onto the space generated by the eigenfunctions associated to the first k eigenvalues of the operator A_0. There exists an integer $k_0 = k_0(\delta_0)$ such that,

$$\|(I - \mathbb{P}_k)\Sigma U_0\|_{L^2}^2 \leq \frac{\delta_0 \nu_h}{16} \quad \forall k \geq k_0. \tag{7.56}$$

If $v_0 \in H^1_{0,\mathrm{per}}(\widetilde{Q})$ satisfies condition (7.50), property (7.56) implies that

$$\|(I - \mathbb{P}_{k_0})v_0\|_{L^2}^2 \leq 2\|(I - \mathbb{P}_{k_0})U_0\|_{L^2}^2 + 2\|(I - \mathbb{P}_{k_0})(U_0 - v_0)\|_{L^2}^2 \leq \frac{\delta_0 \nu_h}{8} + 2\eta^2.$$

On the other hand, the following obvious estimate holds,

$$\|\mathbb{P}_{k_0}v_0\|_{H^1(\widetilde{Q})}^2 \leq \lambda_{k_0+1}\|v_0\|_{L^2}^2 \leq \lambda_{k_0+1}(2\|U_0\|_{L^2}^2 + 2\eta^2). \tag{7.57}$$

We next decompose v_ε into the sum $v_\varepsilon = v_{1,\varepsilon} + v_{2,\varepsilon}$ where $v_{1,\varepsilon}$ is the solution of the Stokes problem

$$(\mathrm{LS}_{1,\varepsilon}) \begin{cases} \partial_t v_{1,\varepsilon} - \nu_h \Delta_h v_{1,\varepsilon} - \varepsilon \partial_{x_3}^2 v_{1,\varepsilon} = -\nabla q_{1,\varepsilon} & \text{in } \widetilde{Q},\ t > 0, \\ \mathrm{div}\, v_{1,\varepsilon} = 0 & \text{in } \widetilde{Q},\ t > 0, \\ v_{1,\varepsilon}|_{\partial\Omega \times (-1,1)} = 0,\ t > 0, \\ v_{1,\varepsilon}(x_h, x_3) = v_{1,\varepsilon}(x_h, x_3 + 2),\ t > 0, \\ v_{1,\varepsilon}|_{t=0} = \mathbb{P}_{k_0}v_0, \end{cases}$$

and $v_{2,\varepsilon}$ is the solution of the Stokes problem

$$(\mathrm{LS}_{2,\varepsilon}) \begin{cases} \partial_t v_{2,\varepsilon} - \nu_h \Delta_h v_{2,\varepsilon} - \varepsilon \partial_{x_3}^2 v_{2,\varepsilon} = -\nabla q_{2,\varepsilon} & \text{in } \widetilde{Q},\ t > 0, \\ \mathrm{div}\, v_{2,\varepsilon} = 0 & \text{in } \widetilde{Q},\ t > 0, \\ v_{2,\varepsilon}|_{\partial\Omega \times (-1,1)} = 0,\ t > 0, \\ v_{2,\varepsilon}(x_h, x_3) = v_{2,\varepsilon}(x_h, x_3 + 2),\ t > 0, \\ v_{2,\varepsilon}|_{t=0} = (I - \mathbb{P}_{k_0})v_0. \end{cases}$$

From Lemma 7.2.6 we deduce that, for $t \geq 0$,

$$\|v_{2,\varepsilon}(t)\|_{L^2}^2 + \nu_h \int_0^t \|\nabla_h v_{2,\varepsilon}(s)\|_{L^2}^2 ds + \varepsilon \int_0^t \|\partial_3 v_{2,\varepsilon}(s)\|_{L^2}^2 ds$$
$$\leq \|(I - \mathbb{P}_{k_0})v_0\|_{L^2}^2 \leq \frac{\delta_0 \nu_h}{8} + 2\eta^2. \tag{7.58}$$

Hence, if $\eta^2 \leq \delta_0 \nu_h / 16$, we obtain, for any $t \geq 0$,

$$\int_0^t \|\nabla_h v_{2,\varepsilon}(s)\|_{L^2}^2 ds \leq \frac{\delta_0}{8} + \frac{2\eta^2}{\nu_h} \leq \frac{\delta_0}{4}. \tag{7.59}$$

In order to get an upper bound of the term $\int_0^t \|\nabla_h v_{1,\varepsilon}(s)\|_{L^2}^2 ds$, we first estimate $\|\nabla_h v_{1,\varepsilon}(s)\|_{L^2}^2$ for any $s \geq 0$. As in the proof of Proposition 7.2.7, we take the inner product in $L^2(\widetilde{Q})^3$ of the first equation of $(LS_{1,\varepsilon})$ with $-P\Delta_h v_{1,\varepsilon}$. Arguing as in the proof of Proposition 7.2.7, we obtain, for any $t \geq 0$,

$$\partial_t \|\nabla_h v_{1,\varepsilon}\|_{L^2}^2 + \nu_h \|P\Delta_h v_{1,\varepsilon}\|_{L^2}^2 + 2\varepsilon \|\nabla_h \partial_3 v_{1,\varepsilon}\|_{L^2}^2 \leq 0.$$

Integrating the above inequality between 0 and t and taking into account the estimate (7.57), we obtain, for $t \geq 0$,

$$\|\nabla_h v_{1,\varepsilon}(t)\|_{L^2}^2 + \nu_h \int_0^t \|P\Delta_h v_{1,\varepsilon}(s)\|_{L^2}^2 ds + 2\varepsilon \int_0^t \|\nabla_h \partial_3 v_{1,\varepsilon}(s)\|_{L^2}^2 ds$$
$$\leq \|\mathbb{P}_{k_0} v_0\|_{H^1(\widetilde{Q})}^2 \leq \lambda_{k_0+1}(2\|U_0\|_{L^2}^2 + 2\eta^2).$$

From the above inequality, we deduce that, for $t \geq 0$,

$$\int_0^t \|\nabla_h v_{1,\varepsilon}(s)\|_{L^2}^2 ds \leq t\lambda_{k_0+1}(2\|U_0\|_{L^2}^2 + 2\eta^2) \leq t\lambda_{k_0+1}(2\|U_0\|_{L^2}^2 + \frac{\delta_0 \nu_h}{8}),$$

and thus, if

$$0 < T_0 \leq \frac{\delta_0}{4\lambda_{k_0+1}}(2\|U_0\|_{L^2}^2 + \frac{\delta_0 \nu_h}{8})^{-1}, \tag{7.60}$$

then

$$\int_0^{T_0} \|\nabla_h v_{1,\varepsilon}(s)\|_{L^2}^2 ds \leq \frac{\delta_0}{4}. \tag{7.61}$$

Inequalities (7.59) and (7.61) imply that, if $\eta^2 \leq \delta_0 \nu_h / 16$ and if the condition (7.60) holds, then

$$\int_0^{T_0} \|\nabla_h v_\varepsilon(s)\|_{L^2}^2 ds \leq \delta_0 \leq \delta. \tag{7.62}$$

It remains to bound the integral $\int_0^{T_0} \|\nabla_h z_\varepsilon(s)\|_{L^2}^2 \mathrm{d}s$. Taking the inner product in $L^2(\widetilde{Q})^3$ of the first equation of System $(\mathbf{Z}_\varepsilon)$ with z_ε, we obtain the equality

$$\frac{1}{2}\partial_t\|z_\varepsilon\|_{L^2}^2 + \nu_h\|\nabla_h z_\varepsilon\|_{L^2}^2 + \varepsilon\|\partial_3 z_\varepsilon\|_{L^2}^2 = -(z_{\varepsilon,3}\partial_3 v_\varepsilon, z_\varepsilon)$$
$$- (z_{\varepsilon,h}\nabla_h v_\varepsilon, z_\varepsilon) - (v_{\varepsilon,3}\partial_3 v_\varepsilon, z_\varepsilon) - (v_{\varepsilon,h}\nabla_h v_\varepsilon, z_\varepsilon). \quad (7.63)$$

We next estimate the four terms of the right-hand side of the equality (7.63). Applying Lemma 7.2.3 and using the fact that $\partial_3 z_{\varepsilon,3} = -\mathrm{div}\,_h z_{\varepsilon,h}$, we obtain, for $0 \le t \le T_\varepsilon$,

$$|(z_{\varepsilon,3}\partial_3 v_\varepsilon, z_\varepsilon)| \le \|z_{\varepsilon,3}\|_{L_v^\infty(L_h^2)}\|\partial_3 v_\varepsilon\|_{L_h^2(L_h^4)}\|z_\varepsilon\|_{L_v^2(L_h^4)}$$
$$\le C_0^3\big(\|z_{\varepsilon,3}\|_{L^2}^{1/2}\|\partial_3 z_{\varepsilon,3}\|_{L^2}^{1/2} + \|z_{\varepsilon,3}\|_{L^2}\big)$$
$$\times \|\partial_3 v_\varepsilon\|_{L^2}^{1/2}\|\partial_3\nabla_h v_\varepsilon\|_{L^2}^{1/2}\|z_\varepsilon\|_{L^2}^{1/2}\|\nabla_h z_\varepsilon\|_{L^2}^{1/2}$$
$$\le C_0^3\big(\|z_{\varepsilon,3}\|_{L^2}^{1/2}\|\nabla_h z_{\varepsilon,h}\|_{L^2}^{1/2} + \|z_{\varepsilon,3}\|_{L^2}\big)$$
$$\times \|\partial_3 v_\varepsilon\|_{L^2}^{1/2}\|\partial_3\nabla_h v_\varepsilon\|_{L^2}^{1/2}\|z_\varepsilon\|_{L^2}^{1/2}\|\nabla_h z_\varepsilon\|_{L^2}^{1/2}.$$

Applying Young's inequality to the above estimate, we get the inequality

$$|(z_{\varepsilon,3}\partial_3 v_\varepsilon, z_\varepsilon)| \le \frac{\nu_h}{8}\|\nabla_h z_\varepsilon\|_{L^2}^2 + \frac{4C_0^6}{\nu_h}\|\partial_3 v_\varepsilon\|_{L^2}\|\partial_3\nabla_h v_\varepsilon\|_{L^2}\|z_\varepsilon\|_{L^2}^2$$
$$+ \frac{3C_0^4}{2\nu_h^{1/3}}\|\partial_3 v_\varepsilon\|_{L^2}^{2/3}\|\partial_3\nabla_h v_\varepsilon\|_{L^2}^{2/3}\|z_\varepsilon\|_{L^2}^2. \quad (7.64)$$

Applying Lemma 7.2.3 and Young's inequality again, we also obtain the following estimate, for $0 \le t \le T_\varepsilon$,

$$|(z_{\varepsilon,h}\nabla_h v_\varepsilon, z_\varepsilon)| \le C_0^2\|\nabla_h v_\varepsilon\|_{L_v^\infty(L_h^2)}\|\nabla_h z_\varepsilon\|_{L^2}\|z_\varepsilon\|_{L^2}$$
$$\le C_0^3\big(\|\nabla_h v_\varepsilon\|_{L^2}^{1/2}\|\nabla_h\partial_3 v_\varepsilon\|_{L^2}^{1/2} + \|\nabla_h v_\varepsilon\|_{L^2}\big)\|\nabla_h z_\varepsilon\|_{L^2}\|z_\varepsilon\|_{L^2}$$
$$\le \frac{\nu_h}{8}\|\nabla_h z_\varepsilon\|_{L^2}^2 + \frac{C_0^6}{\nu_h}\|z_\varepsilon\|_{L^2}^2\big(9\|\nabla_h v_\varepsilon\|_{L^2}^2 + \|\nabla_h\partial_3 v_\varepsilon\|_{L^2}^2\big). \quad (7.65)$$

Applying Lemma 7.2.3 again, we can write, for $0 \le t \le T_\varepsilon$,

$$|(v_{\varepsilon,3}\partial_3 v_\varepsilon, z_\varepsilon)| \le \|v_{\varepsilon,3}\|_{L_v^\infty(L_h^2)}\|\partial_3 v_\varepsilon\|_{L_v^2(L_h^4)}\|z_\varepsilon\|_{L_v^2(L_h^4)}$$
$$\le C_0^3\big(\|v_{\varepsilon,3}\|_{L^2}^{1/2}\|\partial_3 v_{\varepsilon,3}\|_{L^2}^{1/2} + \|v_{\varepsilon,3}\|_{L^2}\big)$$
$$\times \|\partial_3 v_\varepsilon\|_{L^2}^{1/2}\|\partial_3\nabla_h v_\varepsilon\|_{L^2}^{1/2}\|z_\varepsilon\|_{L^2}^{1/2}\|\nabla_h z_\varepsilon\|_{L^2}^{1/2}.$$

Using Young's inequality several times, we deduce from the above estimate that

$$
|(v_{\varepsilon,3}\partial_3 v_\varepsilon, z_\varepsilon)|
$$
$$
\leq \frac{3C_0^4}{2\nu_h^{1/3}}\left(\|v_\varepsilon\|_{L^2}^{2/3}\|\partial_3 v_\varepsilon\|_{L^2}^{2/3} + \|v_\varepsilon\|_{L^2}^{4/3}\right)\|\partial_3 v_\varepsilon\|_{L^2}^{2/3}\|\partial_3\nabla_h v_\varepsilon\|_{L^2}^{2/3}\|z_\varepsilon\|_{L^2}^{2/3}
$$
$$
+ \frac{\nu_h}{8}\|\nabla_h z_\varepsilon\|_{L^2}^2,
$$

and so

$$
|(v_{\varepsilon,3}\partial_3 v_\varepsilon, z_\varepsilon)| \leq \frac{\nu_h}{8}\|\nabla_h z_\varepsilon\|_{L^2}^2 + \frac{C_0^4}{\nu_h}\|\partial_3\nabla_h v_\varepsilon\|_{L^2}^2\|z_\varepsilon\|_{L^2}^2
$$
$$
+ C_0^4\|v_\varepsilon\|_{L^2}^2\|\partial_3 v_\varepsilon\|_{L^2}^2 + C_0^4\|v_\varepsilon\|_{L^2}\|\partial_3 v_\varepsilon\|_{L^2}^2. \tag{7.66}
$$

Finally, arguing as above by applying Lemma 7.2.3 and Young's inequality, we get the estimate

$$
|(v_{\varepsilon,h}\nabla_h v_\varepsilon, z_\varepsilon)| \leq C_0^2\|\nabla_h v_\varepsilon\|_{L_v^\infty(L_h^2)}\|\nabla_h v_\varepsilon\|_{L^2}^{1/2}\|v_\varepsilon\|_{L^2}^{1/2}\|\nabla_h z_\varepsilon\|_{L^2}^{1/2}\|z_\varepsilon\|_{L^2}^{1/2}
$$
$$
\leq C_0^3\left(\|\nabla_h v_\varepsilon\|_{L^2}^{1/2}\|\nabla_h\partial_3 v_\varepsilon\|_{L^2}^{1/2} + \|\nabla_h v_\varepsilon\|_{L^2}\right)
$$
$$
\times \|\nabla_h v_\varepsilon\|_{L^2}^{1/2}\|v_\varepsilon\|_{L^2}^{1/2}\|\nabla_h z_\varepsilon\|_{L^2}^{1/2}\|z_\varepsilon\|_{L^2}^{1/2}
$$
$$
\leq \frac{\nu_h}{8}\|\nabla_h z_\varepsilon\|_{L^2}^2 + 2C_0^4\|\nabla_h v_\varepsilon\|_{L^2}^2\|v_\varepsilon\|_{L^2}
$$
$$
+ \frac{C_0^4}{2\nu_h}\|z_\varepsilon\|_{L^2}^2\left(\|\nabla_h v_\varepsilon\|_{L^2}^2 + \|\nabla_h\partial_3 v_\varepsilon\|_{L^2}^2\right). \tag{7.67}
$$

Integrating the equality (7.63) from 0 to t, taking into account the estimates (7.64) to (7.67) and, applying Gronwall's lemma yields, for $0 \leq t \leq T_\varepsilon$,

$$
\|z_\varepsilon(t)\|_{L^2}^2 + \nu_h\int_0^t\|\nabla_h z_\varepsilon(s)\|_{L^2}^2 ds + 2\varepsilon\int_0^t\|\partial_3 z_\varepsilon(s)\|_{L^2}^2 ds
$$
$$
\leq \int_0^t B_1(s)\|z_\varepsilon(s)\|_{L^2}^2 ds + \int_0^t B_2(s)ds,
$$

and

$$\|z_\varepsilon(t)\|_{L^2}^2 + \nu_h \int_0^t \|\nabla_h z_\varepsilon(s)\|_{L^2}^2 ds \leq 2 \int_0^t B_2(s) ds \left(\exp 2 \int_0^t B_1(s) ds \right), \tag{7.68}$$

where

$$B_1(s) = 2 \Big[\frac{C_0^4}{\nu_h} (3C_0^2 + 2) \|\nabla_h \partial_3 v_\varepsilon(s)\|_{L^2}^2 + \frac{C_0^4}{2\nu_h} (18 C_0^2 + 1) \|\nabla_h v_\varepsilon(s)\|_{L^2}^2$$

$$+ \frac{2C_0^6}{\nu_h} \|\partial_3 v_\varepsilon(s)\|_{L^2}^2 + C_0^4 \|\partial_3 v_\varepsilon(s)\|_{L^2} \Big]$$

$$B_2(s) = 2C_0^4 \Big(\|v_\varepsilon(s)\|_{L^2}^2 \|\partial_3 v_\varepsilon(s)\|_{L^2} + \|v_\varepsilon(s)\|_{L^2} \|\partial_3 v_\varepsilon(s)\|_{L^2}^2$$

$$+ 2\|v_\varepsilon(s)\|_{L^2} \|\nabla_h v_\varepsilon(s)\|_{L^2}^2 \Big). \tag{7.69}$$

The inequalities (7.68) and (7.69) and the estimates (7.53), (7.54), and (7.62) imply that, for $0 \leq t \leq T_\varepsilon$,

$$\int_0^t \|\nabla_h z_\varepsilon(s)\|_{L^2}^2 ds \leq \frac{c_1}{\nu_h} \|v_0\|_{L^2} \Big(t\|v_0\|_{L^2} \|\partial_3 v_0\|_{L^2} + t\|\partial_3 v_0\|_{L^2}^2 + \delta_0 \Big)$$

$$\times \exp c_2 \Big(t\|\partial_3 v_0\|_{L^2} + \nu_h^{-1} \|\partial_3 v_0\|_{L^2}^2 (t + \nu_h^{-1}) + \nu_h^{-1} \delta_0 \Big), \tag{7.70}$$

where c_1 and c_2 are two positive constants independent of v_0 and ε. Since

$$\|v_0\|_{H^{0,1}(\tilde{Q})} \leq \|\Sigma U_0\|_{H^{0,1}(\tilde{Q})} + \eta \leq \|\Sigma U_0\|_{H^{0,1}(\tilde{Q})} + \frac{(\delta_0 \nu_h)^{1/2}}{4},$$

inequality (7.70) shows that we can choose $\delta_0 > 0$ and $T_0 > 0$ independent of ε and v_0 such that

$$\frac{c_1}{\nu_h} \|v_0\|_{L^2} \Big(T_0 \|v_0\|_{L^2} \|\partial_3 v_0\|_{L^2} + T_0 \|\partial_3 v_0\|_{L^2}^2 + \delta_0 \Big)$$

$$\times \exp c_2 \Big(T_0 \|\partial_3 v_0\|_{L^2} + \nu_h^{-1} \|\partial_3 v_0\|_{L^2}^2 (T_0 + \nu_h^{-1}) + \nu_h^{-1} \delta_0 \Big) \leq \frac{\delta}{2}. \tag{7.71}$$

As we have explained at the beginning of the proof, properties (7.70) and (7.71) imply that the maximal time T_ε of existence of z_ε and of u_ε is larger than T_0 and that

$$\int_0^{T_0} \|\nabla_h z_\varepsilon(s)\|_{L^2}^2 ds \leq \frac{\delta}{2}.$$

Thus the inequalities (7.55) are proved, which concludes the proof of the theorem. □

Notice that the classical approach to showing the local-in-time existence result for large initial data, consisting in the decomposition of the problem into a large data linear problem and a small data, perturbed nonlinear problem, does not work here, since we cannot prove that, for initial data U_0 in $\tilde{H}^{0,1}(Q)$, the quantity $\|(I - \mathbb{P}_{k_0})v_0\|_{H^{0,1}(\tilde{Q})}$ is small. In the above proof, the decomposition of the linear system into two systems, one with smooth initial data $\mathbb{P}_{k_0}v_0$ and the other one with small initial data $(I - \mathbb{P}_{k_0})v_0$ avoids this difficulty. Indeed, in the estimates (7.58) and (7.59), we only need to know that $\|(I - \mathbb{P}_{k_0})v_0\|_{L^2(\tilde{Q})}$ is small.

References

Chemin, J.Y., Desjardins, B., Gallagher, I., & Grenier, E. (2000) Fluids with anisotropic viscosity. *M2AN. Mathematical Modelling and Numerical Analysis* **34**, no. 2, 315–335.

Chemin, J.Y., Desjardins, B., Gallagher, I., & Grenier, E. (2006) *Mathematical Geophysics: An introduction to rotating fluids and to the Navier–Stokes equations*. Oxford University Press.

Chemin, J.Y. & Zhang, P. (2007) On the global wellposedness to the 3-D incompressible anisotropic Navier–Stokes equations. *Comm. Math. Phys.* **272**, 529–566.

Constantin, P. & Foias, C. (1988) *Navier–Stokes equations*. University of Chicago Press, Chicago.

Fujita, H. & Kato, T. (1964) On the Navier–Stokes initial value problem I. *Archive for Rational Mechanic and Analysis* **16**, 269–315.

Grenier, E. & Masmoudi, N. (1997) Ekman layers of rotating fluids, the case of well prepared initial data. *Commun. in Partial Differential Equations* **22**, 953–975.

Iftimie, D. (1999) The resolution of the Navier–Stokes equations in anisotropic spaces. *Revista Matemática Iberoamericana* **15**, 1–36.

Iftimie, D. (2002) A uniqueness result for the Navier–Stokes equations with vanishing vertical viscosity. *SIAM Journal on Mathematical Analysis* **33**, no. 6, 1483–1493.

Iftimie, D. & Planas, G. (2006) Inviscid limits for the Navier–Stokes equations with Navier friction boundary conditions. *Nonlinearity* **19**, 899–918.

Paicu, M. (2004) Étude asymptotique pour les fluides anisotropes en rotation rapide dans le cas périodique. *Journal de Mathématiques Pures et Appliquées* **83**, no. 2, 163–242.

Paicu, M. (2005a) Équation périodique de Navier–Stokes sans viscosité dans une direction. *Communications in Partial Differential Equations* **30**, no. 7-9, 1107–1140.

Paicu, M. (2005b) Équation anisotrope de Navier–Stokes dans des espaces critiques. *Revista Matemática Iberoamericana* **21**, no. 1, 179–235.

Pedlovsky, J. (1979) *Geophysical fluid dynamics*. Springer-Verlag.

Solonnikov, V.A. & Ščadilov, V.E. (1973) On a boundary value problem for a stationary system of Navier–Stokes equations. *Trudy Mat. Inst. Steklov* **125**, 186–199.

Taylor, G.I. (1923) Experiments on the motion of solid bodies in rotating fluids. *Proc. Roy. Soc. A* **104**, 213–218.

Temam, R. (1979) *Navier–Stokes Equations*, Revised Edition. Studies in Mathematics and its Applications **2**, North-Holland.

8

The regularity problem for the three-dimensional Navier–Stokes equations

James C. Robinson

Mathematics Institute, University of Warwick,
Coventry, CV4 7AL. U.K.
J.C.Robinson@warwick.ac.uk

Witold Sadowski

Faculty of Mathematics, Informatics and Mechanics,
University of Warsaw, Banacha 2,
02-097 Warszawa. Poland.
witeks@hydra.minuw.edu.pl

Abstract

This paper gives a brief summary of some of the main results concerning the regularity of solutions of the three-dimensional Navier–Stokes equations. We then outline the basis of a numerical algorithm that, at least in theory, can verify regularity for all initial conditions in any bounded subset of H^1.

8.1 Introduction

The aim of this paper is to present some partial results concerning the problem of regularity of global solutions of the three-dimensional Navier-Stokes equations. Since these equations form the fundamental model of hydrodynamics it is a matter of great importance whether or not they can be uniquely solved. However, one hundred and fifty years after the Navier–Stokes model was presented for the first time, we still lack an existence and uniqueness theorem, and the most significant contributions to the subject remain those of Leray (1934) and Hopf (1951).

Nevertheless, there have been many advances since their work, and it would not be possible to give an exhaustive presentation of these in a short article. We give a brief overview of some of the main results, and then concentrate on one specific and in some ways non-standard approach to the problem, with a discussion of the feasibility of testing for regularity via numerical computations following Chernysehnko et al. (2007), Dashti & Robinson (2008), and Robinson & Sadowski (2008).

Published in *Partial Differential Equations and Fluid Mechanics*, edited by James C. Robinson and José L. Rodrigo. © Cambridge University Press 2009.

In some ways this contribution can be viewed as a companion to the introductory review by Robinson (2006).

The plan of this paper is as follows. First we formulate the main problem and define the function spaces that provide us with a convenient framework in which to present the modern research in this field. Then we formulate classical theorems due to Leray and Hopf and sketch some state-of-the-art results concerning sufficient conditions for regularity ('conditional regularity') and restrictions on singularities ('partial regularity'). We then move on to the main part of the paper and describe some recent results that show that in theory one can verify numerically the regularity of solutions of the Navier–Stokes equations arising from large sets of initial conditions.

8.2 Formulation of the problem

The Navier–Stokes equations that model the motion of an incompressible fluid in three dimensions are

$$\frac{\partial u}{\partial t} - \nu \Delta u + (u \cdot \nabla)u + \nabla p = f \qquad (8.1)$$

(momentum conservation) and the incompressibility condition

$$\nabla \cdot u = 0.$$

We consider the problem with the initial condition $u(x,0) = u_0(x)$.

The unknowns are the three-component velocity $u(x,t)$ and the scalar pressure $p(x,t)$. The coefficient $\nu > 0$ is the kinematic viscosity of the fluid and $f(x,t)$ denotes a body force applied (by some means) internally. Throughout what follows we take $\nu = 1$, and generally we will treat the case of unforced 'decaying turbulence', i.e. we take $f = 0$ and consider

$$\frac{\partial u}{\partial t} - \Delta u + (u \cdot \nabla)u + \nabla p = 0 \qquad \nabla \cdot u = 0. \qquad (8.2)$$

We will consider the flow of the fluid in a three-dimensional torus or in the whole of $\mathbb{R}^3$. While these cases are less physically relevant than that of a bounded domain, they avoid technical difficulties that arise from the presence of boundaries while retaining the vortex stretching mechanism which arises from the nonlinear term and appears to be the main factor obstructing a proof of existence and uniqueness of solutions.

Indeed, when the Clay Institute chose to include the question of the regularity of Navier–Stokes solutions as one of their seven Millennium Problems, it is the boundary-free cases that were listed in the official description (Fefferman, 2000).

8.3 Classical results of Leray and Hopf

8.3.1 Function spaces

The formulation of the problem in terms of smooth functions is not mathematically convenient and it is very useful to recast it in a form that allows for solutions in a larger function space.

We take $\Omega = [0, 2\pi]^3$ to be the domain of the flow, and let $\mathcal{C}$ denote the space of all divergence-free smooth periodic three-component functions with zero average on Ω (the zero average condition corresponds to zero total momentum). We introduce a collection of function spaces that arise as the closure of $\mathcal{C}$ with respect to various norms.

The space H, which consists essentially of allowable (divergence-free) fields with finite kinetic energy, is the closure of $\mathcal{C}$ in the L^2 norm

$$\|u\| = \|u\|_{[L^2(\Omega)]^3} = \left(\sum_{i=1}^{3} \int_\Omega |u_i(x)|^2 \, dx \right)^{1/2}.$$

We denote by P the orthogonal projector from $[L^2(\Omega)]^3$ onto H.

The enstrophy space V is the closure of $\mathcal{C}$ in the norm

$$\|Du\| = \left(\int_\Omega |Du(x)|^2 \, dx \right)^{1/2}, \quad \text{where} \quad |Du(x)|^2 = \sum_{i,j=1}^{3} \left| \frac{\partial u_i(x)}{\partial x_j} \right|^2,$$

which, since u has zero average, is equivalent to the standard H^1 norm. (The quantity $\|Du\|^2$ is equal to the square integral of the vorticity (curl u), a quantity that is commonly referred to as the enstrophy.)

Finally, by V^2 we denote the closure of $\mathcal{C}$ in the norm

$$\|\Delta u\| = \left(\int_\Omega |\Delta u(x)|^2 \, dx \right)^{1/2},$$

which is equivalent to the H^2 norm in this periodic setting (again we use the fact that u has zero average).

Since we are working in a periodic domain we can characterize the spaces V^s in a very simple way, in terms of the eigenfunctions of the Stokes operator A, defined as $Au = -P(\Delta u)$. The periodic geometry simplifies calculations greatly, since in this case if $u \in V^2$ then $Au = -\Delta u$.

The Stokes operator is self-adjoint with compact inverse; it follows that there exists an orthonormal basis $\{w_k\}$ for H consisting of eigenfunctions of A:

$$Aw_k = -\Delta w_k = \lambda_k w_k$$

where

$$0 < \lambda_1 \leq \lambda_2 \leq \lambda_3 \leq \cdots$$

are the eigenvalues of the Stokes operator corresponding to the eigen-functions $w_1, w_2, w_3, \ldots$, with $\lambda_1 = 1$ and $\lambda_k \to \infty$ as $k \to \infty$.

The spaces V^s ($s = 0, 1, 2$) consist of all those u that can be obtained as

$$u = \sum_{k=1}^{\infty} c_k w_k,$$

with coefficients $\{c_k\}$ for which

$$\|u\|_s := \left(\sum_{k=1}^{\infty} \lambda_k^{2s} c_k^2 \right)^{1/2}$$

is finite. This expression in fact coincides with the norms defined above (identifying V^0 with H and V^1 with V). However, note that there is no reason to restrict to integer values of s in this definition, so in this way we can obtain a family of spaces V^s for any $s \in \mathbb{R}$, which have the property that $V^0 = H$, $V^1 = V$, and (since $\lambda_k \geq 1$ for all k) that

$$\|u\|_s \leq \|u\|_{s'} \qquad \text{if} \qquad s \leq s'. \tag{8.3}$$

We adopt the more compact notation $\|u\|_s$ (as opposed to $\|u\|_{V^s}$) from now on.

As a final piece of notation, we use $B(u, v)$ to denote the bilinear form defined by

$$B(u, v) = P[(u \cdot \nabla)v].$$

8.3.2 Weak solutions

Now we are in a position to define weak and strong solutions of the Navier–Stokes equations. First let us notice that taking the inner product of both sides of (8.1) with a function $\varphi(x, t) \in \mathcal{C}$ and then integrating by parts we obtain (for $\nu = 1$ and $f \equiv 0$)

$$-\int_0^T \int_\Omega u \frac{\mathrm{d}\varphi}{\mathrm{d}t} + \int_0^T \int_\Omega u \, A\varphi - \int_0^T \int_\Omega u B(u, \varphi) = 0. \tag{8.4}$$

Thus any smooth solution of the Navier–Stokes equations has to satisfy (8.4) for every choice of φ. However, it is not *a priori* excluded that a function u satisfying (8.4) for all such φ may not satisfy (8.1). This leads us to the following notion of a weak solution.

Definition 8.3.1 *Let* $u \in L^\infty(0, T; H) \cap L^2(0, T; V)$ *satisfy the equation*

$$\frac{\mathrm{d}u}{\mathrm{d}t} + A(u) + B(u, u) = 0 \qquad (8.5)$$

in a weak sense, i.e. for every smooth divergence-free function $\varphi(x, t)$ *(with* $\varphi(\cdot, t) \in C$ *for each t) the equation (8.4) holds. Then the function* *u is called a* weak solution *of the Navier–Stokes equations. If in addition* $u \in L^\infty(0, T; V) \cap L^2(0, T; V^2)$ *then u is called a* strong solution *of the Navier–Stokes equations.*

Since the works of Leray (1934) and Hopf (1951) it has been known that given an initial condition in H at least one weak solution exists for all time. Indeed, using the modern language of functional analysis this can be shown relatively easily. The main idea is to construct Galerkin approximations of the solution u: for each $k \in \mathbb{N}$ we find smooth functions $b_i^{(k)}(t)$ $(i = 1, \ldots, k)$ such that

$$u_k(t) = \sum_{i=1}^{k} b_i^{(k)}(t) \, w_i$$

satisfies the equation

$$\frac{\mathrm{d}u_k}{\mathrm{d}t} + Au_k + P_k B(u_k, u_k) = 0, \qquad (8.6)$$

where P_k denotes the orthogonal projection from $[L^2(\Omega)]^3$ onto the span of $\{w_1, \ldots, w_k\}$. Such functions – which are all smooth since the eigenfunctions w_k are smooth – exist due to basic theory of ordinary differential equations. Then one proves estimates on u_k that are uniform (with respect to k) in various spaces, and uses compactness theorems to show that, for some appropriate subsequence (which we relabel here), $u_k \to u$ strongly in $L^2(0, T; H)$, weakly-* in $L^\infty(0, T; H)$, and weakly in $L^2(0, T; V)$; the limit function u is an element of both $L^\infty(0, T; H)$ and $L^2(0, T; V)$, and satisfies (8.4), i.e. is a weak solution. (Note that although each u_k is smooth, the limiting function u may be much less regular.) For more detail see Constantin & Foias (1988), Doering & Gibbon (1995), Robinson (2001), or Temam (2001).

Strong solutions are the key to the regularity problem. While uniqueness of weak solutions is not known, strong solutions are unique in the larger class of weak solutions (see, for example, Temam, 2001). Furthermore, it is relatively straightforward to show that a strong solution is automatically smooth for $t > 0$ using the following bounds on the nonlinear term, which are proved in Constantin & Foias (1988):

$$|(B(w,v), A^m w)| \le c_m \|v\|_{m+1} \|w\|_m^2 \qquad m \ge 2, \qquad (8.7)$$

$$|(B(v,w), A^m w)| \le c_m \|v\|_m \|w\|_m^2 \qquad m \ge 3. \qquad (8.8)$$

Theorem 8.3.2 (cf. Constantin & Foias, 1988, Theorem 10.6)
Let $u \in L^\infty(0,T;V) \cap L^2(0,T;V^2)$ be a strong solution of (8.2) with $u_0 \in V^m$. Then in fact $u \in L^\infty(0,T;V^m) \cap L^2(0,T;V^{m+1})$.

Proof We give a formal argument which can be made rigorous using the Galerkin procedure outlined above. The proof is inductive, supposing initially that $u \in L^2(0,T;V^k)$ for some $k \le m$. Taking the inner product of equation (8.2) with $A^k u$ we obtain

$$\frac{1}{2} \frac{\mathrm{d}}{\mathrm{d}t} \|u\|_k^2 + \|u\|_{k+1}^2 \le |(B(u,u), A^k u)|$$

and so using (8.7) (valid here for $k \ge 2$)

$$\frac{1}{2} \frac{\mathrm{d}}{\mathrm{d}t} \|u\|_k^2 + \|u\|_{k+1}^2 \le c_k \|u\|_k^2 \|u\|_{k+1}.$$

Therefore

$$\frac{\mathrm{d}}{\mathrm{d}t} \|u\|_k^2 + \|u\|_{k+1}^2 \le c_k^2 \|u\|_k^4. \qquad (8.9)$$

Dropping the term $\|u\|_{k+1}^2$ we have

$$\frac{\mathrm{d}}{\mathrm{d}t} \|u\|_k^2 \le \left(c_k^2 \|u\|_k^2 \right) \|u\|_k^2.$$

It now follows from the Gronwall inequality that our inductive assumption $u \in L^2(0,T;V^k)$ implies that $u \in L^\infty(0,T;V^k)$.
Returning to (8.9) and integrating between 0 and T we obtain

$$\int_0^T \|u(s)\|_{k+1}^2 \,\mathrm{d}s \le \|u(0)\|_k^2 + c_k^2 \int_0^T \|u(s)\|_k^4 \,\mathrm{d}s,$$

which shows in turn that $u \in L^2(0,T;V^{k+1})$.
Since by assumption $u \in L^2(0,T;V^2)$, the first use of the induction requires $k = 2$, for which inequality (8.7) is valid: we can therefore conclude by induction that $u \in L^\infty(0,T;V^m) \cap L^2(0,T;V^{m+1})$. $\qquad \square$

It follows from this theorem that strong solutions (whatever the regularity of the initial condition) immediately become smooth: given any $T > 0$ and $k \in \mathbb{N}$, one can easily show that $u(T) \in V^k$. Indeed, if u is a strong solution then $u \in L^2(0,T;V^2)$; it follows that $u(t) \in V^2$ for almost every $t \in (0, T)$, and in particular $u(t_2) \in V^2$ for some $t_2 < T$. So $u \in L^2(t_2, T; V^3)$, which gives $u(t_3) \in V^3$ for some $t_3 \in (t_2, T)$; one can proceed inductively to show that $u(T) \in V^k$. Thus $u(t) \in C^\infty$ for every $t > 0$. Given this, we also refer to a strong solution as a 'regular' solution.

8.3.3 Energy and enstrophy inequalities

Taking the inner product of both sides of (8.6) with u_k and integrating over Ω we obtain

$$\frac{1}{2}\frac{d}{dt}\|u_k\|^2 + \|Du_k\|^2 = 0,$$

since the nonlinear term vanishes (this can be checked by a direct computation). Now, integration with respect to time gives us

$$\|u_k(T)\|^2 + \int_0^T \|Du_k(s)\|^2 \, ds = \|u_k(0)\|^2,$$

the energy equality for the Galerkin approximations. Since

$$\|u_k(0)\| \le \|u(0)\|$$

we also have the energy *inequality* which is satisfied by the function u that is the weak limit of u_k:

$$\|u(T)\|^2 + \int_0^T \|Du(s)\|^2 \, ds \le \|u(0)\|^2. \tag{8.10}$$

Note that this energy inequality is derived here as a consequence of the Galerkin procedure for constructing weak solutions. While it is not known if any weak solution must satisfy the energy inequality (see Cheskidov, Friedlander, & Shvydkoy (2008) for recent work on this problem), this shows that there exists at least one solution that satisfies (8.10). A weak solution that satisfies the energy inequality is called a Leray–Hopf weak solution, and we concentrate on this class of weak solutions from now on.

In a similar way we can obtain an enstrophy inequality. To this end, we take the inner product of (8.5) with $-\Delta u$, integrate over Ω and get

$$\frac{1}{2}\frac{d}{dt}\|Du\|^2 + \|\Delta u\|^2 - \int_\Omega B(u, u) \cdot (\Delta u) = 0.$$

This time the nonlinear term does not vanish, but standard inequalities (Hölder and interpolation inequalities) allow us to estimate the last term on the left-hand side as

$$\left| \int_{\Omega} B(u, u) \cdot (\Delta u) \right| \leq C_B \|Du\|^{3/2} \|\Delta u\|^{3/2}, \tag{8.11}$$

to obtain

$$2\frac{\mathrm{d}}{\mathrm{d}t}\|Du\|^2 + \|\Delta u\|^2 \leq C_B^4 \|Du\|^6. \tag{8.12}$$

8.3.4 Regularity of solutions for small times

We will now investigate consequences of the energy and enstrophy inequalities. Dropping the second term on the left-hand side of (8.12) we obtain

$$2\frac{\mathrm{d}X}{\mathrm{d}t} \leq C_B^4 X^3,$$

where $X = \|Du\|^2$. Since this implies that for $t \geq s$

$$\|Du(t)\|^2 \leq \frac{\|Du(s)\|^2}{\sqrt{1 - 2C_B^4(t - s)\|Du(s)\|^4}}, \tag{8.13}$$

it follows that if the enstrophy of an initial condition is bounded then it cannot blow-up immediately and the minimal time T in which a blow-up could potentially occur can be estimated from below.

Theorem 8.3.3 *Let $u(t)$ be a Leray–Hopf weak solution of the Navier–Stokes equations arising from initial condition $u_0 \in V$. Then there exists a time T^*,*

$$T^* = \frac{1}{2C_B^4 \|Du(0)\|^4}, \tag{8.14}$$

such that for each $T < T^*$

$$u \in L^\infty(0, T; V) \cap L^2(0, T; V^2).$$

Therefore weak solutions obtained from the Galerkin approximation and arising from sufficiently smooth initial conditions remain regular for sufficiently small times. However, what happens after time T^* is unknown.

8.3.5 Regularity of solutions arising from a bounded set of initial conditions

Another consequence of (8.12) is that if the initial enstrophy is sufficiently small then the corresponding solution cannot blow up at all.

Indeed, it follows from (8.3) and (8.12) that

$$2\frac{\mathrm{d}}{\mathrm{d}t}\|Du\|^2 \le C_B^4 \|Du\|^6 - \|Du\|^2. \tag{8.15}$$

Therefore if

$$\|Du\| < C_B^{-1} \tag{8.16}$$

the right-hand side of (8.15) is negative and as a consequence the enstrophy $\|Du\|^2$ is decreasing in time. We can rephrase this as follows:

Theorem 8.3.4 *There exists an $R_0 > 0$ such that every initial condition with $\|Du_0\| \le R_0$ gives rise to a strong solution that exists for all $t \ge 0$.*

The value of R_0 depends only on absolute constants involved in certain Sobolev embedding results (recall that we have set $L = 2\pi$, $\nu = 1$, and $f = 0$), and one can find an explicit bound on this quantity. For our parameter values, one has

$$R_0 \approx 0.00008$$

(see Dashti & Robinson (2008) for a derivation of (8.11) that keeps careful track of the constants involved in estimating C_B).

8.3.6 Regularity of solutions for large times

It turns out that even if the initial enstrophy of an unforced flow ($f = 0$) is large then nevertheless any weak solution arising from this initial condition has to be strong eventually. Indeed, it follows from the energy inequality (8.10) that

$$\int_0^T \|Du(s)\|^2 \, \mathrm{d}s \le \|u(0)\|^2, \tag{8.17}$$

which shows that if $T > T^{**}$, where

$$T^{**} = \frac{\|u(0)\|^2}{C_B^{-2}}, \tag{8.18}$$

then there must be some time $s \in (0, T)$ such that $\|Du(s)\| < C_B^{-1}$. By (8.16) this implies that $u(t)$ is regular for all $t > s$, and so in particular any such u is regular for all $t \ge T^{**}$.

Theorem 8.3.5 *If $u(t)$ is a Leray–Hopf weak solution of the unforced Navier–Stokes equations on $\Omega = [0, 2\pi]^3$ then it is a strong solution for all times $t > T^{**}$, where T^{**} is defined by (8.18).*

8.4 Modern approach

At present there are only partial results concerning the existence of regular solutions of the Navier–Stokes equations, which can roughly be divided into three groups: results that give sufficient conditions for regularity; results that restrict the occurrence of possible singularities; and results that provide large sets of forces or initial conditions that give rise to regular solutions. The main part of this paper is devoted to presenting some of the results from the last group, but first we briefly review conditional and partial regularity results.

8.4.1 Conditional regularity results

In this section we state some of the current results (which mainly concern flows in $\mathbb{R}^3$) that give sufficient conditions for regularity of weak solutions. We state them as simply as possible, and skip any technical assumptions: an interested reader should consult the relevant papers for details.

The first example provides regularity criteria for the velocity field that ensure the existence and uniqueness of smooth solutions (Serrin, 1962; Fabes, Jones, & Riviere, 1972; Escauriaza, Seregin, & Šverák, 2003): if a weak Leray–Hopf solution u of the Navier–Stokes equations satisfies

$$u \in L^r(0,T;L^s(\mathbb{R}^3)) \quad \text{for some } \frac{2}{r} + \frac{3}{s} \leq 1, \ 3 \leq s \leq \infty \qquad (8.19)$$

then u is regular.

The second example involves an analogous criterion for the pressure (Chae & Lee, 2001; Berselli & Galdi, 2002; Zhou, 2006): if the pressure p satisfies

$$p \in L^r(0,T;L^s) \quad \text{for some } \frac{2}{r} + \frac{3}{s} \leq 2, \ s > \frac{3}{2} \qquad (8.20)$$

or

$$\nabla p \in L^r(0,T;L^s) \quad \text{for some } \frac{2}{r} + \frac{3}{s} \leq 3, \ 1 \leq s \leq \infty \qquad (8.21)$$

then the corresponding velocity u is regular.

It is also possible to state a sufficient condition for regularity in terms of the vorticity (da Veiga, 1995): if the vorticity $\omega = \text{curl } u$ of a Leray–Hopf weak solution u belongs to the space

$$L^r(0,T;L^s) \quad \text{for some } \frac{2}{r} + \frac{3}{s} \leq 2, \ s > 1 \qquad (8.22)$$

then the corresponding velocity u is regular. Note that this generalizes the well-known criterion due to Beale, Kato, & Majda (1984) that a solution cannot blow up as $t \to T$ if $\omega \in L^1(0, T; L^\infty)$.

8.4.2 Restrictions on singularities

One of the two most celebrated results that places restrictions on singularities is the following theorem of Scheffer (1976), which treats the set of singular times:

Theorem 8.4.1 (Scheffer, 1976) *If u is a Leray–Hopf weak solution of the Navier–Stokes equations then the $1/2$-dimensional Hausdorff measure of the set*

$$X = \{t : \|Du(t)\| = \infty\}$$

of singular times of the flow is zero.

Here we give a very simple proof of a related result due to Robinson & Sadowski (2007), which guarantees that the upper box-counting dimension of X is no greater than $1/2$. This notion of dimension (which provides an upper bound on the Hausdorff dimension) is usually defined as

$$d_{\mathrm{box}}(X) = \limsup_{\epsilon \to 0} \frac{\log N(X, \epsilon)}{-\log \epsilon},$$

where $N(X, \epsilon)$ is the minimum number of balls of radius ϵ required to cover X. However, one can also take $N(X, \epsilon)$ to be the maximum number of disjoint balls of radius ϵ with centres in X, a variant of the definition that will be useful here. (For more details about this definition, see Falconer, 1990.)

Theorem 8.4.2 (Robinson & Sadowski, 2007) *If X denotes the set of singular times of a Leray–Hopf weak solution then $d_{\mathrm{box}}(X) \le \frac{1}{2}$.*

As remarked above, this implies that the Hausdorff dimension of X is no larger than one half, but not that its $1/2$-dimensional Hausdorff measure is zero. However, there are subsets of $\mathbb{R}$ (e.g. the rationals) that have Hausdorff dimension zero but box-counting dimension one.

Proof The proof relies on the observation that (8.13) implies a lower bound on singular solutions: if $\|Du(t)\| = +\infty$ then

$$\|Du(s)\|^2 \ge c(t - s)^{-1/2} \qquad \text{for all} \qquad s \le t$$

(this observation goes back to Leray, 1934). Suppose that $d_{\text{box}}(X) = d > 1/2$. Then for some δ with $1/2 < \delta < d$ there exists a sequence $\epsilon_j \to 0$ such that $N_j = N(X, \epsilon_j) > \epsilon_j^{-\delta}$. If the centres of these N_j balls are at t_n, $1 \leq n \leq N_j$, then since the balls are disjoint

$$\int_0^1 \|Du(s)\|^2 \, ds \geq \sum_{n=1}^{N_j} \int_{t_n - \epsilon_j}^{t_n + \epsilon_j} \|Du(s)\|^2 \, ds > \sum_{n=1}^{N_j} \int_{t_n - \epsilon_j}^{t_n} \|Du(s)\|^2 \, ds.$$

Since $\|Du(t_n)\| = +\infty$, $\|Du(s)\|^2 \geq c(t_n - s)^{-1/2}$, and so

$$\int_0^1 \|Du(s)\|^2 \, ds \geq \frac{1}{2c} \sum_{i=1}^{N_j} \sqrt{c\epsilon_j} \geq \frac{1}{2\sqrt{c}} \epsilon_j^{(1/2)-\delta}.$$

The right-hand side tends to infinity as $j \to \infty$, but we know that the left-hand side is finite by (8.17); so we must have $d_{\text{box}}(X) \leq 1/2$. $\qquad\square$

The other well-known result is due to Caffarelli, Kohn, & Nirenberg (1982), which considers the points of singularity in space-time:

Theorem 8.4.3 (Caffarelli et al., 1982) *Let $u(t)$ be a suitable weak solution[1] of the Navier–Stokes equations and let S be the set of all points (x, t) for which there is no $M > 0$ such that $|u(y, s)| < M$ for almost all (y, s) in some neighbourhood of (x, t). Then the one-dimensional parabolic Hausdorff measure of S is zero.*

The statement that the one-dimensional parabolic Hausdorff measure of S is zero is equivalent to the statement that for any $\epsilon > 0$, S can be covered by a (perhaps countable) collection of 'parabolic cylinders' $B(x_j, r_j) \times [t_j - r_j^2, t_j + r_j^2]$ such that $\sum_j r_j < \epsilon$.

Finally, it should be mentioned that some other restrictions on the singular behaviour of the Navier–Stokes flows are known. For example, it has been proved that if the vorticity blows up in a finite time, then at least two of its components must blow up simultaneously (Chae & Choe, 1999).

8.5 Regularity for a dense set of forces

Another approach to the problem of global regularity of weak solutions focuses on proving regularity of solutions for large sets of initial conditions or forces. Much effort has been made to prove regularity of solutions

---

[1] One has to assume, in addition to the usual definition of a weak solution, that $p \in L^{5/4}((0, T) \times \Omega)$ and that a local form of the energy inequality holds. It can be shown that at least one such suitable weak solution exists.

arising from initial conditions that are in some sense 'small'. For example, flows in thin domains have been investigated in detail and many interesting results have been proved in this direction (see Raugel & Sell, 1993, or Temam & Ziane, 1996, for example). However, we will not discuss these problems here and we refer the reader directly to the relevant research papers.

Instead we will concentrate on the following two problems, of which the second appears to be much harder:

Problem 1. For a given initial condition u_0, find a large set $\mathcal{F}$ of forces such that u_0 gives rise to a strong solution of the Navier–Stokes equations for any forcing $f \in \mathcal{F}$.

Problem 2. For a given force f find large set $\mathcal{U}$ of initial conditions such that any $u_0 \in \mathcal{U}$ gives rise to a strong solution of the Navier–Stokes equations with forcing f.

In what follows we restrict ourselves to a periodic domain with $L = 2\pi$ and $\nu = 1$ as above.

First we will address Problem 1. A slightly naive question is whether for a given initial condition u_0 we can always find *at least one* forcing $f(t)$ for which u_0 gives rise to the strong solution of the Navier–Stokes equations. An easy answer is provided by the following lemma, which although elementary contains an idea that will be significant later.

Lemma 8.5.1 *Each $u_0 \in V$ gives rise to a strong solution of the Navier–Stokes equations for some forcing $f \in L^2(0, T; H)$.*

Proof Let $u(t) = e^{-At}u_0 \in C^0([0, T]; V)$. Then since u is the unique solution of

$$\frac{du}{dt} = -Au \qquad u(0) = u_0$$

it follows that $u \in L^2(0, T; V^2)$ and $du/dt \in L^2(0, T; H)$. In addition, therefore, $B(u, u) \in L^2(0, T; H)$. Now define

$$\hat{f} = \frac{du}{dt} + Au + B(u, u);$$

we have shown that $\hat{f} \in L^2(0, T; H)$, and clearly u is the unique strong solution of the Navier–Stokes equations with forcing $\hat{f}$. ☐

A great improvement on this result is the following theorem:

Theorem 8.5.2 (Fursikov, 1980) *For a given initial condition $u_0 \in V^{1/2}$ there exists an open set $\mathcal{F} \subset L^2(0, T; V^{-1/2})$ such that for every function $f \in \mathcal{F}$ there exists a unique solution of the Navier–Stokes equations on the time interval $(0, T)$. Moreover, the set $\mathcal{F}$ is dense in the topology of $L^p(0, T; V^{-l})$, where $1 \le p < 4/(5 - 2l)$ for $1/2 < l \le 3/2$, and $1 \le p \le 2$ for $l > 3/2$.*

As we can see, for a given initial condition we can find a very large set $\mathcal{F}$ of forces that induce regularity of weak solutions arising from u_0.

Proof We sketch the idea of the density part of the proof, following the presentation in Fursikov (1980).

Step 1. Let P_k be the orthogonal projector from H onto the k-dimensional space spanned by the first k eigenvectors of the Stokes operator A. For a given initial condition $u_0 \in V^{1/2}$ and a given $f \in L^2(0, T; V^{-1/2})$ we consider $y_k = z_k + v_k$, where z_k is a solution of the Galerkin system

$$\frac{\mathrm{d}z_k}{\mathrm{d}t} + Az_k + P_k B(z_k, z_k) = P_k f \qquad z_k(0) = P_k u_0$$

and v_k is the solution of a linear problem

$$\frac{\mathrm{d}v_k}{\mathrm{d}t} + Av_k = 0 \qquad v_k(0) = u_0 - P_k u_0.$$

Step 2. We show that y_k is the solution of the Navier–Stokes equations for a forcing $f + g_k$, where

$$\|g_k\|_{L^2(0,T;V^{-l})} \to 0 \quad \text{as} \ k \to \infty,$$

which finishes the proof.   □

However, it should be noticed that the density of $\mathcal{F}$ is proved in a weaker topology than the openness property. So the set of forces is not 'generic' in the standard sense. Note also that for a given forcing f, Fursikov's theorem says nothing about the sets of initial conditions that give rise to strong solutions.

8.6 Robustness of regularity under perturbation

In this section we consider two results that show that regularity is stable 'under perturbation'. First we present a result that requires $u_0 \in V^m$

with $m \geq 3$, which is also valid for the Euler equations ($\nu = 0$), and then a result valid for standard 'strong solutions' with $u_0 \in V$.

8.6.1 Robustness of regularity for $u_0 \in V^m$, $m \geq 3$

The next theorem, due to Chernysehnko et al. (2007), addresses the problem of regularity both in the space of initial conditions and of forces, showing that regularity is stable under perturbations:

Theorem 8.6.1 (Chernysehnko et al., 2007) *Let $m \geq 3$. Assume that $v_0 \in V^m$ gives rise to a strong solution $v(t)$ on $[0, T]$ when the forcing is $g(t) \in L^\infty(0, T; V^{m-1}) \cap L^1(0, T; V^m)$. Then so does any initial condition $u_0 \in V^m$ and forcing $f(t) \in L^2(0, T; V^{m-1})$ for which*

$$\|v_0 - u_0\|_m + \int_0^T \|g(s) - f(s)\|_m \, \mathrm{d}s \leq P(v), \qquad (8.23)$$

where

$$P(v) = (c_m T)^{-1} \exp\left(-c_m \int_0^T \|v(s)\|_m + \|v(s)\|_{m+1} \, \mathrm{d}s\right). \qquad (8.24)$$

Here $\|\cdot\|_m$ denotes the norm in V^m, and c_m is an absolute constant related to certain Sobolev embedding results.

(We have fixed $L = 2\pi$ and $\nu = 1$, so the statement of this theorem here hides the fact that $P(v)$ does not depend on ν; with some extra work the same result can be shown to hold for the Euler equations, see Chernysehnko et al., 2007, for details.)

Proof (Sketch) First, using inequalities (8.7) and (8.8) for the nonlinear term, we prove that $w = u - v$ satisfies

$$\frac{\mathrm{d}}{\mathrm{d}t} \|w\|_m \leq \|f - g\|_m + c(\|v\|_m + \|v\|_{m+1})\|w\|_m + \|w\|_m^2.$$

Multiplication by $\exp(-c \int_0^T \|v(s)\|_m + \|v(s)\|_{m+1} \, \mathrm{d}s)$ gives

$$\frac{\mathrm{d}y}{\mathrm{d}t} \leq \|f - g\|_m + \alpha y^2,$$

where

$$y(t) = \|w(t)\|_m \exp\left(-c \int_0^T \|v(s)\|_m + \|v(s)\|_{m+1} \, \mathrm{d}s\right)$$

and $y(0) = \|u_0 - v_0\|_m$. One then applies a simple ODE result due to Constantin (1986): if

$$\frac{dy}{dt} \le \delta(t) + \alpha y^2 \qquad y(0) = y_0 \ge 0, \qquad t \in [0, T],$$

where $\alpha > 0$ and $\delta(t) \ge 0$, then

$$y(t) \le \frac{\eta}{1 - \alpha \eta t}$$

for all $\alpha \eta t < 1$, where

$$\eta = y_0 + \int_0^T \delta(s)\, ds. \qquad\qquad \square$$

8.6.2 Robustness of regularity for $u_0 \in V$

In what follows we will frequently use a similar robustness result that requires less regularity for the initial condition:

Theorem 8.6.2 (Dashti & Robinson, 2008) *Assume that $v_0 \in V$ gives rise to a strong solution $v(t)$ on $[0, T^*]$ with forcing $\widetilde{f}(t)$. Then so does any initial condition $u_0 \in V$ and forcing $f(t)$ with*

$$\|D(v_0 - u_0)\| + \int_0^{T^*} \|D\widetilde{f}(s) - Df(s)\|\, ds \le W(v), \qquad (8.25)$$

where

$$W(v) = (cT^*)^{-1/4} \exp\left(-c \int_0^{T^*} \|Dv(s)\|^4 + \|Dv(s)\|\, \|Av(s)\|\, ds\right) \tag{8.26}$$

and c is an absolute constant related to certain Sobolev embedding results.

The proof is as above, but is based on the differential inequality

$$\frac{d}{dt}\|Dw\| \le (c_1\|Du\|^4 + c_2\|Du\|\,\|Au\|)\|Dw\| + c_1\|Dw\|^5 + \|D(f - g)\|$$

where $w = u - v$, and properties of functions $y(t)$ that satisfy

$$\frac{dy}{dt} \le \delta(t) + \alpha y^5 \qquad \alpha \ge 0, \ \delta(t) \ge 0.$$

8.7 Numerical verification of regularity

We now discuss how this robustness result can be used to provide a method of (at least theoretical) numerical verification of regularity: first for a single initial condition, then for any bounded set in V^2, and finally for any bounded set in V.

8.7.1 Verification for a single initial condition in V

The following result is a simple consequence of Theorem 8.6.2 when it is combined with the idea used in Lemma 8.5.1 that if both v and $f := \mathrm{d}v/\mathrm{d}t + Av + B(v, v)$ are sufficiently regular then v is the unique strong solution of the Navier–Stokes equation $\mathrm{d}u/\mathrm{d}t + Au + B(u, u) = f$.

Corollary 8.7.1 (Dashti & Robinson, 2008) *Suppose that the function $v \in L^\infty(0, T; V) \cap L^2(0, T; V^2)$ satisfies*

$$\frac{\mathrm{d}v}{\mathrm{d}t} + Av + B(v, v) \in L^1(0, T; V) \cap L^2(0, T; V)$$

and

$$\|Dv(0) - Du_0\| + \int_0^T \left\| \frac{\mathrm{d}v}{\mathrm{d}t}(s) + Av(s) + B(v(s), v(s)) \right\|_1 \mathrm{d}s \leq W(v),$$
$$(8.27)$$

where $W(v)$ is defined by (8.26). Then u is a regular solution of the Navier–Stokes equations with $u \in L^\infty(0, T; V) \cap L^2(0, T; V^2)$.

This corollary tells us that we can obtain a purely theoretical result (regularity of the solution arising from a given initial condition) from approximate numerical computations. (Of course, a numerical solution will be given only at discrete time points but this is not an obstacle since one can, for example, construct u via linear interpolation between these points.)

In fact, it is possible to verify regularity corresponding to a particular initial condition u_0 in a finite time: if u_0 does give rise to a regular solution on $[0, T]$, finite-dimensional Galerkin approximations (based on the eigenfunctions of the Stokes operator) can be guaranteed to converge to $u(t)$ (see Dashti & Robinson, 2008, for details). It follows that if u_0 leads to such a regular solution, one can take v to be the result of successively larger-dimensional Galerkin approximations until (8.27) is satisfied. On the other hand, if a weak solution arising from u_0 evolves to produce a singularity then the numerical computation will not be able to show this. In other words, we can verify regularity of a strong solution, but we cannot obtain in this way any proof of the hypothetical breakdown of regularity of a weak solution.

8.7.2 Verification for a ball in V^2

Since it is possible to use numerical computations to check regularity of a single given initial condition it is sensible to ask whether numerical computations may improve the theoretical result given in Theorem 8.3.4. The answer is yes. However, before we prove this we will investigate an auxiliary problem. Instead of considering a set of initial conditions in the space V we will, for the moment, restrict our attention to a set of more regular initial conditions that is a ball in V^2. Our aim is to present a method of verifying the following statement, for some fixed $S > 0$:

Statement 1. Every initial condition with $\|Au_0\| \leq S$ gives rise to a strong solution that exists for all $t \geq 0$.

Following Robinson & Sadowski (2008) we will sketch the proof of the following theorem.

Theorem 8.7.2 (Robinson & Sadowski, 2008) *Assume that Statement 1 is true. Then it is possible to verify it numerically in a finite time.*

Proof (Sketch)
Step 1. We have already remarked (Theorem 8.3.5) that any weak solution is strong for $t \geq T^{**}$, where T^{**} depends only on the norm of the initial data in V. So we need only verify regularity on the bounded interval $[0, T^{**}]$.
Step 2. We assume that Statement 1 is true. Following Constantin, Foias, & Temam (1985) one can prove, under this assumption, that for any $T > 0$ there exist constants $D_S(T)$ and $E_S(T)$ such that every initial condition u_0 with $\|Au_0\| \leq S$ gives rise to a solution u satisfying

$$\sup_{0 \leq t \leq T} \|Du(s)\| \leq D_S \quad \text{and} \quad \int_0^T \|Au(s)\|^2 \, ds \leq E_S.$$

Step 3. Using these bounds one can deduce that for any such a solution u the following integral is uniformly bounded:

$$\int_0^{T^{**}} \|Du(s)\|^4 + \|Du(s)\| \, \|Au(s)\| \, ds < Q.$$

It follows that there exists a $\delta = \delta(S)$ such that if $\|Au_0\| \leq S$ then $W(u) > \delta$. Theorem 8.6.2 implies that every initial condition $v_0 \in V$ with

$$\|Du_0 - Dv_0\| < \delta$$

also gives rise to a strong solution on $[0, T^{**}]$.

Step 4. One can find an explicit finite collection of balls in V of radius δ that cover the ball in V^2 (centred at the origin) of radius S. One can then check numerically, using Corollary 8.7.1, that the centres of each of these balls are initial conditions giving rise to strong solutions on $[0, T^{**}]$. $\square$

Observe that the proof is based only on an assumption of regularity of solutions arising from initial conditions with $\|Au_0\| \leq S$. Whether or not singularities may occur in solutions arising from initial conditions u_0 with $\|Au_0\| > S$ has no effect on the possibility of numerical verification of regularity of solutions with $\|Au_0\| < S$.

8.7.3 Verification for a ball in V

Our main aim now is to show that we can verify numerically the regularity of solutions arising from initial conditions in some ball in V. More precisely, we wish to verify the following statement for some fixed $R > 0$:

Statement 2. Every initial condition with $\|Du_0\| \leq R$ gives rise to a strong solution that exists for all $t \geq 0$.

In order to verify this we have to make the stronger assumption that the Navier–Stokes equations are in fact regular, i.e. that Statement 2 holds for any value of $R > 0$.

Theorem 8.7.3 (Robinson & Sadowski, 2008) *Assume that the Navier–Stokes equations are regular. Then it is possible to verify Statement 2 numerically in a finite time.*

Proof (Sketch)

Step 1. We find time $T_R > 0$ such that every solution with initial condition $\|Du_0\| \leq R$ is regular on the interval $[0, T_R]$ and has an explicit bound on $\|Au(T_R)\|$. We can do this using a result due to Foias & Temam (1989) on the Gevrey regularity of solutions, a consequence of which is that for

$$T_R = \frac{1}{K(1 + R^2)}$$

we have

$$\|Au(T_R)\|^2 \le K^2(1 + R^2)^{5/2},$$

where $K \le 3266$.

Step 2. If an initial condition with $\|Du_0\| \le R$ leads to a singularity, then so does some initial condition $v_0 = u(T_R)$ with $\|Av_0\| \le S$, where $S = S(R) = K_1(1 + R^2)^{5/2}$. So it suffices to verify the regularity of all initial conditions within this ball in V^2, which we know that we can do. □

Observe that in order to prove Statement 2 we need to assume that *all* solutions of the Navier–Stokes equations with initial conditions in V are regular. Indeed, we cannot exclude the possibility that there is an initial condition with $\|Au_0\| \le S$ that gives rise to a solution developing a singularity, while initial conditions with $\|Du_0\| \le R$ all give rise to strong solutions, as illustrated below.

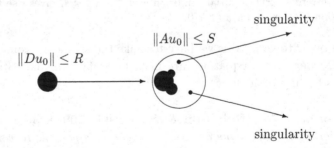

Of course, in this case one would not be able to "verify" numerically a false statement about the regularity of solutions arising from initial conditions with $\|Au_0\| \le S$, even though it would be true that every solution with $\|Du_0\| \le R$ leads to a regular solution.

8.8 Acknowledgements

JCR was partially supported by a Royal Society University Research Fellowship and the Leverhulme Trust; WS was partially supported by Polish Government Grant 1 P03A 017 30.

References

Beale, J.T., Kato, T., & Majda, A.J. (1984) Remarks on the breakdown of smooth solutions for the 3-D Euler equations. *Comm. Math. Phys.* **94**, 61–66.

Berselli, L.C. & Galdi, G.P. (2002) Regularity criterion involving the pressure for the weak solutions to the Navier–Stokes equations. *Proc. Amer. Math. Soc.* **130**, 3585–3595.

Caffarelli, L., Kohn, R., & Nirenberg, L. (1982) Partial regularity of suitable weak solutions of the Navier–Stokes equations. *Comm. Pure Appl. Math.* **35**, 771–831.

Chae, D. & Choe, H.J. (1999) Regularity of solutions to the Navier–Stokes equation. *Electronic Journal of Differential Equations* 5, 1–7.

Chae, D. & Lee, J. (2001) Regularity criterion in terms of pressure for the Navier–Stokes equations. *Nonlinear Anal.* **46**, 727–735.

Chernysehnko, S.I., Constantin, P., Robinson, J.C., & Titi, E.S. (2007) A posteriori regularity of the three-dimensional Navier–Stokes equations from numerical computations. *J. Math. Phys.* **48**, 1–15.

Cheskidov, A., Friedlander, S., & Shvydkoy, R. (2008) On the energy equality for weak solutions of the 3D Navier–Stokes equations, submitted. arXiv:0704.2089.

Constantin, P. (1986) Note on loss of regularity for solutions of the 3D incompressible Euler and related equations. *Comm. Math. Phys.* **104**, 311–326

Constantin, P. & Foias, C. (1988) *Navier–Stokes Equations.* University of Chicago Press, Chicago.

Constantin, P., Foias, C., & Temam, R. (1985) Attractors representing turbulent flows. *Memoirs of the Amer. Math. Soc.* **53** (314), Providence, Rhode Island.

Dashti, M. & Robinson, J.C. (2008) An a posteriori condition on the numerical approximations of the Navier–Stokes equations for the existence of a strong solution. *SIAM J. Numer. Anal.* **46**, 3136–3150.

Doering, C.R. & Gibbon, J.D. (1995) *Applied Analysis of the Navier–Stokes Equations.* Cambridge Texts in Applied Mathematics, Cambridge University Press, Cambridge.

Escauriaza, L., Seregin, G.A., & Šverák, V. (2003) $L_{3,\infty}$-solutions of Navier–Stokes equations and backward uniqueness. *Russian Math. Surveys* **58**, 211–250.

Fabes, E., Jones, B., & Riviere, N. (1972) The initial value problem for the Navier–Stokes equations with data in L^p. *Arch. Rat. Mech. Anal.* **45**, 222–248.

Falconer, K. (1990) *Fractal Geometry.* Wiley, Chichester.

Fefferman, C.L. (2000) Existence and smoothness of the Navier–Stokes equation. http://www.claymath.org/millennium/Navier-Stokes_Equations/navierstokes.pdf.

Foias, C. & Temam, R. (1989) Gevrey class of regularity for the solutions of the Navier–Stokes equations. *J. Funct. Anal.* **87**, 359–369.

Fursikov, A.V. (1980) On some control problems and results concerning the unique solvability of a mixed boundary value problem for the three dimensional Navier–Stokes and Euler systems. *Soviet Math. Dokl.* **21**, 181–188.

Hopf, E. (1951) Über die Anfangswertaufgabe für die hydrodynamischen Grundgleichungen. *Math. Nachr.* **4**, 213–231.

Leray, J. (1934) Essai sur le mouvement d'un fluide visqueux emplissant l'espace. *Acta Math.* **63**, 193–248.

Raugel, G. & Sell, G.R. (1993) Navier–Stokes equations on thin 3D domains. I: global attractors and global regularity of solutions. *Journal of the Amer. Math. Soc.* **6**, 503–568.

Robinson, J.C. (2001) *Infinite-dimensional dynamical systems.* Cambridge Texts in Applied Mathematics, Cambridge University Press, Cambridge.

Robinson, J.C. (2006) Regularity and singularity in the three-dimensional Navier–Stokes equations. *Boletín de la Sociedad Española de Matemática Aplicada* **35**, 43–71.

Robinson, J.C. & Sadowski, W. (2007) Decay of weak solutions and the singular set of the three-dimensional Navier–Stokes equations. *Nonlinearity* **20**, 1185–1191.

Robinson, J.C. & Sadowski, W. (2008) Numerical verification of regularity in the three-dimensional Navier–Stokes equations for bounded sets of initial data. *Asymptot. Anal.* **59**, 39–50.

Scheffer, V. (1976) Turbulence and Hausdorff dimension, in *Turbulence and Navier–Stokes equations*, Orsay 1975, Springer LNM **565**, 174–183, Springer-Verlag, Berlin.

Serrin, J. (1962) On the interior regularity of weak solutions of the Navier–Stokes equations. *Arch. Rat. Mech. Anal.* **9**, 187–191.

Temam, R. (2001) *Navier–Stokes Equations.* North Holland, Amsterdam, 1977. Reprinted by AMS Chelsea.

Temam, R. & Ziane, M. (1996) Navier–Stokes equations in three-dimensional thin domains with various boundary conditions. *Adv. in Differential Equations* **1**, 499–546.

da Veiga, H.B. (1995) Concerning the regularity problem for the solutions of the Navier–Stokes equations. *C. R. Acad. Sci. Paris I* **321**, 405–408.

Zhou, Y. (2006) On the regularity criteria in terms of pressure of the Navier–Stokes equations in $\mathbb{R}^3$. *Proc. Amer. Math. Soc.* **134**, 149–156.

9

Contour dynamics for the surface quasi-geostrophic equation

José L. Rodrigo

Mathematics Institute, University of Warwick,
Coventry, CV4 7AL. UK.
J.Rodrigo@warwick.ac.uk

Abstract

We review recent progress on the evolution of sharp fronts for the surface quasi-geostrophic equations and related problems, with special emphasis on techniques that can be extended to the study of vortex dynamics for the 3D Euler equations.

9.1 Introduction

In these notes we will review a series of problems related to contour dynamics for two-dimensional active-scalar equations. More precisely, the central problem we will concentrate on is the evolution of sharp fronts for the surface quasi-geostrophic equation, obtained by considering the evolution of initial data given by the indicator of a smooth, open set by the equations

$$\frac{D\theta}{Dt} := \frac{\partial \theta}{\partial t} + u \cdot \nabla \theta = 0, \tag{9.1}$$

$$u = \nabla^{\perp} \psi = (-\partial_{x_2} \psi, \partial_{x_1} \psi), \tag{9.2}$$

$$(-\Delta)^{\frac{1}{2}} \psi = \theta. \tag{9.3}$$

Here the variable θ represents a two-dimensional active scalar (potential temperature in physical terms) advected by the velocity field u and ψ is the stream function. Finally, we define the fractional Laplacian in terms of the Fourier transform,

$$\widehat{(-\Delta)^{\frac{1}{2}} \psi}(\xi) = |\xi| \hat{\psi}(\xi).$$

We will refer to the system (9.1)–(9.3) as SQG.

The term 'sharp front', to denote the boundary of the set whose indicator we are considering, stems from the fact that the SQG system was originally introduced as a model for atmospheric turbulence, where the curve represents an abrupt change in temperature, called a sharp front.

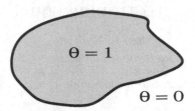

Fig. 9.1. Sharp Front

It is easy to see that if the initial condition is the indicator of an open set with C^1 boundary then the solution remains of the same form, reducing the problem to understanding the evolution of the boundary (hence the name contour dynamics).

The surface quasi-geostrophic equation has been studied extensively, both from the geophysical and mathematical point of view. A derivation of the system of equations can be found in Pedlosky (1987), for the evolution of the temperature on the 2D boundary of a half-space with small Rossby and Ekman numbers and constant potential vorticity. See Held, Pierrehumbert, & Swanson (1994) for an analysis of the statistical turbulence theory for the equation and Garner et al. (1995) for a qualitative analysis of the solutions. We will not make a systematic review of the literature concerning SQG here. Concerning the question that we consider here, that is, the evolution of sharp fronts, one of the most active questions about SQG is the study of the frontogenesis, precisely the formation of a discontinuous temperature front in finite time. We refer the reader to Constantin, Majda, & Tabak (1994a,b), Córdoba (1998), Ohkitani & Yamada (1997), Constantin, Nie, & Schörghofer (1998, 1999), Rodrigo (2004, 2005), Córdoba, Fefferman, & Rodrigo (2004), Córdoba et al. (2005) and Gancedo (2008) for more details on the questions of the frontogenesis and the evolution of sharp fronts.

The main mathematical interest in SQG arises from the analogies of this system with the 3D Euler equations (described in detail in the next section) and it is this connection that motivates our approach to the problems considered. More precisely, we attempt to consider only techniques that can be extended to the study of the analogous problems in 3D Euler, that is, as we will see in next section, the study of vortex lines for 3D Euler.

9.2 SQG and 3D Euler

As mentioned before, the main mathematical interest in the surface quasi-geostrophic system lies in its strong analogies with the 3D Euler equation. SQG presents a two-dimensional equation that contains many of the features of 3D Euler. We will present a very brief review of the analogies between these two equations. These similarities are best established when considering 3D Euler in vorticity form. Recall that in terms of the vorticity $\omega = \operatorname{curl} u$, 3D Euler becomes

$$\frac{D\omega}{Dt} = (\nabla u)\omega,$$

where $u = (u_1, u_2, u_3)$ is the 3D velocity satisfying $div\, u = 0$.

In particular, we observe that by differentiating equation (9.1) we obtain

$$\frac{D(\nabla^\perp \theta)}{Dt} = (\nabla u)\nabla^\perp \theta.$$

This shows a very strong analogy between the 3D Euler equation and QG, where $\nabla^\perp \theta$ plays the role of ω. We briefly recall some further analogies:

- The velocity is recovered via the formulas

$$u(x) = \int_{\mathbb{R}^3} K_3(y)\omega(x+y)\mathrm{d}y \qquad u(x) = \int_{\mathbb{R}^2} K_2(y)\nabla^\perp \theta(x+y)\mathrm{d}y,$$

  where the kernels K_d, $d = 2, 3$ are homogeneous of degree $1 - d$. Additionally, the strain matrix (the symmetric part of the gradient of the velocity) can be recovered via a singular integral operator, given by kernels of degree $-d$.
- Both systems have conserved energy.
- Both $|\omega|$ and $|\nabla^\perp \theta|$ evolve according to the same type of equation. ($|\omega|$ measures the infinitesimal length of a vortex line).
- Both systems have analogous conditions for a break up of a solution, that is, the well-known criterion due to Beale, Kato, & Majda (1984) translates directly to SQG with $\nabla^\perp \theta$ in place of ω.
- The integral curves of ω, and of $\nabla^\perp \theta$ move with the fluid.

These analogies were first noticed by Constantin, Majda, & Tabak. We refer the reader to Constantin et al. (1994a,b) and Majda & Tabak (1996) for a complete presentation. Another detailed exposition is found in Majda & Bertozzi (2002).

It is the last analogy in the previous list that motivates the work presented in these notes. We will start by reviewing some results about vortex lines.

9.2.1 Sharp Fronts and Vortex Lines

An outstanding open problem in fluid dynamics is the evolution of a single (idealized) vortex line. In general, a vortex line is an integral curve of the vorticity field, but the problem we want to understand here is the solution of 3D Euler when the vorticity ω is supported on a curve Γ, and has the form

$$\omega = |\omega|\, \delta_\Gamma\, \boldsymbol{T}, \tag{9.4}$$

where $\boldsymbol{T}$ is the tangent to the curve (see Figure 9.2). One can consider this problem as the evolution of an idealized vortex tube of thickness zero. It is not known whether solutions of 3D Euler of this form actually exist, and if they do what the evolution equation for the curve Γ is.

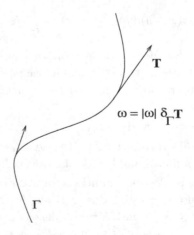

Fig. 9.2. Vortex Line

The main difficulty in the study the evolution of a vortex line is that we need to understand the velocity field (given by the Biot–Savart Law)

$$u(x,t) = \frac{1}{4\pi} \int \frac{x-y}{|x-y|^3} \times \omega(y,t)\,dy, \qquad (9.5)$$

which appears in the Euler equation in vorticity form

$$\frac{D\omega}{Dt} = (\nabla u)\omega.$$

Equation (9.5) is obtained by inverting the *curl* operator under reasonable assumptions on the regularity and decay of ω that, of course, are not satisfied by a vortex line as defined in (9.4). In the case of a vortex line, the velocity is of order

$$u(x) \approx \frac{1}{\text{distance to the curve}},$$

which means that u is not in $L^2(\mathbb{R}^3)$. This means that we cannot use any of the various definitions of solution for 3D Euler to understand a vortex line.

No rigorous derivation of an equation for the evolution of a vortex line is known but a well-known approximation for its evolution is given by the equation

$$\partial_t \gamma = \kappa \boldsymbol{B}, \qquad (9.6)$$

where κ is the curvature of γ and $\boldsymbol{B}$ is the binormal. Equation (9.6) is known as the LIA equation (Locally Induced Approximation), and was first introduced by da Rios in 1906. We refer the reader to Ricca (1996) for an excellent review of the work of da Rios and the derivation of the LIA equation.

Equation (9.6) can be converted into a nonlinear Schrödinger equation using the Hasimoto transformation. Using this formulation the existence of self-similar singular solutions has been proved (Vega, 2003; Gutiérrez, Rivas, & Vega, 2003; Gutiérrez & Vega, 2004). We also remark that modifications of equation (9.6) have been studied for example by Klein & Majda (1991a,b). The main reason for introducing these modified equations is to obtain an equation that contains self-stretching, since the length of any curve evolving under (9.6) remains constant in time.

Postponing the details until the next section, we compare in the picture below the velocity fields for both problems, to show that sharp fronts for SQG retain many of the features of vortex lines, while being a more tractable problem. In particular, the velocity field for SQG while still singular, contains only a logarithmic singularity (see Figure 9.3).

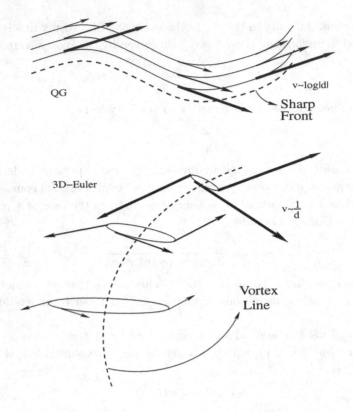

Fig. 9.3. Velocity for a Sharp Front ($v \sim |\log d|$) and a Vortex Line ($v \sim \frac{1}{d}$)

9.2.2 The vortex patch problem: 2D Euler and SQG

We have motivated the study of sharp fronts by considering the evolution of vortex lines for 3D Euler. But the evolution of sharp fronts for SQG also has an analogous problem in 2D Euler. We notice that 2D Euler, in its vorticity formulation, provides us with a scalar equation that presents a very similar analytical structure to SQG. The 2D Euler equation reads

$$\frac{D\omega}{Dt} = 0, \tag{9.7}$$

where

$$(u_1, u_2) = (-\frac{\partial \psi}{\partial y}, \frac{\partial \psi}{\partial x}), \tag{9.8}$$

and

$$-\Delta \psi = \omega. \tag{9.9}$$

Observe that the above system is very similar to (9.1)–(9.3) except for the relationship between the stream function and the active scalar. In this case the analogy takes place at the level of ω and θ, not at the level of ω and $\nabla^\perp \theta$ as before. In the case of SQG the fractional power of the Laplacian makes the equation more singular than 2D Euler.

For 2D Euler, the evolution of the indicator of a set is known as the vortex patch problem, and in this case, the evolution of the boundary curve is well understood. In particular, the derivation of the equation presents no problems (the velocity is not singular) and the global regularity of the vortex patches was proved by Chemin (1993) using paradifferential calculus. A simpler proof can be found in Bertozzi & Constantin (1993) and Majda & Bertozzi (2002).

It is natural to introduce the following family of interpolating models between 2D Euler (9.7)–(9.9) and SQG (9.1)–(9.3), that we will refer to as the α-models:

$$\frac{D\theta_\alpha}{Dt} = 0,$$

$$u = \nabla^\perp \psi, \qquad \text{and}$$

$$(-\Delta)^{1-\frac{\alpha}{2}} \psi = \theta_\alpha,$$

where $0 < \alpha < 1$. They were first introduced in Córdoba et al. (2005) in the context of sharp fronts and vortex patches. Notice that when $\alpha = 0$ we recover 2D Euler, and when $\alpha = 1$ we obtain SQG. Also, the larger the parameter α the more complicated the problem becomes. For all this family of equations we could consider the evolution of the indicator of a smooth set (see Figure 9.1). We remark that the global regularity of the evolution of the boundary of the set is only known when $\alpha = 0$, precisely when we are back in the case of 2D Euler and so considering the evolution of a vortex patch. We will return to this family later in these notes.

9.3 Evolution equation

In order to simplify the presentation we will consider the evolution of sharp fronts that are periodic in one of the space variables.

We consider the front originally given by the curve $y = \varphi_0(x)$ (see Figure 9.4), a smooth periodic function, and assume that the solution to the system (9.1)–(9.3) is of the same form and is given by $\varphi(x, t)$, a

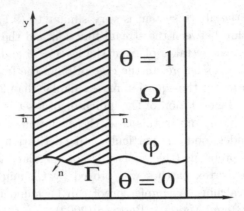

Fig. 9.4. Periodic sharp front

smooth periodic function. This means that the scalar function $\theta(x, y, t)$ is given by

$$\begin{cases} \theta(x, y, t) = 1 & y \geq \varphi(x, t) \\ \theta(x, y, t) = 0 & y < \varphi(x, t). \end{cases} \tag{9.10}$$

The first derivation we will consider is purely formal but it shows interesting features of the velocity field, and how it affects the evolution of the front.

9.3.1 First approach: redefining the velocity

We will start by eliminating the stream function ψ from the system (9.1)–(9.3), by using the second and third equations. The operator $(-\Delta_{x,y})^{-\frac{1}{2}}$ in the cylinder is given by a convolution with the kernel

$$K(x, y) = \frac{\chi(x, y)}{(x^2 + y^2)^{\frac{1}{2}}} + \eta(x, y), \tag{9.11}$$

for (x, y) in $[-\frac{1}{2}, \frac{1}{2}] \times \mathbb{R}$, and defined in the rest of the plane by extending it periodically in x. Here, χ is a smooth function with compact support that satisfies

$$\chi(x, y) = 1 \text{ if } |x - y| \leq r \quad \text{and} \quad \operatorname{supp} \chi \subset \{|x - y| \leq R\},$$

where $0 < r < R < \frac{1}{2}$ are positive numbers to be chosen later. Also η is smooth with compact support[1] and satisfies $\eta(0,0) = 0$.

Observe that both χ and η can be taken to be even functions. Also, changing the value of r and R does not affect the structure of K given by (9.11), since the difference in the function χ created by changing r and R can be absorbed by the correction term η.

And so, by inverting the fractional Laplacian we have (for a point (x, y) not in the front)

$$\psi(x,y,t) = \int_{\mathbb{R} \times \mathbb{R}/\mathbb{Z}} \frac{\theta(\widetilde{x}, \widetilde{y}, t)\chi(x - \widetilde{x}, y - \widetilde{y})}{[(x - \widetilde{x})^2 + (y - \widetilde{y})^2]^{\frac{1}{2}}} + \theta(\widetilde{x}, \widetilde{y}, t)\eta(x - \widetilde{x}, y - \widetilde{y}) \, d\widetilde{x} \, d\widetilde{y}$$

and since $u = \nabla^\perp \psi$ we obtain

$$u(x, y, t) = \int_{\mathbb{R} \times \mathbb{R}/\mathbb{Z}} \frac{\nabla^\perp_{\widetilde{x}, \widetilde{y}} \theta(\widetilde{x}, \widetilde{y}, t)\chi(x - \widetilde{x}, y - \widetilde{y})}{[(x - \widetilde{x})^2 + (y - \widetilde{y})^2]^{\frac{1}{2}}} \, d\widetilde{x} \, d\widetilde{y}$$

$$+ \int_{\mathbb{R} \times \mathbb{R}/\mathbb{Z}} \nabla^\perp_{\widetilde{x}, \widetilde{y}} \theta(\widetilde{x}, \widetilde{y}, t)\eta(x - \widetilde{x}, y - \widetilde{y}) \, d\widetilde{x} \, d\widetilde{y}. \quad (9.12)$$

A simple calculation yields

$$\nabla^\perp \theta(x, y, t) = (-1, -\frac{\partial \varphi}{\partial x}(x, t))\delta(y - \varphi(x, t)). \quad (9.13)$$

Plugging this expression into (9.12), and carrying out the integration with respect to $\widetilde{y}$, we obtain

$$u(x, y, t) = -\int_{\mathbb{R}/\mathbb{Z}} (1, \frac{\partial \varphi}{\partial \widetilde{x}}(\widetilde{x}, t))\frac{\chi(x - \widetilde{x}, y - \varphi(\widetilde{x}, t))}{[(x - \widetilde{x})^2 + (y - \varphi(\widetilde{x}, t))^2]^{\frac{1}{2}}} \, d\widetilde{x}$$

$$- \int_{\mathbb{R}/\mathbb{Z}} (1, \frac{\partial \varphi}{\partial \widetilde{x}}(\widetilde{x}, t))\eta(x - \widetilde{x}, y - \varphi(\widetilde{x}, t)) \, d\widetilde{x}. \quad (9.14)$$

Notice that the first integral (9.14) is divergent as we approach the front, i.e. as $y \to \varphi(x, t)$. We look more closely at the original equation (9.1) to redefine u as we approach the front. We use the following simple observation: if θ solves equation (9.1) then it also solves

$$(\partial_t + [u + h\nabla^\perp \theta] \cdot \nabla_{x,y})\theta = 0$$

for any smooth periodic function h.

We want to use this observation to correct the singularity of u in equation (9.14). Since the direction of $\nabla^\perp \theta$ (see (9.13)) is the same as

---

[1] In order to avoid irrelevant considerations at ∞ we will consider the correcting function η to be compactly supported. This has the effect of modifying $(-\Delta_{x,y})^{-1/2}$ by adding a smoothing operator.

the tangent to the curve, given by $(1, \frac{\partial \varphi}{\partial x}(x,t))$, we redefine u by adding the term $h \nabla^\perp \theta$ given by

$$(1, \frac{\partial \varphi}{\partial x}(x,t)) \int_{\mathbb{R}/\mathbb{Z}} \frac{\chi(x - \tilde{x}, y - \varphi(\tilde{x}, t))}{[(x - \tilde{x})^2 + (y - \varphi(\tilde{x}, t))^2]^{\frac{1}{2}}} \, d\tilde{x}$$

$$+(1, \frac{\partial \varphi}{\partial x}(x,t)) \int_{\mathbb{R}/\mathbb{Z}} \eta(x - \tilde{x}, y - \varphi(\tilde{x}, t)) \, d\tilde{x}.$$

We obtain

$$u(x, y, t) = \int_{\mathbb{R}/\mathbb{Z}} (0, \frac{\partial \varphi}{\partial x}(x,t) - \frac{\partial \varphi}{\partial \tilde{x}}(\tilde{x}, t)) \frac{\chi(x - \tilde{x}, y - \varphi(\tilde{x}, t))}{[(x - \tilde{x})^2 + (y - \varphi(\tilde{x}, t))^2]^{\frac{1}{2}}} \, d\tilde{x} +$$

$$+ \int_{\mathbb{R}/\mathbb{Z}} (0, \frac{\partial \varphi}{\partial x}(x,t) - \frac{\partial \varphi}{\partial \tilde{x}}(\tilde{x}, t)) \eta(x - \tilde{x}, y - \varphi(\tilde{x}, t)) \, d\tilde{x}.$$

Notice that now we can pass to the limit when (x, y) approaches the front, i.e. as (x, y) approaches $(x, \varphi(x, t))$. We obtain

$$u(x, \varphi(x, t), t) =$$

$$= \int_{\mathbb{R}/\mathbb{Z}} (0, \frac{\partial \varphi}{\partial x}(x,t) - \frac{\partial \varphi}{\partial \tilde{x}}(\tilde{x}, t)) \frac{\chi(x - \tilde{x}, \varphi(x,t) - \varphi(\tilde{x}, t))}{[(x - \tilde{x})^2 + (\varphi(x,t) - \varphi(\tilde{x}, t))^2]^{\frac{1}{2}}} \, d\tilde{x}$$

$$+ \int_{\mathbb{R}/\mathbb{Z}} (0, \frac{\partial \varphi}{\partial x}(x,t) - \frac{\partial \varphi}{\partial \tilde{x}}(\tilde{x}, t)) \eta(x - \tilde{x}, \varphi(x,t) - \varphi(\tilde{x}, t)) \, d\tilde{x}.$$

Since u is now purely vertical, the fact that $\Omega = \{y \geq \varphi(x,t)\}$ is convected by u we obtain the evolution equation we were looking for:

$$\frac{\partial \varphi}{\partial t}(x,t) =$$

$$= \int_{\mathbb{R}/\mathbb{Z}} \frac{\frac{\partial \varphi}{\partial x}(x,t) - \frac{\partial \varphi}{\partial \tilde{x}}(\tilde{x}, t)}{[(x - \tilde{x})^2 + (\varphi(x,t) - \varphi(\tilde{x}, t))^2]^{\frac{1}{2}}} \chi(x - \tilde{x}, \varphi(x,t) - \varphi(\tilde{x}, t)) \, d\tilde{x}$$

$$+ \int_{\mathbb{R}/\mathbb{Z}} \left[\frac{\partial \varphi}{\partial x}(x,t) - \frac{\partial \varphi}{\partial \tilde{x}}(\tilde{x}, t) \right] \eta(x - \tilde{x}, \varphi(x,t) - \varphi(\tilde{x}, t)) \, d\tilde{x},$$

with initial data $\varphi(x, 0) = \varphi_0(x)$.

9.3.2 *Rigorous derivation: using weak solutions*

The above derivation is purely formal, but it clearly shows the main features of the velocity. These notes are designed with special emphasis on the connections with vortex dynamics. Before we present a rigorous derivation that could be applied to 3D Euler we briefly include a derivation using weak solutions (an idea that cannot be used for vortex lines since $u \notin L^2$). We start with the definition of a weak solution for SQG.

Definition 9.3.1 *A bounded function θ is a weak solution of SQG if for any function $\phi \in C_0^\infty(\mathbb{R}/\mathbb{Z} \times \mathbb{R} \times [0, \varepsilon])$ we have*

$$\int_{\mathbb{R}^+ \times \mathbb{R}/\mathbb{Z} \times \mathbb{R}} \theta(x, y, t) \, \partial_t \phi(x, y, t) + \theta(x, y, t) u(x, y, t) \cdot \nabla \phi(x, y, t) \, dy \, dx \, dt = 0.$$

We state the following two theorems that justify the previous formal derivation (details can be found in Rodrigo, 2004).

Theorem 9.3.2 *If θ is a weak solution of SQG of the form described in (9.10), then the function φ satisfies the equation*

$$\frac{\partial \varphi}{\partial t}(x, t) =$$

$$+ \int_{\mathbb{R}/\mathbb{Z}} \frac{\dfrac{\partial \varphi}{\partial x}(x, t) - \dfrac{\partial \varphi}{\partial y}(y, t)}{[(x-y)^2 + (\varphi(x, t) - \varphi(y, t))^2]^{\frac{1}{2}}} \chi(x - y, \varphi(x, t) - \varphi(y, t)) \, dy$$

$$+ \int_{\mathbb{R}/\mathbb{Z}} \left[\frac{\partial \varphi}{\partial x}(x, t) - \frac{\partial \varphi}{\partial y}(y, t) \right] \eta(x - y, \varphi(x, t) - \varphi(y, t)) \, dy. \qquad (9.15)$$

Theorem 9.3.3 *Given any periodic, smooth function $\varphi_0(x)$ the initial value problem determined by the equation (9.15) with initial data $\varphi(x, 0) = \varphi_0(x)$ has a unique smooth solution for a small time, determined by the initial data φ. Moreover the function θ defined by (9.10) is a weak solution of the SQG equation.*

In the last section of these notes we will present an additional derivation that does not use the fact that sharp fronts are weak solutions for SQG. The reason for looking for a new approach is that, as mentioned before, vortex lines are not weak solutions for 3D Euler, and we would like to obtain a derivation that only requires techniques available for vortex lines. The derivation will use a family of almost sharp fronts (the analogue of vortex tubes for 3D Euler). Before we go into this construction we will briefly review the existence theory for equation (9.15).

9.4 Existence Results

Concerning the existence of solutions for equation (9.15) the following local existence theorem was proved by Rodrigo (2005).

Theorem 9.4.1 (SQG, $\alpha = 1$) *Given any periodic, smooth function $\varphi_0(x)$ the equation (9.15) for the evolution of sharp fronts has a unique smooth solution for a small time, determined by the initial data φ_0.*

For the case $0 < \alpha < 1$, which we have not explicitly considered here, the following result was obtained in Córdoba et al. (2005).

Theorem 9.4.2 ($0 < \alpha < 1$) *Given φ_0 a smooth, periodic curve, the equation obtained for the evolution of α-patches is locally well posed (has a unique solution on a short time interval).*

Remark 9.4.3 *We have presented the results for the periodic version (in one variable), requiring smooth initial conditions. Gancedo (2008) has obtained local existence results for closed curves in $\mathbb{R}^2$ with initial data in Sobolev spaces. This improvement is due to the presence of an extra cancellation that does not appear in the periodic case.*

We briefly review some features of the proof of Theorem 9.4.1. We refer the reader to Rodrigo (2005) for more details. The analysis of Theorem 9.4.2 is similar (and simpler). Details can be found in Córdoba et al. (2005).

First, we observe that equation (9.15) is a nonlinear version of

$$\frac{\partial \varphi}{\partial t} = \underbrace{\left[\frac{\partial}{\partial x} \log \left(\left| \frac{\partial}{\partial x} \right| \right) \right]}_{i\,k \log |k|} \varphi, \tag{9.16}$$

where the notation in the above expression means that the operator in the right hand side is given by the Fourier multiplier $ik \log |k|$.

The main tool used is a Nash–Moser implicit function theorem. As part of the iterative argument, in the analysis of the linearization, a series of integrating factors and canonical transformations are used to simplify the structure of the singular terms. In particular it is necessary to deal with terms presenting singularities of the form $\log |k|$ and $k \log |k|$. After these series of transformations the linearization becomes

$$\frac{\partial f}{\partial t}(x,t) = \int_{|y|<\frac{1}{2}} \frac{\frac{\partial f}{\partial x}(x,t) - \frac{\partial f}{\partial y}(y,t)}{|y-x|} \, dy + \text{"smooth bounded terms"}$$

In this form, it is now possible to complete a Banach fixed point argument in various Sobolev spaces. In obtaining energy estimates we exploit the fact that the most singular term in the linearization is skew symmetric. We remark that this is reasonable, since the most singular term in the right-hand side of (9.16) is skew symmetric, since it can be easily seen that it is given by a purely imaginary multiplier, a fact that the linearization also shares.

9.4.1 About Global Existence

The equation for the evolution of sharp fronts is believed to be ill-posed. Numerical work (Córdoba et al., 2005) shows the existence of a self-similar blow up for SQG, and the α systems, at least for $\alpha \geq \frac{1}{2}$.

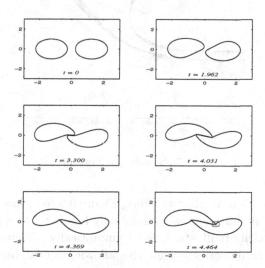

Fig. 9.5. Singularity for sharp fronts

We present in bullet form the main results obtained in the numerical work, and reproduce some of pictures obtained for SQG ($\alpha = 1$) when we consider two circular patches. Figure 9.5 corresponds to the evolution

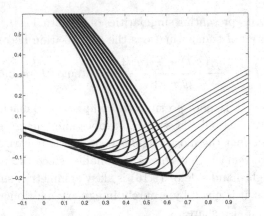

Fig. 9.6. Close up near the singularity

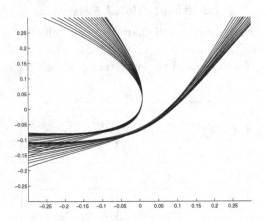

Fig. 9.7. Self-similar blow up

of the two patches at 6 different times, Figure 9.6 to a close up near the
singularity for 10 different times, and finally Figure 9.7 to a rescaled plot
near the singularity to illustrate the asymptotically self-similar blow-up.
In summary, the numerics suggest the following conclusions:

(i) There exists a singularity in finite time, say T.

(ii) The singularity is point-like.

(iii) The singularity is self-similar $(T - t)^{-\frac{1}{\alpha}}$, for the corresponding
α-model (including SQG).

(iv) The minimum distance d between the curves tends to 0 like

$$d \sim C_1 (T-t)^{\frac{1}{\alpha}}.$$

(v) The maximum curvature κ tends to ∞ like

$$\kappa \sim \frac{C_2}{(T-t)^{\frac{1}{\alpha}}}.$$

9.5 Almost Sharp Fronts

As indicated before it would be desirable to have a derivation of the equation for a sharp front that does not use the fact that the sharp front itself is a weak solution. This can be accomplished by considering almost sharp fronts. These are weak solutions of the equation with large gradient ($\sim \delta^{-1}$, where 2δ is the thickness of the transition layer for θ).

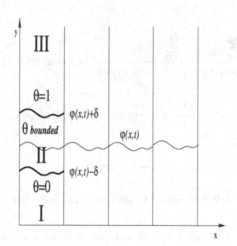

Fig. 9.8. Almost Sharp Front

More precisely, we consider θ of the following form

$$\theta = 0 \text{ if } \quad y \le \varphi(x,t) - \delta$$

$$\theta \text{ bounded if } \quad |\varphi(x,t) - y| \le \delta$$

$$\theta = 1 \text{ if } \quad y \ge \varphi(x,t) + \delta$$

where φ is a smooth periodic function and $0 < \delta < \frac{1}{2}$. We will denote the three regions determining the structure of θ by I, II and III respectively (see Figure 9.8).

This type of solution can be thought of as the analogue of a vortex tube for 3D Euler, i.e. a regularized vortex line.

9.5.1 Stability and derivation

Almost-sharp fronts can be used to obtain a derivation of equation (9.15) using the following theorem (see Córdoba et al., 2004).

Theorem 9.5.1 *If θ is an almost-sharp front and is a weak solution of SQG, then φ satisfies the equation*

$$\frac{\partial \varphi}{\partial t}(x,t) =$$

$$= \int_{\mathbb{R}/\mathbb{Z}} \frac{\frac{\partial \varphi}{\partial x}(x,t) - \frac{\partial \varphi}{\partial \widetilde{x}}(\widetilde{x},t)}{[(x-\widetilde{x})^2 + (\varphi(x,t) - \varphi(\widetilde{x},t))^2]^{\frac{1}{2}}} \chi(x-\widetilde{x}, \varphi(x,t) - \varphi(\widetilde{x},t)) \, d\widetilde{x}$$

$$+ \int_{\mathbb{R}/\mathbb{Z}} \left[\frac{\partial \varphi}{\partial x}(x,t) - \frac{\partial \varphi}{\partial \widetilde{x}}(\widetilde{x},t) \right] \eta(x-\widetilde{x}, \varphi(x,t) - \varphi(\widetilde{x},t)) \, d\widetilde{x}$$

$$+ \mathcal{E}rror,$$

with $|\mathcal{E}rror| \leq C\,\delta|\log \delta|$ where C depends only on $\|\theta\|_{L^\infty}$ and $\|\nabla \varphi\|_{L^\infty}$.

Remark 9.5.2 *Note that an almost-sharp front specifies the function φ up to an error of order δ. The above theorem provides an evolution equation for the function φ up to an error of order $\delta|\log \delta|$.*

Proof We briefly sketch the main ideas in the proof. Recall the definition of a weak solution

$$\int_{\mathbb{R}^+ \times \mathbb{R}/\mathbb{Z} \times \mathbb{R}} \theta(x,y,t)\, \partial_t \phi(x,y,t) + \theta(x,y,t)u(x,y,t) \cdot \nabla \phi(x,y,t) \, dy \, dx \, dt = 0.$$

We need to study the above equation in the three regions of θ. Notice that region I does not contribute since $\theta = 0$. We will concentrate on region II, where θ is simply bounded. There we have

$$\int_{II \times \mathbb{R}^+} \theta(x, y, t) \partial_t \phi(x, y, t) \, dx \, dy \, dt = O(\delta)$$

since $\theta = O(1)$, and area(II) $= O(\delta)$.

We still have to estimate $\int_{\mathbb{R}^2} u \cdot (\mathbb{1}_{II} \theta \nabla \phi) \, dx dy$. Recall that the velocity for SQG is given by

$$u = \nabla^\perp((-\Delta)^{-\frac{1}{2}} \theta) = K * \theta,$$

where K looks locally like the orthogonal of the Riesz transform.

In order to complete the estimate all that is required are two classical results in harmonic analysis. Namely, the fact that the Riesz transform maps bounded functions into functions of exponential class and the fact that the integral of a function of exponential class on a domain of area δ is of order $\delta \log \delta$. The first guarantees the fact that u is of exponential class, and hence the whole integrand in the expression below, while the second directly leads to

$$\int_{\mathbb{R}^2} u \cdot (\mathbb{1}_{II} \theta \nabla \phi) \, dx \, dy = O(\delta \log \delta).$$

Region III can be handled in a similar manner to obtain more error terms of order $\delta \log \delta$ and the terms from the equation of a sharp front in a similar manner to Theorem 9.3.2.   □

9.5.2 Construction of Almost-Sharp Fronts

In order to use Theorem 9.5.1 to find a derivation of the equation, we need to prove the existence of almost-sharp fronts of thickness δ for some small time independent of δ. Notice that only local existence results are know for SQG and so the time of existence for a family of almost-sharp fronts is not uniform with respect to δ, but rather goes to zero as δ goes to zero. In this section we present some of the ideas involved in the constructions of some families of solutions such that for $\delta \leq \delta_0$ the solutions θ_δ exist for at least time T, independent of δ. In particular we concentrate on the crucial problem of finding a limit equation when δ goes to zero. This corresponds to upcoming joint work with Charles Fefferman.

As a first approach, we look for solutions of the form

$$\theta_\delta(x,y,t) = \Omega(x, \frac{y - \varphi(x,t)}{\delta}, t),$$

where Ω is smooth and satisfies

$$\begin{cases} \Omega(x,\xi,t) = \frac{1}{2} & \xi > 1 \\ \Omega(x,\xi,t) \text{ smooth} & |\xi| \le 1 \\ \Omega(x,\xi,t) = -\frac{1}{2} & \xi < -1. \end{cases} \qquad (9.17)$$

In order to simplify the presentation, we will use the following expression for the velocity

$$u(x,y,t) = \int \frac{(-\partial_{\widetilde{y}}\theta(\widetilde{x},\widetilde{y},t), \partial_{\widetilde{x}}\theta(\widetilde{x},\widetilde{y},t))}{[(x - \widetilde{x})^2 + (y - \widetilde{y})^2]^{\frac{1}{2}}} \, d\widetilde{x} \, d\widetilde{y},$$

which amounts to taking the cut-off function χ as 1 and the correction term η as 0, as can be seen by comparing the above formula with the true expression for u (see (9.12)).

Once the equation is rewritten in terms of Ω, x, ξ we expect, since the logarithmic singularity in the velocity for a sharp front does not affect the shape of the front, to obtain an equation of the form

$$\frac{1}{\delta}(\text{Sharp Front Equation for } \varphi) + \text{terms of order 1 depending on } \Omega, x, \xi$$

so that if the curve φ solves the sharp front equation we could take a limit in the equation. Unfortunately that is not the case, and while we do obtain the term of order $\frac{1}{\delta}$ there are a priori harmless terms for which we cannot take a limit. Namely, among the terms arising from $u \cdot \nabla\theta$ we have the term

$$\int\int \frac{\Omega_{\widetilde{\xi}}(\widetilde{x},\widetilde{\xi},t)\Omega_x(x,\xi,t) - \Omega_{\widetilde{x}}(\widetilde{x},\widetilde{\xi},t)\Omega_\xi(x,\xi,t)}{[(x-\widetilde{x})^2 + (\varphi(x,t) - \varphi(\widetilde{x},t) + \delta(\xi - \widetilde{\xi}))^2]^{\frac{1}{2}}} \, d\widetilde{x} \, d\widetilde{\xi}.$$

The problem with this term is that when $\delta = 0$ the kernel looses the dependence on $\widetilde{\xi}$ making the integral singular! We have

$$\int\int \frac{\Omega_{\widetilde{\xi}}(\widetilde{x},\widetilde{\xi},t)\Omega_x(x,\xi,t) - \Omega_{\widetilde{x}}(\widetilde{x},\widetilde{\xi},t)\Omega_\xi(x,\xi,t)}{[(x-\widetilde{x})^2 + (\varphi(x,t) - \varphi(\widetilde{x},t))^2]^{\frac{1}{2}}} \, d\widetilde{x} \, d\widetilde{\xi} =$$

$$= \int \frac{\Omega_x(x,\xi,t) - \int \Omega_{\widetilde{x}}(\widetilde{x},\widetilde{\xi},t)d\widetilde{\xi} \, \Omega_\xi(x,\xi,t)}{[(x-\widetilde{x})^2 + (\varphi(x,t) - \varphi(\widetilde{x},t))^2]^{\frac{1}{2}}} \, d\widetilde{x}$$

since $\int \Omega_{\widetilde{\xi}}(\widetilde{x},\widetilde{\xi},t)\, d\widetilde{\xi} = 1$ due to the form of Ω (see (9.17)).

At this formal level, having made $\delta = 0$ we observe that an integration of the equation with respect to ξ makes the singular term convergent

$$\int \frac{\int \Omega_x(x, \xi, t)\, d\xi - \int \Omega_{\tilde{x}}(\tilde{x}, \tilde{\xi}, t)\, d\tilde{\xi}}{[(x - \tilde{x})^2 + (\varphi(x, t) - \varphi(\tilde{x}, t))^2]^{\frac{1}{2}}}\, d\tilde{x}.$$

Remark 9.5.3
The above formal calculation suggests that

(i) $\int \Omega(x, \xi, t)\, d\xi =: h(x, t)$ *satisfies a much better equation.*

(ii) *If the velocity can be expressed in terms of h rather than in terms of Ω, then we can convert the initial equation into a transport-like equation with coefficients that tend to ∞ as $\delta \to 0$.*

The main reason for the presence of these unexpected singular terms can be found in the expression for the velocity. We know that it is singular and of order $\log \delta$ and we have seen in the initial derivation of the equation that it does not influence the evolution of the curve, but there is a nonsingular term, which will lead to a singularity in the term $u \cdot \nabla \theta$. We have, for $\delta > 0$, an expression of the form

$$u(x, \xi, t) \approx \log \delta\, (1, \varphi(x, t)) + \delta \log \delta \text{ term} + \cdots$$

or more precisely

$$u(x, t) = (\log \delta)\, \alpha(x, t)(1, \varphi(x, t))$$
$$- \delta \log \delta \Big\{ \alpha_x(x, t)\xi + \alpha_x(x, t)h(x, t) + \alpha(x, t)h_x(x, t) \Big\}(0, 1) + \cdots,$$

where α only depends on φ and $h(x, t) = \int \Omega(x, \xi, t)\, d\xi$. Observe that as remarked before, the velocity can be expressed in terms of $h(x, t)$.

Notice that since the gradient of θ is of order δ^{-1}, the term of order $\delta \log \delta$ will generate a singularity. In more detail, we have

$$\nabla \theta = \Omega_x(x, \xi, t)(1, 0) + (-\varphi(x, t), 1)\frac{1}{\delta}\Omega_\xi(x, \xi, t)$$

and so

$$u \cdot \nabla \theta = \log \delta \Big\{ \alpha\Omega_x - [\alpha_x\, \xi + \alpha_x h + \alpha h_x]\Omega_\xi \Big\} + \cdots$$

Notice that integrating in the vertical direction, that is with respect to ξ, has the same regularizing effect, since a simple calculation shows that

$$\int \alpha\Omega_x - [\alpha_x\, \xi + \alpha_x h + \alpha h_x]\Omega_\xi\, d\xi = 0.$$

The advantage of integrating the equation with respect to ξ is that it allows us to obtain a "simple" equation for $h(x,t) = \int \Omega(x,\xi,t)\mathrm{d}\xi$, that actually has a limit as δ goes to 0.

We use its solution in the original equation for Ω,

$$\Omega_t + \log \delta\Big\{\alpha\Omega_x - \big[\alpha_x\,\xi + \alpha_x h + \alpha h_x\big]\Omega_\xi\Big\} + \cdots = 0, \qquad (9.18)$$

which is now a transport-like equation since h is known.

While equation (9.18) does not have a limiting form when δ goes to zero, all we need to do is integrate the following vector fields

$$\left\{ \begin{aligned} \frac{\mathrm{d}x}{\mathrm{d}t} &= \alpha(x,t) \\ \frac{\mathrm{d}\xi}{\mathrm{d}t} &= -[\alpha_x\,\xi + \alpha_x h + \alpha h_x], \end{aligned} \right.$$

with initial conditions $x(0) = x_0$ and $\xi(0) = \xi_0$.

The process of integrating the system of ODEs and rewriting the equation in terms of x_0 and ξ_0 corresponds to unwinding the logarithmic divergence that appears in $u \cdot \nabla\theta$, before considering the map

$$(x,y,t) \longmapsto (x, \frac{y - \varphi(x,t)}{\delta}, t).$$

It is this unwinding process that makes it possible to obtain a limiting equation for Ω, in terms of the Lagrangian coordinates x_0 and ξ_0.

Acknowledgements

The author was partially supported by MTM2005–05980, Ministerio de Educación y Ciencia (Spain) and MTM2008–03754, Ministerio de Ciencia e Innovación (Spain). The author wishes to acknowledge A.M. Mancho for producing Figures 9.5, 9.6, and 9.7.

References

Beale, J.T., Kato, T., & Majda, A.J. (1984) Remarks on the breakdown of smooth solutions for the 3-D Euler equations. *Comm. Math. Phys.* **94**, 61–66.

Bertozzi, A.L. & Constantin, P. (1993) Global regularity for vortex patches. *Comm. Math. Phys.* **152**, no. 1, 19–28.

Chemin, J.Y. (1993) Persistance de structures géométriques dans les fluides incompressibles bidimensionnels. *Ann. Ec. Norm. Supér.* **26**, no. 4, 1–16.

Constantin, P., Majda, A., & Tabak, E. (1994a) Singular front formation in a model for quasigesotrophic flow. *Phys. Fluids* **6**, no. 1, 9–11.

Constantin, P., Majda, A., & Tabak, E. (1994b) Formation of strong fronts in the $2-D$ quasigeostrophic thermal active scalar. *Nonlinearity* **7**, no. 6, 1495–1533.

Constantin, P., Nie, Q., & Schörghofer, N. (1998) Nonsingular surface quasi-geostrophic flow. *Phys. Lett. A* **241**, no. 3, 168–172.

Constantin, P., Nie, Q., & Schörghofer, N. (1999) Front formation in an active scalar equation. *Phys. Rev. E (3)* **60**, no. 3, 2858–2863.

Córdoba, D. (1998) Nonexistence of simple hyperbolic blow-up for the quasi-geostrophic equation. *Ann. of Math. (2)* **148**, no. 3, 1135–1152.

Córdoba, D., Fefferman, C., & Rodrigo, J.L. (2004) Almost sharp fronts for the surface Quasi-Geostrophic equation. *Proc. Natl. Acad. Sci. USA* **101**, no. 9, 2487–2491.

Córdoba, D., Fontelos, M.A., Mancho, A.M., & Rodrigo, J.L. (2005) Evidence of singularities for a family of contour dynamics equations. *Proc. Natl. Acad. Sci. USA* **102**, no. 17, 5949–5952.

da Rios, L.S. (1906) Sul moto d'un liquido indefinito con un filetto vorticoso di forma qualunque. *Rend. Circ. Mat. Palermo* **22**, 117–135.

Gancedo, F. (2008) Existence for the α-patch model and the QG sharp front in Sobolev spaces. *Adv. Math.* **217**, no. 6, 2569–2598.

Garner, S.T., Held, I.M., Pierrehumbert, R.T., & Swanson, K. (1995) Surface quasi-geostrophic dynamics. *J. Fluid Mech.* **282**, 1–20.

Gutiérrez, S. & Vega, L. (2004) Self-similar solutions of the localized induction approximation: singularity formation. *Nonlinearity* **17**, no. 6, 2091–2136.

Gutiérrez, S., Rivas, J., & Vega, L. (2003) Formation of singularities and self-similar vortex motion under the localized induction approximation. *Comm. Partial Differential Equations* **28**, no. 5-6, 927–968.

Held, I.M., Pierrchumbert, R.T., & Swanson. K. (1994) Spectra of local and nonlocal two dimensional turbulence. *Chaos, Solitons and Fractals* **4**, 1111–1116.

Klein, R. & Majda, A.J. (1991a) Self-stretching of a perturbed vortex filament. I. The asymptotic equation for deviations from a straight line. *Phys. D* **49**, no. 3, 323–352.

Klein, R. & Majda, A.J. (1991b) Self-stretching of perturbed vortex filaments. II. Structure of solutions. *Phys. D* **53**, no. 2-4, 267–294.

Majda, A.J. & Bertozzi, A. (2002) *Vorticity and incompressible flow.* Cambridge Texts in applied Mathematics, **27**.

Madja, A.J. & Tabak, E. (1996) A two-dimensional model for quasigeostrophic flow: comparison with the two-dimensional Euler flow. *Physica D* **98**, no. 2-4, 515–522.

Ohkitani, K. & Yamada, M. (1997) Inviscid and inviscid-limit behavior of a surface quasigeostrophic flow. *Phys. Fluids* **9**, no. 4, 876–882.

Pedlosky, J. (1987) *Geophysical Fluid Dynamics.* Springer-Verlag, 345–368 and 653–670, New York.

Ricca, R.L. (1996) The contributions of da Rios and Levi-Civita to asymptotic potential theory and vortex filament dynamics. *Fluid Dynam. Res.* **18**, no. 5, 245–268.

Rodrigo, J.L. (2004) The vortex patch problem for the Quasi-Geostrophic equation. *PNAS* **101**, no. 9, 2484–2486.

Rodrigo, J.L. (2005) On the evolution of sharp fronts for the quasi-geostrophic equation. *Comm. Pure Appl. Math.* **58**, no. 6, 821–866.

Vega, L. (2003) Kink solutions of the binormal flow. *Journées Équations aux Dérivées Partielles*, **XIV**, Univ. Nantes, Nantes.

10

Theory and applications of statistical solutions of the Navier–Stokes equations

Ricardo M.S. Rosa

Instituto de Matemática,
Universidade Federal do Rio de Janeiro,
Caixa Postal 68530 Ilha do Fundão, Rio de Janeiro,
RJ 21945-970. Brazil.
`rrosa@ufrj.br`

Abstract

Since the 1970s the use of statistical solutions of the Navier–Stokes equations has led to a number of rigorous results for turbulent flows. This paper reviews the concept of a statistical solution, its role in the mathematical foundation of the theory of turbulence, some of its successes, and the theoretical and applied challenges that still remain. The theory is illustrated in detail for the particular case of a two-dimensional flow driven by a uniform pressure gradient.

10.1 Introduction

It is believed that turbulent fluid motions are well modelled by the Navier–Stokes equations. However, due to the complicated nature of these equations, most of our understanding of turbulence relies to a great extent on laboratory experiments and on heuristic and phenomenological arguments. Nevertheless, a number of rigorous mathematical results have been obtained directly from the Navier–Stokes equations, particularly in the last two decades.

Of great interest in turbulence theory are mean quantities, which are in general well behaved, in contrast to the corresponding instantaneous values, which tend to vary quite dramatically in time. The treatment of mean values, however, is a delicate problem, as remarked by Monin & Yaglom (1975). In practice time and space averages are the most generally used, while in theory averages with respect to a large ensemble of flows avoid some analytical difficulties and have a more universal character. In the conventional theory of turbulence, all three types of average are taken to agree by invoking some sort of ergodic property, although no general result in this direction exists.

Published in *Partial Differential Equations and Fluid Mechanics*, edited by James C. Robinson and José L. Rodrigo. © Cambridge University Press 2009.

Rigorous results for time and space averages can be obtained from individual weak solutions of the Navier–Stokes equations, while rigorous results for ensemble averages are obtained via the concept of statistical solutions.

Probability theory has been used to formalize the concept of an ensemble average since very early in the development of the conventional theory. Later on, Hopf (1952) published one of the first works with a more mathematical flavour. A rigorous mathematical foundation for this formalization, however, came only with the introduction of statistical solutions by Foias (1972, 1973). Another rigorous framework was introduced a few years later by Vishik & Fursikov (1977, 1988). Both frameworks involve measures on function spaces and require a number of deep results in analysis. Just as for individual solutions of the Navier–Stokes equations, there are various challenging problems regarding the regularity of such statistical solutions.

Statistical solutions can be time-dependent or stationary in time. The time-dependent case is suitable, for instance, for the study of decaying turbulence, while the stationary case applies to turbulence in statistical equilibrium in time.

The stationary statistical solution is a generalization of the notion of an invariant measure for a semigroup. The major difficulty here lies in the fact that the three-dimensional Navier–Stokes equations are not know to generate a well-defined semigroup. An invariant measure in this case does not make sense, and the statistical solution allows one to address this issue. In the two-dimensional case, in which the associated semigroup is well defined, both notions coincide.

Beyond the theoretical issues permeating the concept of statistical solutions, a number of applications have been given yielding rigorous bounds for characteristic physical quantities taken in the mean with respect to such solutions. Estimates have been obtained involving the mean energy dissipation rate, mean enstrophy dissipation rate, mean skin-drag coefficient, mean kinetic-energy flux between different scales of motions, and so on. A number of such rigorous results are related to important properties of turbulent flows as derived in the conventional theory of turbulence via heuristic or phenomenological arguments, and yield rigorous results such as estimates for the number of degrees of freedom, estimates related to the Kolmogorov energy dissipation law, exponential decay of the power spectrum in the two-dimensional case, conditions for the existence of an energy cascade in the three-dimensional case, conditions for the existence of inverse energy and direct enstrophy cascades in the two-dimensional case, etc.

While the results for time averages and ensemble averages have recently paralleled one another, our main concern in this article is on ensemble averages. We recall the definition of a statistical solution, discuss some of the associated delicate theoretical problems, and mention, without proof, a number of rigorous estimates obtained recently. We also include, for the sake of illustration, a proof of two estimates pertaining to a two-dimensional channel flow driven by a uniform pressure gradient. More precisely, we prove the curious fact that, for this geometry, the plane Poiseuille flow minimizes the mean rate of enstrophy dissipation and maximizes the mean rate of energy dissipation. These minimization and maximization properties are with respect to invariant measures of the semigroup generated by the two-dimensional Navier–Stokes equations; a Dirac measure concentrated on the velocity field for the Poiseuille flow is one such measure. The minimization part is new while the maximization is an adaptation of a corresponding recent result for the three-dimensional channel problem treated by Ramos, Rosa, & Temam (2008). Although this example is for a two-dimensional problem, it has many similarities with the corresponding estimates for the three-dimensional problem and so serves as a useful illustration.

10.2 The incompressible Navier–Stokes equations

We consider the incompressible Navier–Stokes equations (NSE) and write them in the form

$$\frac{\partial \mathbf{u}}{\partial t} - \nu \Delta \mathbf{u} + (\mathbf{u} \cdot \boldsymbol{\nabla})\mathbf{u} + \boldsymbol{\nabla} p = \boldsymbol{f}, \quad \boldsymbol{\nabla} \cdot \mathbf{u} = 0, \tag{10.1}$$

where $\mathbf{u} = \mathbf{u}(\mathbf{x}, t) = (u_1, u_2, u_3)$ denotes the three-component velocity field, $\mathbf{x} = (x_1, x_2, x_3)$ is the space variable; t is the time variable; ν is the kinematic viscosity; $p = p(\mathbf{x}, t)$ is the kinematic pressure; and $\boldsymbol{f} = \boldsymbol{f}(\mathbf{x}, t) = (f_1, f_2, f_3)$ is the density of volume forces.

For the classical mathematical theory of the NSE the reader is referred to Ladyzhenskaya (1963), Temam (1984), and Constantin & Foias (1988), which are based on the earlier works of Leray (1933, 1934a,b) and Hopf (1951).

The spatial domain is denoted by $\Omega \subset \mathbb{R}^3$. The boundary conditions are assumed to be homogeneous and may be no-slip, periodic, or combinations of these, as in the case of a periodic channel flow. In the fully-periodic case, the velocity field and the forcing term can be assumed, without loss of generality, to have zero space average over the periodic domain, while in cases involving no-slip boundary conditions, the domain is assumed to be either bounded or to have finite width in one direction. In all these cases, the Stokes operator (the linear part of the time-independent problem), which will be denoted by A, is strictly positive.

The NSE can be written in the form

$$\frac{d\mathbf{u}}{dt} = \mathbf{F}(\mathbf{u}),$$

in a suitable space H of square-integrable divergence-free vector fields with the appropriate boundary conditions, and with an appropriate function $\mathbf{F} : V \to V'$, with $V \subseteq H \subseteq V'$. Here, V is the space of divergence-free vector fields that are square integrable along with their first order partial derivatives, equipped with the appropriate boundary conditions, and V' is its dual. See, for instance, Section 6 for a precise definition of these spaces in the case of a two-dimensional channel. The pressure can, in principle, be determined from the vector field $\mathbf{u}$.

The L^2-like norm and inner product in H are denoted by $|\cdot|$ and $(\cdot, \cdot)$, while the H_0^1-like norm and inner product in V are denoted by $\|\cdot\|$ and $((\cdot, \cdot))$. For the sake of simplicity, the duality product between V and V' is also denoted by $(\cdot, \cdot)$. These inner products can be written explicitly as

$$(\boldsymbol{u}, \boldsymbol{v}) = \int_\Omega \sum_{i=1}^3 u_i v_i \, d\boldsymbol{x} \quad \text{and} \quad ((\boldsymbol{u}, \boldsymbol{v})) = \int_\Omega \sum_{i,j=1}^3 \frac{\partial u_i}{\partial x_j} \frac{\partial v_i}{\partial x_j} \, d\boldsymbol{x}.$$

The inner products and duality are such that $(A\boldsymbol{u}, \boldsymbol{v}) = ((\boldsymbol{u}, \boldsymbol{v}))$ for all $\boldsymbol{u}, \boldsymbol{v} \in V$. Under the assumptions made on the domain Ω, there is a largest positive number $\lambda_1 > 0$ such that $(A\boldsymbol{u}, \boldsymbol{u}) = \|\boldsymbol{u}\|^2 \geq \lambda_1 |\boldsymbol{u}|^2$, for all $\boldsymbol{u} \in V$.

The nonlinear term gives rise to a bilinear form $B : V \times V \to V'$, which can be defined by duality through

$$(B(\mathbf{u}, \mathbf{v}), \mathbf{w}) = b(\mathbf{u}, \mathbf{v}, \mathbf{w}),$$

where the trilinear term $b : V \times V \times V \to \mathbb{R}$ is defined by

$$b(\mathbf{u}, \mathbf{v}, \mathbf{w}) = \int_\Omega ((\mathbf{u} \cdot \nabla)\mathbf{v}) \cdot \mathbf{w} \, dx. \qquad (10.2)$$

In the three-dimensional case and with typical boundary conditions (periodic, no-slip, etc.) it is well-known that given an initial velocity field

u_0 in H, there exists a weak solution (in a suitable sense) $u = u(t)$ that is defined for all $t \geq 0$ and satisfies $u(0) = u_0$. In fact, there exists a weak solution satisfying, in addition, a certain energy-type inequality, and this solution is called a Leray–Hopf weak solution. But the regularity obtained for this solution is still not sufficient to guarantee its uniqueness. If the initial condition is more regular, however, there exists a time interval, $0 \leq t < \delta$, with δ depending on the size of the initial condition in an appropriate sense, in which the solution is regular and unique. Therefore, we have local unique regular solutions and global Leray–Hopf weak solutions.

In the two-dimensional case, the system is more amenable to analysis, and the existence and uniqueness of global regular solutions are known for arbitrary initial conditions in H; the system is well-posed and generates a continuous nonlinear semigroup $\{S(t)\}_{t \geq 0}$ in H.

10.3 Mean quantities and statistical solutions

As mentioned in the introduction, there is much interest in mean quantities, whether space averaged, time averaged, or with respect to ensemble averages. One is typically interested in relating different quantities in the hope of finding universal properties of turbulent flows (Kolmogorov, 1941; Batchelor, 1953; Monin & Yaglom, 1975; Hinze, 1975; Tennekes & Lumley, 1972; Lesieur, 1997; Frisch, 1995; Foias et al., 2001a; Rosa, 2002, 2006).

Types of averages

Time averages can be easily defined for a given individual solution $u(t)$ of the NSE through the relation

$$\frac{1}{T} \int_t^{t+T} \varphi(u(s)) \, ds,$$

where $T > 0$ and $\varphi : H \to \mathbb{R}$ is associated with some physical quantity one wants to consider (energy, rate of energy dissipation, etc.). Space averages can be defined in a similar way, although one must be careful with the boundary conditions.

In situations where a statistical equilibrium is of interest (the instantaneous quantities still vary widely in time but average quantities seem stationary), an "infinite-time" average is often more meaningful and in this case the limit as $T \to \infty$ is taken in the definition of the average. One may assume that the limit exists, although there is no general result about this, or one may consider superior or inferior limits, or generalized limits.

In the case of ensemble averages, one may think of a number of experiments yielding velocity fields $\mathbf{u}^{(n)}(t)$, $t \geq 0$, $n = 1, \ldots, N$, and take the average over this ensemble of experiments:

$$\frac{1}{N} \sum_{n=1}^{N} \varphi(\mathbf{u}^{(n)}(t)).$$

In a more general and rigorous way, one considers a family of probability measures $\{\mu_t\}_{t \geq 0}$ defined on H, representing the probability distribution of the velocity field, and takes the average

$$\int_H \varphi(\mathbf{v}) \, d\mu_t(\mathbf{v}).$$

Note that in this expression the quantity $\mathbf{v}$ is simply a dummy variable in the integration. The important quantity is the family of measures.

Time-dependent statistical solutions

In relation with the ensemble averages, for a family $\{\mu_t\}_{t \geq 0}$ of measures to be meaningful for the flow, it must satisfy an equation analogous to the Liouville equation in Statistical Mechanics, namely

$$\frac{d}{dt} \int_H \Phi(\mathbf{v}) \, d\mu_t(\mathbf{v}) = \int_H (\mathbf{F}(\mathbf{v}), \Phi'(\mathbf{v})) \, d\mu_t(\mathbf{v}), \qquad (10.3)$$

for all "test" functions Φ. Note that since $\mathbf{F} : V \to V'$, the measures μ_t are expected to be carried by V, while the derivative $\Phi'(\mathbf{v})$ of the test function is assumed to belong to V. More general test functions $\varphi : H \to \mathbb{R}$ can be evaluated in the mean, but for the Liouville equation, a more restricted set of test functions Φ is considered, which explains the difference in notation.

The reason behind the form of the equation (10.3) can be seen from the following derivation, for a family of measures

$$\mu_t = \frac{1}{N} \sum_{i=1}^{N} \delta_{\mathbf{u}^{(n)}(t)}$$

obtained as a convex combination of Dirac measures concentrated on N weak solutions $\mathbf{u}^{(n)}(t)$ with equal probability $1/N$:

$$\frac{\mathrm{d}}{\mathrm{d}t} \int_H \Phi(\mathbf{v}) \, \mathrm{d}\mu_t(\mathbf{v}) = \frac{\mathrm{d}}{\mathrm{d}t} \frac{1}{N} \sum_{n=1}^N \Phi(\mathbf{u}^{(n)}(t)) = \frac{1}{N} \sum_{n=1}^N \frac{\mathrm{d}}{\mathrm{d}t} \Phi(\mathbf{u}^{(n)}(t))$$

$$= \frac{1}{N} \sum_{n=1}^N \Phi'(\mathbf{u}^{(n)}(t)) \circ \frac{\mathrm{d}}{\mathrm{d}t} \mathbf{u}^{(n)}(t) = \frac{1}{N} \sum_{n=1}^N \Phi'(\mathbf{u}^{(n)}(t)) \circ \mathbf{F}(\mathbf{u}^{(n)}(t))$$

$$= \frac{1}{N} \sum_{n=1}^N (\mathbf{F}(\mathbf{u}^{(n)}(t)), \Phi'(\mathbf{u}^{(n)})) = \int_H (\mathbf{F}(\mathbf{v}), \Phi'(\mathbf{v})) \, \mathrm{d}\mu_t(\mathbf{v}).$$

Other conditions are necessary for the problem to make sense. For instance, concerning generalized moments, one asks that

$$t \mapsto \int_H \varphi(\mathbf{v}) \, \mathrm{d}\mu_t(\mathbf{v}) \text{ is measurable on } [0, \infty), \qquad (10.4)$$

for every bounded and continuous real-valued function φ on H.

From the physical point of view, it is also natural to assume that the mean kinetic energy per unit mass and the mean rate of enstrophy dissipation per unit time per unit mass are finite, at least almost everywhere. In fact, it is assumed that

$$t \mapsto \int_H |\mathbf{v}|^2 \, \mathrm{d}\mu_t(\mathbf{v}) \in L^\infty_{\mathrm{loc}}(0, \infty), \qquad t \mapsto \int_H \|\mathbf{v}\|^2 \, \mathrm{d}\mu_t(\mathbf{v}) \in L^2_{\mathrm{loc}}(0, \infty),$$
$$(10.5)$$

as obtained for individual weak solutions.

Finally, a mean energy inequality for the statistical solutions is also assumed, analogous to the energy inequality for Leray–Hopf weak solutions:

$$\frac{1}{2} \int_H |\mathbf{v}|^2 \, \mathrm{d}\mu_t(\mathbf{v}) + \nu \int_{t'}^t \int_H \|\mathbf{v}\|^2 \, \mathrm{d}\mu_s(\mathbf{v}) \, \mathrm{d}s$$

$$\leq \frac{1}{2} \int_H |\mathbf{v}|^2 \, \mathrm{d}\mu_{t'}(\mathbf{v}) + \int_{t'}^t \int_H (\boldsymbol{f}, \mathbf{v}) \, \mathrm{d}\mu_s(\mathbf{v}) \, \mathrm{d}s,$$

for almost all $t' \geq 0$, and all $t \geq t'$, with $t' = 0$ included.

A stricter definition requires the validity of the following strengthened form of energy inequality:

$$\frac{1}{2} \int_H \psi\left(|\mathbf{v}|^2\right) \, \mathrm{d}\mu_t(\mathbf{v}) + \nu \int_{t'}^t \int_H \psi'\left(|\mathbf{v}|^2\right) \|\mathbf{v}\|^2 \, \mathrm{d}\mu_s(\mathbf{v}) \, \mathrm{d}s$$

$$\leq \frac{1}{2} \int_H \psi\left(|\mathbf{v}|^2\right) \, \mathrm{d}\mu_{t'}(\mathbf{v}) + \int_{t'}^t \int_H \psi'\left(|\mathbf{v}|^2\right) (\boldsymbol{f}, \mathbf{v}) \, \mathrm{d}\mu_s(\mathbf{v}) \, \mathrm{d}s, \quad (10.6)$$

for almost all $t' \geq 0$, and all $t \geq t'$, with $t' = 0$ included, and for arbitrary non-negative, non-decreasing, continuously-differentiable real-valued function ψ with bounded derivative.

For us here, a family $\{\mu_t\}_{t \geq 0}$ of Borel probability measures on H satisfying (10.4), (10.5), (10.3) (for suitable "test functions"), and the strengthened energy inequality (10.6), is then called a *statistical solution* on the time-interval $[0, \infty)$.

This fundamental definition of statistical solution, which makes rigorous the concept of ensemble average which is ubiquitous in turbulence theory, was introduced by Foias (1972, 1973) (see also Hopf, 1952; Vishik & Fursikov, 1977, 1988; Foias et al., 2001a) and has been extensively exploited in recent years.

As with the theory for individual solutions, there are results on the global existence of statistical solutions. The typical assumption on the initial measure μ_0 is that it has finite mean kinetic energy:

$$\int_H |\mathbf{v}|^2 \, d\mu_0(\mathbf{v}) < \infty.$$

Stationary statistical solutions

In the case of turbulence in statistical equilibrium in time, the mean quantities are time independent and a single measure μ represents the statistics of the flow. This measure satisfies the stationary Liouville-type equation,

$$\int_H (\mathbf{F}(\mathbf{v}), \Phi'(\mathbf{v})) \, d\mu(\mathbf{v}) = 0, \tag{10.7}$$

and is called a *stationary statistical solution* of the NSE.

These stationary statistical solutions, first defined by Foias (1973), are generalizations of invariant measures, which do not make sense in the three-dimensional case due to the lack of a global well-posedness result. Foias's concept of a statistical solution was inspired by the two unpublished pioneering works of Prodi (1960, 1961), who insisted on the idea of working with measures that should be invariant in some suitable sense.

In the two-dimensional case, on the other hand, global well-posedness has been established and ensemble averages can be interpreted as averages with respect to invariant measures. We recall here that an invariant Borel probability measure for the semigroup $\{S(t)\}_{t \geq 0}$ generated by the 2D Navier–Stokes equations is a Borel probability measure μ on H such that

$$\mu(S(t)^{-1}E) = \mu(E),$$

for all Borel sets E in H. An analogous definition of a stationary statistical solution can be given in 2D and it turns out to be equivalent to the definition of an invariant measure (Foias, 1973; Foias et al., 2001a).

Time-average stationary statistical solutions

A particular class of stationary statistical solution is obtained via a generalized limit of time averages of weak solutions, as in the Krylov–Bogoliubov procedure (Krylov & Bogoliubov, 1937). More precisely, given a global weak solution $\mathbf{u} = \mathbf{u}(t)$ defined for all $t \geq 0$ and given a function $\varphi : H \to \mathbb{R}$ which is weakly continuous on H, the generalized Banach limit

$$\varphi \mapsto \underset{T \to \infty}{\mathrm{LIM}} \frac{1}{T} \int_0^T \varphi(\mathbf{u}(t)) \, dt$$

exists and defines a linear functional with respect to φ. It can be proved (Bercovici et al., 1995; Foias et al., 2001a) that there exists a measure $\mu_{\mathbf{u}}$ associated with this weak solution for which

$$\underset{T \to \infty}{\mathrm{LIM}} \frac{1}{T} \int_0^T \varphi(\mathbf{u}(t)) \, dt = \int_H \varphi(\mathbf{v}) \, d\mu_{\mathbf{u}}(\mathbf{v})$$

for all such φ, and that this measure is a stationary statistical solution.

Mean quantities

A common notation for the ensemble average is $\langle \cdot \rangle$. We use this notation here in the case of stationary statistical turbulence, writing

$$\langle \varphi \rangle = \int_H \varphi(\mathbf{u}) \, d\mu(\mathbf{u}),$$

for a given functional $\varphi : H \to \mathbb{R}$ and a given stationary statistical solution μ on H. The quantity $\langle \varphi \rangle$ is called a *generalized moment*.

Taking κ_0 to denote a macro-scale wavenumber (e.g. $\kappa_0 = \lambda_1^{1/2}$), two important examples of ensemble-averaged mean quantities in the three-dimensional case are the mean kinetic energy per unit mass e and the mean energy dissipation rate per unit time per unit mass ϵ, defined respectively as

$$e = \frac{\kappa_0^3}{2} \left\langle |\mathbf{u}|^2 \right\rangle, \qquad \epsilon = \kappa_0^3 \nu \left\langle \|\mathbf{u}\|^2 \right\rangle, \tag{10.8}$$

which are obtained with the choices

$$\varphi(\mathbf{u}) = \frac{\kappa_0^3}{2}|\mathbf{u}|^2, \qquad \varphi(\mathbf{u}) = \kappa_0^3 \nu \|\mathbf{u}\|^2,$$

respectively.

Several other mean quantities can be considered, such as mean velocity field, Reynolds number, Taylor wavenumber, structure functions, skin-friction coefficient in the case of a channel, and so on (see Section 5).

In the two-dimensional case, two important physical quantities are the mean enstrophy per unit mass and the mean enstrophy dissipation rate per unit time and unit mass, given respectively by

$$E = \frac{\kappa_0^2}{2} \langle \|\mathbf{u}\|^2 \rangle, \qquad \eta = \nu \kappa_0^2 \langle |A\mathbf{u}|^2 \rangle. \tag{10.9}$$

10.4 Mathematical aspects of the theory of statistical solutions

As introduced above a statistical solution is a family $\{\mu_t\}_{t \geq 0}$ of probability measures μ_t on the phase space H satisfying the Liouville-type equation (10.3) along with appropriate regularity conditions.

A delicate issue when working with such a statistical solution is the measurability of certain dynamic sets with respect to the measures μ_t. The lack of a well-posedness result for the 3D NSE makes this a nontrivial and intriguing issue.

There are also a number of interesting regularity problems for the statistical solutions. One type of regularity problem is related to the localization of the carriers of the measures. Let us just recall that a carrier for a measure is any set of full measure, i.e. for which the measure of the complement of the carrier is null. From the definition of the statistical solutions (the assumption that enstrophy is finite almost always), it follows immediately that μ_t is carried by the set V for almost every $t \geq 0$. This is important, in particular, for the definition of the mean rate of energy dissipation. More regular statistical solutions exists for which μ_t is carried by the domain of the Stokes operator $D(A)$.

A related "asymptotic regularity" problem is whether a stationary statistical solution is carried by a set in H for which all weak solutions are in fact global strong solutions. This is a major and challenging open problem akin to the global well-posedness problem for the 3D NSE. The *Prodi conjecture* is that this asymptotic regularity holds true.

Another problem relates to a different notion of statistical solution due to Vishik & Fursikov (1988). They considered a measure ρ defined on the space of trajectories $Z = L^2(0, T; H) \cap \mathcal{C}([0, T], V^{-s})$, for a given $s \geq 2$, where V^{-s} is the dual of the space $V^s = D(A^{s/2})$ defined in terms of the domain of powers of the Stokes operator. The assumption that makes this measure relevant to the NSE is that this measure be carried by the space of weak solutions (not necessarily of Leray–Hopf type, in their framework, although a certain form of mean energy estimate is assumed).

A slightly different definition in the spirit of Vishik & Fursikov is to consider a Borel probability measure on the space $\mathcal{C}([0, T], H_w)$ that is carried by the set of Leray–Hopf weak solutions, where H_w denotes the space H endowed with its weak topology. Under natural integrability and continuity assumptions, the family of projections in H, at each time t, of this measure, gives rise to a statistical solution in the sense given earlier and that we term a *Vishik–Fursikov statistical solution*. With this definition, any Vishik–Fursikov statistical solution is a statistical solution as defined earlier.

The statistical solutions obtained as projections of measures in some suitable trajectory space have a few additional properties which make them much more amenable to analysis. Another important open problem is then the converse statement, characterizing under which conditions a statistical solution is a Vishik–Fursikov statistical solution.

Similar regularity problems appear in the two-dimensional case. Note, for instance, that an important quantity is the mean rate of enstrophy dissipation per unit time per unit mass, associated with the choice of the generalized moment $\varphi(\mathbf{u}) = \nu \kappa_0^2 |A\mathbf{u}|^2$. It is important to know whether this choice is allowed in the expression for $\langle \varphi \rangle$. It can be proved that if f belongs to H then any invariant measure μ for the 2D NSE is carried by $D(A)$ and $\langle |A\mathbf{u}|^2 \rangle$ is finite (Bercovici et al., 1995; Foias et al., 2001a). If the forcing is more regular, so is the invariant measure. For example, if f is in some Gevrey space, any invariant measure is carried by an associated Gevrey space. This has important consequences for the exponential decay of the energy spectrum, for instance.

10.5 Applications to turbulent flows

A number of rigorous results related to the conventional theory of turbulence have been obtained in different contexts. We briefly mention some of them in what follows.

The Reynolds equations

One of the pioneering works in turbulence to address statistical properties of turbulent flows is due to Reynolds (1895), who proposed decomposing the flow into a regular mean part and an irregular fluctuating part. The mean flow was found to satisfy what are now called the *Reynolds equations*.

In the framework of statistical solutions, for a family $\{\mu_t\}_{t\geq 0}$, a mean velocity field $\mathbf{U}(t)$ can be defined by duality,

$$(\mathbf{U}(t), \mathbf{v}) = \int_H (\mathbf{u}, \mathbf{v}) \, d\mu_t,$$

and it follows that $\mathbf{U} \in L^\infty_{\mathrm{loc}}(0, \infty; H) \cap L^2_{\mathrm{loc}}(0, \infty; V)$. Thus the Reynolds equations can be written rigorously as a functional equation in the space $L^{4/3}_{\mathrm{loc}}(\mathbb{R}, V')$:

$$\mathbf{U}_t + \nu A\mathbf{U} + B(\mathbf{U}, \mathbf{U}) = \boldsymbol{f} + \langle B(\mathbf{u} - \mathbf{U}, \mathbf{u} - \mathbf{U}) \rangle,$$

where the second term in the right-hand side is due to the fluctuations from the mean and is related to the Reynolds stress tensor. The mean pressure term P, the Reynolds stress tensor $\langle \boldsymbol{u}' \otimes \boldsymbol{u}' \rangle$, where $\boldsymbol{u}' = \boldsymbol{u} - \boldsymbol{U}$, and the usual form of the Reynolds equations,

$$\mathbf{U}_t + \nu \Delta \mathbf{U} + (\mathbf{U} \cdot \nabla)\mathbf{U} + \nabla P = \boldsymbol{f} + \nabla \cdot \langle \boldsymbol{u}' \otimes \boldsymbol{u}' \rangle,$$

can be recovered just as in the classical mathematical theory of the Navier–Stokes equations. The Reynolds stress tensor

$$\langle \boldsymbol{u}' \otimes \boldsymbol{u}' \rangle = \left(\int_H u'_i u'_j \, d\mu(\mathbf{u}) \right)_{i,j=1,2,3}$$

belongs to $L^{4/3}_{\mathrm{loc}}(\mathbb{R}, L^2(\Omega)^{3\times 3})$, for instance.

For more details, the reader is referred to Foias (1974). See also Foias et al. (2001a) and Vishik & Fursikov (1988).

The closure problem

The mean flow is one part of the puzzle. Notice that the Reynolds equations are not closed since they involve a term that depends on the fluctuations. One way to look at this problem is to notice that the mean velocity field is a first-order moment, and the extra term in the Reynolds equations involves second-order moments. Equations for the second-order moments can be derived but they depend on third-order moments. This process can be continued indefinitely, leading to an infinite system of equations, with the equations for the nth order moment depending on the moments of order $n + 1$. This is called the *Friedmann-Keller system* of equations.

Keller & Friedman (1925) stated that the statistical moments are the fundamental characteristic of turbulent flows. Here, it must be recalled that a result in Probability Theory, the so-called *moment problem*, guarantees that the distribution function of a probability measure is uniquely determined by the corresponding infinite set of moments, provided that a summability criterion holds. Attempts to truncate the sequence to a finite system lead to the so-called *closure models*. The moment problem has been addressed by Vishik & Fursikov in a number of papers (e.g. Vishik & Fursikov, 1988, and Fursikov, 1999).

The Hopf equation

Hopf (1952) studied the evolution of the probability measure of a flow and arrived at an equation with infinitely many variables for the corresponding characteristic functions. This equation is now known as the *Hopf equation*.

Two decades later, Foias (1974) recast this problem in a rigorous form as a functional equation for the characteristic functions $\chi(t, \mathbf{g})$ associated with a statistical solution $\{\mu_t\}_t$ and a test function $\mathbf{g} \in H$, with

$$\chi(t, \mathbf{g}) = \int_H e^{i(\mathbf{u}, \mathbf{g})} \, d\mu_t(\mathbf{u}).$$

The Hopf equation has also been addressed by Vishik & Fursikov (1988) within their rigorous framework.

The energy cascade

The cascade of energy is a fundamental prediction of the conventional theory of turbulence associated with the famous $\kappa^{-5/3}$ Kolmogorov spectrum. It is based on the phenomenological eddy cascade mechanism devised by Richardson (1922). It says that, for a homogeneous turbulent flow, driven by large scale forces, there exists a certain range of scales, termed the inertial range, within which the transfer of energy from larger to smaller scales is constant and equal to the mean rate of energy dissipation. The main idea is that energy is injected into the system at the largest scales and dissipated through viscosity only at the very smallest scales, leaving a range of scales in between in which the energy is simply transferred to smaller and smaller scales.

This idealized process is usually modelled by a periodic flow and the energy in different scales can be interpreted by the Fourier components in different Fourier wavenumber modes. One expands the velocity field in Fourier modes and writes the energy-budget equation for different wavenumbers. We denote the wavenumbers by κ. The smallest wavenumber κ_0 is the square-root of the first eigenvalue of the Stokes operator and is positive. The energy inequality for a given weak solution in the range of modes larger than a certain κ can then be written as

$$\frac{\kappa_0^3}{2}\frac{d}{dt}|\mathbf{u}_{\kappa,\infty}|^2 + \nu\kappa_0^3\|\mathbf{u}_{\kappa,\infty}\|^2 \leq \kappa_0^3(\mathbf{f}, \mathbf{u}_{\kappa,\infty}) + \kappa_0^3 e_\kappa(\mathbf{u}),$$

where $\mathbf{u}_{\kappa,\infty}$ represents the part of the velocity field with modes higher than κ. The term $\kappa_0^3 e_\kappa(\mathbf{u}) = -\kappa_0^3 b(\mathbf{u}, \mathbf{u}, \mathbf{u}_{\kappa,\infty})$ is interpreted as the energy flux per unit time per unit mass to modes with wavenumber higher than κ (recall that b is defined in (10.2)). The inequality comes from a possible lack of regularity for the weak solution.

In statistical equilibrium in time, the corresponding mean energy inequality takes the form

$$\nu\kappa_0^3 \left\langle \|\mathbf{u}_{\kappa,\infty}\|^2 \right\rangle \leq \kappa_0^3 \left\langle (\mathbf{f}, \mathbf{u}_{\kappa,\infty}) \right\rangle + \kappa_0^3 \left\langle e_\kappa(\mathbf{u}) \right\rangle.$$

The assumption that the energy injection is concentrated on the large scales is typically modelled by assuming that the forcing term $\mathbf{f}$ is only active in lower modes, i.e. $\mathbf{f}_{\kappa,\infty} = 0$ for sufficiently large wavenumber κ, say for $\kappa \geq \bar{\kappa}_f$. Then, it can be show that for $\kappa \geq \bar{\kappa}_f$, $\kappa_0^3 \left\langle e_\kappa(\mathbf{u}) \right\rangle$ is non-negative and decreasing, and the limit

$$\langle e(\mathbf{u}) \rangle_\infty = \lim_{\kappa \to \infty} \langle e_\kappa(\mathbf{u}) \rangle$$

exists. Defining the *restricted energy flux* through wavenumber κ by

$$\langle e_\kappa^*(\mathbf{u}) \rangle = \langle e_\kappa(\mathbf{u}) \rangle - \langle e(\mathbf{u}) \rangle_\infty,$$

one recovers an energy-budget equation:

$$\nu\kappa_0^3 \left\langle \|\mathbf{u}_{\kappa,\infty}\|^2 \right\rangle = \kappa_0^3 \left\langle (\mathbf{f}, \mathbf{u}_{\kappa,\infty}) \right\rangle + \kappa_0^3 \left\langle e_\kappa^*(\mathbf{u}) \right\rangle.$$

Then, the following estimate can be rigorously proved:

$$1 - \left(\frac{\kappa}{\kappa_\tau}\right)^2 \leq \frac{\left\langle \kappa_0^3 e_\kappa^*(\mathbf{u}) \right\rangle}{\epsilon} \leq 1,$$

for $\kappa \geq \bar{\kappa}_f$, where

$$\kappa_\tau = \left(\frac{\langle \|\mathbf{u}\|^2 \rangle}{\langle |\mathbf{u}|^2 \rangle}\right)^{1/2}.$$

is a wavenumber associated with the Taylor wavenumber from the conventional theory. Therefore, if $\kappa_\tau \gg \bar{\kappa}_f$, as expected for turbulent flows, we find the energy cascade

$$\kappa_0^3 \langle \mathbf{e}_\kappa^*(\mathbf{u}) \rangle \approx \epsilon,$$

for a range of wavenumbers κ with $\bar{\kappa}_f < \kappa \ll \kappa_\tau$. Therefore, proving the energy cascade mechanism has been reduced to proving that the Taylor-like wavenumber κ_τ is much larger than the largest wavenumber $\bar{\kappa}_f$ active in the forcing term. More details can be found in Foias et al. (2001a,b), Foias et al. (2001c), and Rosa (2002). The corresponding result for finite-time averages can be found in Foias et al. (2005b).

A similar result has been given by Foias et al. (2002) for the existence of an enstrophy cascade in the two-dimensional case. See also Foias, Jolly, & Manley (2005a) for the corresponding results for finite-time averages, and Rosa (2002) for discussions about the inverse energy cascade.

The energy dissipation law

The energy dissipation law in the conventional theory of turbulence says that for a homogeneous turbulent flow the mean rate of energy dissipation per unit time per unit mass ϵ is related to a characteristic macro-scale length ℓ and a characteristic mean macro-scale velocity U by

$$\epsilon \sim \frac{U^3}{\ell}.$$

In the case of shear-driven flow (with the flow confined between two parallel plates separated by a distance ℓ, with one plate fixed and the other sliding with a longitudinal velocity U), Doering & Constantin (1992) proved the estimate

$$\epsilon \le \frac{1}{8\sqrt{2}} \frac{U^3}{\ell},$$

for $\mathrm{Re} = U\ell/\nu \ge 8\sqrt{2}$.

A number of similar rigorous results of the form

$$\epsilon \le c \frac{U^3}{\ell},$$

for a suitable constant c, have also been proved in a number of other geometries and with various forcing terms or inhomogeneous boundary conditions; see e.g. Constantin & Doering (1994), Foias (1997),

Wang (1997), Nicodemus, Grossman, & Holthaus (1997, 1998), Kerswell (1998), Wang (2000), Foias et al. (2001a), Foias et al. (2001c), Childress, Kerswell, & Gilbert (2001), Doering & Foias (2002), Doering, Eckhardt, & Schumacher (2003), and Foias et al. (2005b). Related estimates for the rate of energy dissipation can also be found in Howard (1972), Busse (1978), Foias, Manley, & Temam (1993), and Constantin & Doering (1995).

Other estimates and characteristic quantities

A number of parameters are important in turbulence theory. Besides the Taylor wavenumber κ_τ given above, another fundamental micro-scale characteristic number is the Kolmogorov wavenumber $\kappa_\epsilon = (\epsilon/\nu^3)^{1/4}$. The conventional theory exploits the Kolmogorov dissipation law to establish a few relations between these quantities and the Reynolds number, namely $\kappa_\epsilon \sim \kappa_0 \mathrm{Re}^{3/4}$, $\kappa_\tau \sim \kappa_0^{1/3} \kappa_\epsilon^{2/3}$, and $\kappa_\tau \sim \kappa_0 \mathrm{Re}^{1/2}$. Likewise, in the periodic or no-slip cases with a steady forcing, the following rigorous results hold for large Reynolds number flows: $\kappa_\epsilon \leq c\kappa_0 \mathrm{Re}^{3/4}$, $\kappa_\tau \leq c\kappa_0^{1/3} \kappa_\epsilon^{2/3}$, and $\kappa_\tau \leq c\kappa_0 \mathrm{Re}^{1/2}$, for a suitable universal constant c (see Foias ct al., 2001c).

Further estimates involve the non-dimensional Grashof number $G^* = |A^{-1/2}\mathbf{f}|/\nu^2\kappa_0^{1/2}$, such as $\mathrm{Re} \leq G^*$, $\kappa_\epsilon \leq \kappa_0 G^{*1/2}$, and so on. This also includes estimates for the number of degrees of freedom $(\kappa_\epsilon/\kappa_0)^3$ of the flow, as derived in the conventional theory of turbulence, which is slightly different from the rigorous estimate obtained for the fractal dimensional of invariant sets in 3D NSE, which involves a Kolmogorov wave number based on a supremum of a certain quantity based on weak solutions (instead of an average for a stationary statistical solution).

For similar estimates in the two-dimensional case, including an improved estimate for the number of degrees of freedom $(\kappa_\eta/\kappa_0)^2$, where $\kappa_\eta = (\eta/\nu^3)^{1/6}$ is the Kraichnan dissipation wave-number, see Foias et al. (2002); Foias et al. (2003).

Homogeneous turbulence

The right framework for treating homogeneous turbulence requires the definition of a homogeneous statistical solution and is a highly non-trivial problem. The case of decaying homogeneous turbulence has been considered by Vishik & Fursikov (1978), Foias & Temam (1980), and more recently by Basson (2006) (see also Dostoglu, Fursikov, & Kahl, 2006). In this framework, a certain self-similar homogeneous statistical solution can also be defined that displays the famous Kolmogorov $\kappa^{-5/3}$ law for the energy spectrum (Foias & Temam, 1983; Foias, Manley, & Temam,

1983). But several basic open problems still persist in this respect, such as the very existence of these self-similar solutions and their relevance to the dynamics of arbitrary homogeneous statistical solutions. Some results for the two-dimensional case can be found in Chae & Foias (1994).

A rigorous framework to treat forced, stationary, locally homogeneous turbulence would also be of crucial importance. In this situation, forcing in the large scales is inhomogeneous but somehow the behaviour at the smaller scales should approach a homogeneous behaviour.

Characterization of "turbulent" statistical solutions

This is one of the most important open problems in the statistical theory of turbulence. Most of the estimates obtained so far are for arbitrary stationary statistical solutions and hold in particular for a Dirac delta measure concentrated on an individual stationary solution of the Navier–Stokes equations. Therefore, the estimates and properties of statistical solutions are not necessarily sharp estimates for turbulent flows since they also include laminar flows.

For instance, while for turbulent flows it is expected that the Kolmogorov dissipation law $\epsilon \sim U^3/\ell$ holds, it has only been proved rigorously that $\epsilon \leq cU^3/\ell$. This means that for some stationary statistical solutions the corresponding value of ϵ may be close to cU^3/ℓ, and these are expected to be associated with turbulent flows, while for other stationary statistical solutions, associated possibly with non-turbulent flows, the equality may be far from being achieved. See for instance the estimates for the skin-friction coefficient in the case of a channel flow driven by a uniform pressure gradient mentioned later in this section, in which this discrepancy is more explicit.

It is therefore of crucial importance to characterize turbulent statistical solutions and obtain sharper results for them.

This is also true for regularity purposes. The idea that mean flows are usually better-behaved than the fluctuations is expected to be reflected by further smoothness of the mean flow. This, however, is not expected to hold in general since a time-dependent statistical solution can be concentrated on an individual weak solution.

Estimates for a channel flow driven by a uniform pressure gradient

In the particular case of a channel flow driven by a uniform pressure gradient, Constantin & Doering (1995) have obtained a lower bound for the mean energy dissipation rate which leads to an important estimate for the skin-friction coefficient. More recently, the same problem was considered by Ramos et al. (2008) and a sharp upper bound for the mean rate of energy dissipation was obtained, corresponding to a sharp lower bound for the skin-friction coefficient. The lower bound for the mean rate of energy dissipation and the corresponding upper bound for the skin-friction coefficient were also slightly improved. All the estimates in Ramos et al. (2008) were obtained for mean quantities averaged with respect to arbitrary stationary statistical solutions.

For the skin-friction coefficient, for example, which is defined by

$$C_f = \frac{Ph}{L_1 U^2},\tag{10.10}$$

where h is the height of the channel, U is the mean longitudinal velocity, and P/L_1 is the imposed pressure gradient, the following estimates, for high-Reynolds-number flows (associated with large pressure gradients), hold:

$$\frac{12}{\text{Re}} \le C_f \le 0.484 + O\left(\frac{1}{\text{Re}}\right),\tag{10.11}$$

where $\text{Re} = hU/\nu$ is the corresponding Reynolds number.

The lower-bound for C_f coincides with the corresponding value of C_f for Poiseuille flow, making this estimate optimal since the estimate is for an arbitrary stationary statistical solution, and a Dirac delta measure concentrated on the Poiseuille flow (which is unstable for high-Reynolds-number flows, but nevertheless exists in a mathematical sense) is an example of a stationary statistical solution. The upper bound, however, might not be optimal since heuristic arguments and flow experiments suggest that $C_f \sim (\ln \text{Re})^{-2}$ for high-Reynolds-number turbulent flows.

The lower-bound for C_f in (10.11) follows more precisely from the following upper bound for the mean rate of energy dissipation per unit time per unit mass:

$$\epsilon \le \epsilon_{\text{Poiseuille}},\tag{10.12}$$

where

$$\epsilon_{\text{Poiseuille}} = \frac{h^2}{12\nu}\left(\frac{P}{L_1}\right)^2.\tag{10.13}$$

is the corresponding rate of energy dissipation for the plane Poiseuille flow. (The definition of ϵ is given in (10.8).)

The estimate related to the Kolmogorov dissipation law in this geometry reads

$$\epsilon \le \left(0.054 + O\left(\frac{1}{\text{Re}^2}\right)\right)\frac{U^3}{h}.\tag{10.14}$$

The upper bound in (10.14) is not related to the upper bound in (10.12). In fact, computing U^3/h for the Poiseuille flow yields $(h^5/27\nu^3)(P/L_1)^3$, which is much larger than $\epsilon_{\text{Poiseuille}}$ for large pressure gradients.

Two-dimensional forced turbulence

In the two-dimensional case, the Kraichnan–Leith–Batchelor theory (Kraichnan, 1967; Leith, 1968; Batchelor, 1969) predicts a direct enstrophy cascade to lower scales and an inverse energy cascade to larger scales, with different power laws for the energy spectrum in each range of scales.

We already mentioned above the result due to Foias et al. (2002) that gives a sufficient condition for the existence of an enstrophy cascade in the two-dimensional case. But this condition is in terms of a parameter that depends on the flow (the Taylor-like wavenumber κ_τ) and is not fully characterized *a priori* from the data of the problem, such as the forcing term.

An important model problem is the Kolmogorov flow, in which the forcing term has only one active mode (associated with an eigenvalue of the Stokes operator). A number of numerical experiments have been performed for this problem in search of the direct and inverse cascades in the two-dimensional case. Many experiments resort to stochastic forcing or a nonlinear feedback-type forcing in the Kolmogorov flow in order to achieve the cascades.

However, no successful experiment has been devised with a steady forcing in a single mode. In fact, it was eventually proved by Constantin, Foias, & Manley (1994) that a single-mode steady forcing is not able to sustain the direct enstrophy cascade (see also Foias et al., 2002, for a slightly different argument). Both works also give necessary conditions for the existence of the direct enstrophy cascade in the case of a two-mode steady forcing. But no proof has been given for the existence of such a forcing term. The existence of a two-mode steady forcing able to sustain the cascades is still an open problem, and it would also be important to characterize such forcings if they exist.

Exponential decay of the power spectrum

A fundamental quantity in the theory of turbulence is the observed power spectrum of the flow. In general, a spectrum is associated with a decomposition of a given quantity with respect to different scales. The *energy spectrum* in turbulence is usually associated with a decomposition of the kinetic energy with respect to different length scales of the flow, while the *power spectrum* is usually associated with the decomposition of the same quantity with respect to different time scales. In practice and in the conventional theory, the *Taylor hypothesis* is usually invoked to relate length-wise and time-wise quantities, in particular the energy and power spectra.

It has been observed that both power and energy spectra decay very fast (with respect to increasing frequency and increasing wave-number, respectively), but how fast is still a matter of debate. One of the first rigorous results using statistical solutions in connection with the conventional theory of turbulence addresses this problem and is due to Bercovici et al. (1995). In this work, the power spectrum is defined in a rigorous way and it is proved that, in the two-dimensional case, the power spectrum decays at least exponentially fast with respect to increasing frequency. The proof is based on the Wiener–Khintchine theory connecting the spectrum with a correlation function through a Fourier transform, and the crucial point guaranteeing the exponential decay is the analyticity in time of the solutions of the two-dimensional Navier–Stokes equations.

The corresponding result for the energy spectrum (exponential decay with respect to wave-number) follows from the analyticity in space, and in this case it follows from assuming that the forcing term is in some Gevrey space; see Foias & Temam (1989) and Foias, Manley, & Sirovich (1989) (see also Doering & Titi, 1995, for a discussion of the three-dimensional case).

It is interesting to notice that these two rigorous mathematical results ignore the Taylor hypothesis and, in fact, need different assumptions: the result for the decay of the power spectrum only assumes that the forcing term belongs to H, while the result for the decay of the energy spectrum depends on the forcing term being analytic in space in some suitable sense.

Inviscid limit

The inviscid limit of the Navier–Stokes equations to the Euler equations ($\nu = 0$ in (10.1)) has been studied in a number of contexts and from different perspectives. Particularly relevant to the conventional theory of turbulence is the limit of the mean energy dissipation rate as the viscosity goes to zero. It is one of the main hypotheses of the Kolmogorov theory of turbulence that this limit is strictly positive, despite the fact that there is no dissipation in the Euler equations. This phenomenon is called *anomalous dissipation*, and its existence (or non-existence) is a major open problem.

The corresponding anomalous dissipation in two-dimensional turbulence was postulated by Kraichnan (1967) and concerns the mean rate of enstrophy dissipation η instead (see (10.9)). However, in this two-dimensional case this has been a controversial issue, and is still an open problem although some partial results have been presented. Most of the results, however, are for finite-time averages, which are not quite the right object to look at in this case since the transient time increases with decreasing viscosity and the long-time behaviour and the associated stationary statistics are not captured. This is one example in which the use of infinite-time averages or, more generally, of stationary statistical solutions is of crucial importance.

One result on the inviscid limit that addresses this long-time, statistical behaviour is due to Constantin & Ramos (2007) (see also Chae, 1991a,b, and Constantin & Wu, 1997) in the context of a damped and driven two-dimensional Navier–Stokes equations on the whole plane. This equation has an extra non-diffusive linear damping term and is known as the Charney–Stommel model of the Gulf Stream. The absence of anomalous dissipation for the Charney–Stommel model had been suggested by Lilly (1972) and Bernard (2000).

Constantin & Ramos (2007) prove that for initial conditions and a forcing term in suitable function spaces, the corresponding time-average stationary statistical solutions are such that their mean rate of enstrophy dissipation vanishes in the inviscid limit, thus proving that in this case there is no anomalous dissipation.

Other fluid-flow problems

A few other rigorous statistical results have been obtained for other fluid-flow problems. See for instance Constantin (1999, 2001) on turbulent convection and transport, Doering & Constantin (2001), and Doering, Otto, & Reznikoff (2006) on infinite Prandtl number convection, and Wang (2008, 2009) on Rayleigh–Bénard convection and some singular perturbation problems.

10.6 Two-dimensional channel flow driven by a uniform pressure gradient

We consider in this section a two-dimensional homogeneous incompressible Newtonian flow confined to a rectangular periodic channel and driven by a uniform pressure gradient. More precisely, the velocity vector field $\mathbf{u} = (u_1, u_2)$ of the fluid satisfies the Navier–Stokes equations

$$\frac{\partial \mathbf{u}}{\partial t} - \nu \Delta \mathbf{u} + (\mathbf{u} \cdot \nabla)\mathbf{u} + \nabla p = \frac{P}{L_1}\mathbf{e}_1, \qquad \nabla \cdot \mathbf{u} = 0,$$

in the domain $\Omega = (0, L_1) \times (0, h)$, $L_1, h > 0$. We denote by $\mathbf{x} = (x, y)$ the space variable; the scalar $p = p(x, y)$ is the kinematic pressure; the boundary conditions are no-slip on the walls $y = 0$ and $y = h$ and periodic in the x direction with period L_1 for both $\mathbf{u}$ and p; the parameter P/L_1 denotes the magnitude of the applied pressure gradient and we assume $P > 0$; the parameter $\nu > 0$ is the kinematic viscosity; and $\mathbf{e}_1$ is the unit vector in the x direction. We sometimes refer to the direction x of the pressure gradient as the *longitudinal direction*.

This problem admits a laminar solution known as the plane Poiseuille flow, for which the velocity field takes the form

$$\mathbf{u}_{\text{Poiseuille}}(x, y) = \frac{P}{2\nu L_1}y(h - y)\mathbf{e}_1.$$

The mathematical formulation of the Navier–Stokes equations in this geometry can be easily adapted from the no-slip and fully-periodic cases, and yields a functional equation for the time-dependent velocity field $\mathbf{u} = \mathbf{u}(t)$ of the form:

$$\frac{d\mathbf{u}}{dt} + \nu A\mathbf{u} + B(\mathbf{u}, \mathbf{u}) = \mathbf{f}_P,$$

where

$$\mathbf{f}_P = \frac{P}{L_1}\mathbf{e}_1.$$

The two fundamental spaces H and V are characterized in this case by

$$H = \left\{ \mathbf{u} = \mathbf{w}|_\Omega; \begin{array}{c} \mathbf{w} \in (L^2_{\text{loc}}(\mathbb{R} \times (0, h)))^2, \ \nabla \cdot \mathbf{w} = 0, \\ \mathbf{w}(x + L_1, y) = \mathbf{w}(x, y), \text{ a.e. } (x, y) \in \mathbb{R} \times (0, h), \\ \mathbf{w}_2(x, 0) = \mathbf{w}_2(x, h) = 0, \text{ a.e. } x \in \mathbb{R}. \end{array} \right\},$$

and

$$V = \left\{ \mathbf{u} = \mathbf{w}|_\Omega; \begin{array}{c} \mathbf{w} \in (H^1_{\text{loc}}(\mathbb{R} \times (0, h)))^2, \ \nabla \cdot \mathbf{w} = 0, \\ \mathbf{w}(x + L_1, y) = \mathbf{w}(x, y), \text{ a.e. } (x, y) \in \mathbb{R} \times (0, h), \\ \mathbf{w}(x, 0) = \mathbf{w}(x, h) = 0, \text{ a.e. } x \in \mathbb{R}. \end{array} \right\}.$$

The Poiseuille flow satisfies

$$\nu A\mathbf{u}_{\text{Poiseuille}} = \mathbf{f}_P$$

and is a stationary solution of the NSE since the nonlinear term vanishes for this flow.

In this two-dimensional channel problem, we are interested in mean quantities averaged with respect to an arbitrary invariant measure μ for the associated semigroup. Our main interest is in the mean enstrophy dissipation rate per unit time and unit mass, given by

$$\eta = \frac{\nu}{L_1 h} \left\langle |A\mathbf{u}|^2 \right\rangle.$$

We also consider the mean energy dissipation rate per unit time per unit mass which in this case takes the form

$$\epsilon = \frac{\nu}{L_1 h} \left\langle \|\mathbf{u}\|^2 \right\rangle.$$

We want to show that the plane Poiseuille flow (or more precisely the invariant measure concentrated on the plane Poiseuille flow) minimizes the mean enstrophy dissipation rate and maximizes the mean energy dissipation rate among all the invariant measures for the system.

Since we assume statistical equilibrium in time, we can recover the stationary form of the Reynolds equations,

$$\nu A \left\langle \mathbf{u} \right\rangle + \left\langle B(\mathbf{u}, \mathbf{u}) \right\rangle = \mathbf{f}_P.$$

In this two-dimensional case, the mean velocity field $\left\langle \mathbf{u} \right\rangle$ belongs to $D(A)$, and the Reynolds equations hold in H.

Using the Reynolds equations, we now prove that for every invariant measure, the corresponding enstrophy dissipation rate satisfies

$$\eta \geq \eta_{\text{Poiseuille}},$$

where

$$\eta_{\text{Poiseuille}} = \frac{\nu}{L_1 h} \left\langle |A\mathbf{u}_{\text{Poiseuille}}|^2 \right\rangle = \frac{1}{\nu} \left(\frac{P}{L_1} \right)^2$$

is the enstrophy dissipation rate for the plane Poiseuille flow.

Subsequently, we will show that

$$\epsilon \leq \epsilon_{\text{Poiseuille}},$$

where

$$\epsilon_{\text{Poiseuille}} = \frac{\nu}{L_1 h} \left\langle \|\mathbf{u}_{\text{Poiseuille}}\|^2 \right\rangle = \frac{h^2}{12\nu} \left(\frac{P}{L_1} \right)^2. \tag{10.15}$$

is the energy dissipation rate for the plane Poiseuille flow.

Note that $\mathbf{f}_P = (1/\nu)P/Le_1$ belongs to H since it is square integrable, divergence free, periodic (in fact constant) in the longitudinal direction, and has zero normal component on the top and bottom walls. Also, any invariant measure in the 2D channel is carried by a bounded set in $D(A)$; this follows from the fact that any invariant measure is carried by the global attractor and that the global attractor in this case is bounded in $D(A)$ (Foias & Temam, 1979; Foias et al., 2001a).

In particular, $A\langle\mathbf{u}\rangle \in H$ and $\langle B(\mathbf{u},\mathbf{u})\rangle \in H$, and hence $A\langle\mathbf{u}\rangle$ and $\langle B(\mathbf{u},\mathbf{u})\rangle$ belong to $L^1(\Omega)^2$. Therefore, the Reynolds equations hold in $L^1(\Omega)^2$ and we are allowed to integrate each term over Ω.

The most notable and important fact in this geometry is that the nonlinear term has zero space average. Let us prove this.

For any smooth vector field $\mathbf{u} = (u,v)$, using that $\mathbf{e}_1 = (1,0)$ belongs to H, we can write the integral of the first component $B(\mathbf{u},\mathbf{u})_1$ of the nonlinear term $B(\mathbf{u},\mathbf{u}) = P(\mathbf{u}\cdot\nabla)\mathbf{u}$ as

$$\int_\Omega B(\mathbf{u},\mathbf{u})_1 \, d\mathbf{x} = (B(\mathbf{u},\mathbf{u}),\mathbf{e}_1) = (P(\mathbf{u}\cdot\nabla\mathbf{u}),\mathbf{e}_1) = ((\mathbf{u}\cdot\nabla\mathbf{u}),\mathbf{e}_1)$$

$$= \int_0^L \int_0^h \left\{ u\frac{\partial u}{\partial x} + v\frac{\partial u}{\partial y} \right\} dx\, dy.$$

Using an integration by parts, the homogeneous no-slip boundary conditions on the walls of the channel, the divergence-free condition, and the periodicity condition in the streamwise direction, we find that

$$\int_\Omega B(\mathbf{u},\mathbf{u})_1 \, d\mathbf{x} = \int_0^L \int_0^h \left\{ u\frac{\partial u}{\partial x} + v\frac{\partial u}{\partial y} \right\} dx\, dy$$

$$= \int_0^L \int_0^h \left\{ u\frac{\partial u}{\partial x} - \frac{\partial v}{\partial y}u \right\} dx\, dy$$

$$\quad + \int_0^L \left\{ v(x,h)u(x,h) - v(x,0)u(x,0) \right\} dx$$

$$= \int_0^L \int_0^h \left\{ u\frac{\partial u}{\partial x} - \frac{\partial v}{\partial y}u \right\} dx\, dy$$

$$= 2\int_0^L \int_0^h u\frac{\partial u}{\partial x} \, dx\, dy$$

$$= \int_0^h \left\{ u(L,y)^2 - u(0,y)^2 \right\} dy$$

$$= 0.$$

(Although the space average of the second component $(\mathbf{u} \cdot \nabla)v$ also vanishes, the divergence-free part of $B(\mathbf{u}, \mathbf{u})$ and the second component of the associated gradient part, of the form p_y, may not. Note, also, that the argument above does not work in this case because $\mathbf{e}_2$ does not belong to H.)

Now, we integrate in space the first component of the Reynolds equations to find that

$$Lh\frac{P}{L} = \nu \int_\Omega \langle A\mathbf{u} \rangle \cdot \mathbf{e}_1 \, d\mathbf{x} \leq \nu L^{1/2} h^{1/2} \left(\int_\Omega |\langle A\mathbf{u} \rangle|^2 \, d\mathbf{x} \right)^{1/2}$$

$$\leq \nu L^{1/2} h^{1/2} \langle |A\mathbf{u}|^2 \rangle^{1/2}.$$

Taking the square of this relation we find

$$\frac{1}{\nu} \frac{P^2}{L^2} \leq \frac{\nu}{Lh} \langle |A\mathbf{u}|^2 \rangle,$$

which is precisely

$$\eta_{\text{Poiseuille}} \leq \eta.$$

Note that in the above we may proceed in a different way to obtain a few interesting exact relations, namely

$$\nu \int_\Omega \langle A\mathbf{u} \rangle \cdot \mathbf{e}_1 \, d\mathbf{x} = \nu \int_\Omega \langle -\Delta\mathbf{u} \rangle \cdot \mathbf{e}_1 \, d\mathbf{x} = -\nu \int_\Omega \langle u_{xx} + u_{yy} \rangle \, d\mathbf{x}$$

$$= -\nu \int_0^h \{u_x(L,y) - u_x(0,y)\} \, dy - \nu \int_\Omega \langle u_{yy} \rangle \, d\mathbf{x} = -\nu \int_\Omega \langle u_{yy} \rangle \, d\mathbf{x},$$

so that

$$-\frac{1}{Lh} \left\langle \int_\Omega u_{yy}(\mathbf{x}) \, d\mathbf{x} \right\rangle = \frac{1}{\nu} \frac{P}{L}.$$

Integrating in y we also obtain

$$-\frac{1}{L} \left\langle \int_0^L \{u_y(x,h) - u_y(x,0)\} \, dx \right\rangle = \frac{h}{\nu} \frac{P}{L}.$$

In terms of the vorticity $\omega = v_x - u_y$ we may rewrite this as

$$\frac{1}{L} \left\langle \int_0^L \{\omega(x,h) - \omega(x,0)\} \, dx \right\rangle = \frac{h}{\nu} \frac{P}{L}.$$

The other remarkable fact is that while the Poiseuille flow minimizes the enstrophy dissipation rate among all the invariant measures, it also maximizes the energy dissipation rate. This follows easily from the energy inequality. Indeed,

$$\nu \langle \|\mathbf{u}\|^2 \rangle \leq \langle (\mathbf{f}_P, \mathbf{u}) \rangle \leq |A^{-1/2}\mathbf{f}_P| \langle \|\mathbf{u}\| \rangle \leq |A^{-1/2}\mathbf{f}_P| \langle \|\mathbf{u}\|^2 \rangle^{1/2},$$

so that

$$\epsilon = \frac{\nu}{L_1 h} \langle \|\mathbf{u}\|^2 \rangle \leq \frac{1}{\nu L_1 h} |A^{-1/2}\mathbf{f}_P|^2.$$

But note that the plane Poiseuille flow is such that

$$A^{-1}\mathbf{f}_P = \nu \mathbf{u}_{\text{Poiseuille}},$$

so that

$$\epsilon \leq \frac{1}{\nu L_1 h} |A^{-1/2}\mathbf{f}_P|^2 = \frac{1}{\nu L_1 h} \|A^{-1}\mathbf{f}_P\|^2 = \frac{\nu}{L_1 h} \|\mathbf{u}_{\text{Poiseuille}}\|^2.$$

The right hand side is exactly the mean rate of energy dissipation per unit time per unit mass, and hence we obtain

$$\epsilon \leq \epsilon_{\text{Poiseuille}}.$$

This mean value can be computed explicitly in terms of the parameters of the problem, as given in (10.15).

Acknowledgements

This work was partially supported by CNPq, Brasília, Brazil, grant 30.7953/2006-8. I would like to thank Ciprian Foias for some comments on the original version of the manuscript. I would also like to thank the editors, James Robinson and José L. Rodrigo, who helped to improve the article and did an impeccable job organizing the Workshop and making my stay in Warwick so pleasant.

References

Basson, A. (2006) Homogeneous statistical solutions and local energy inequality for 3D Navier–Stokes equations. *Comm. Math. Phys.* **266**, 17–35.

Batchelor, G.K. (1953) *The Theory of Homogeneous Turbulence*. Cambridge University Press, Cambridge.

Batchelor, G.K. (1969) Computation of the energy spectrum in homogeneous two-dimensional turbulence. *Phys. Fluids Suppl. II* **12**, 233–239.

Bernard, D. (2000) Influence of friction on the direct cascade of 2D forced turbulence. *Europhys. Lett.* **50**, 333–339.

Bercovici, H., Constantin, P., Foias, C., & Manley, O.P. (1995) Exponential decay of the power spectrum of turbulence. *J. Stat. Phys.* **80**, 579–602.

Busse, F.H. (1978) The optimum theory of turbulence. *Adv. Appl. Mech.* **18**, 77–121.

Chae, D. (1991a) The vanishing viscosity limit of statistical solutions of the Navier–Stokes equations. I. 2-D periodic case. *J. Math. Anal. Appl.* **155**, 437–459.

Chae, D. (1991b) The vanishing viscosity limit of statistical solutions of the Navier–Stokes equations. II. The general case. *J. Math. Anal. Appl.* **155**, 460–484.

Chae, D. & Foias, C. (1994) On the homogeneous statistical solutions of the 2-D Navier–Stokes equations. *Indiana Univ. Math. J.* **43**, 177–185.

Childress, S., Kerswell, R.R., & Gilbert, A.D. (2001) Bounds on dissipation for Navier–Stokes flow with Kolmogorov forcing. *Physica D* **158**, 105–128.

Constantin, P. (1999) Variational bounds in turbulent convection, in *Nonlinear partial differential equations (Evanston, IL, 1998). Contemp. Math.* **238**, 77–88. Amer. Math. Soc., Providence, RI.

Constantin, P. (2001) Bounds for turbulent transport, in *IUTAM Symposium on Geometry and Statistics of Turbulence (Hayama, 1999). Fluid Mech. Appl.* **59**, 23–31. Kluwer Acad. Publ., Dordrecht.

Constantin, P. & Doering, C.R. (1994) Variational bounds on energy dissipation in incompressible flows: shear flow. *Phys. Rev. E* **49**, 4087–4099.

Constantin, P. & Doering. C.R. (1995) Variational bounds on energy dissipation in incompressible flows II: channel flow. *Phys. Rev. E* **51**, 3192–319.

Constantin, P. & Foias, C. (1988) *Navier–Stokes Equations.* University of Chicago Press, Chicago.

Constantin, P. & Ramos, R. (2007) Inviscid limit for damped and driven incompressible Navier–Stokes Equations in $\mathbb{R}^2$. *Comm. Math. Phys.* **275**, 529–551.

Constantin, P. & Wu, J. (1997) Statistical solutions of the Navier–Stokes equations on the phase space of vorticity and the inviscid limits. *J. Math. Phys.* **38**, 3031–3045.

Constantin, P., Foias, C., & Manley, O.P. (1994) Effects of the forcing function on the energy spectrum in 2-D turbulence. *Phys. Fluids* **6**, 427–429.

Doering, C.R. & Constantin, P. (1992) Energy dissipation in shear driven turbulence. *Phys. Rev. Lett.* **69**, 1648–1651. Erratum: *Phys. Rev. Lett.* **69**, 3000.

Doering, C.R. & Constantin, P. (2001) On upper bounds for infinite Prandtl number convection with or without rotation. *J. Math. Phys.* **42**, 784–795.

Doering, C.R., Eckhardt, B., & Schumacher, J. (2003) Energy dissipation in body-forced plane shear flow. *J. Fluid Mech.* **494**, 275–284.

Doering, C.R. & Foias, C. (2002) Energy dissipation in body-forced turbulence. *J. Fluid Mech.* **467**, 289–306.

Doering, C. & Titi, E.S. (1995) Exponential decay rate of the power spectrum for solutions of the Navier–Stokes equations. *Phys. Fluids* **7**, 1384–1390.

Doering, C., Otto, F., & Reznikoff, M.G. (2006) Bounds on vertical heat transport for infinite-Prandtl-number Rayleigh-Bénard convection. *J. Fluid Mech.* **560**, 229–241.

Dostoglou, S., Fursikov, A.V., & Kahl, J.D. (2006) Homogeneous and isotropic statistical solutions of the Navier–Stokes equations. *Math. Phys. Electron. J.* **12** (2006), Paper 2.

Foias, C. (1972) Statistical study of Navier–Stokes equations I. *Rend. Sem. Mat. Univ. Padova* **48**, 219–348.

Foias, C. (1973) Statistical study of Navier–Stokes equations II. *Rend. Sem. Mat. Univ. Padova* **49**, 9–123.

Foias, C. (1974) A functional approach to turbulence. *Russian Math. Survey* **29**, 293–326.

Foias, C. (1997) What do the Navier–Stokes equations tell us about turbulence? in *Harmonic Analysis & Nonlinear Differential Equations* (Riverside, CA, 1995). *Contemp. Math.* **208**, 151–180.

Foias, C. & Temam, R. (1979) Some analytic and geometric properties of the solutions of the evolution Navier–Stokes equations. *J. Math. Pures Appl.* **58**, 339–368.

Foias, C. & Temam, R. (1980) Homogeneous statistical solutions of Navier–Stokes equations. *Indiana Univ. Math. J.* **29**, 913–957.

Foias, C. & Temam, R. (1983) Self-similar universal homogeneous statistical solutions of the Navier–Stokes equations. *Comm. Math. Phys.* **90**, no. 2, 187–206.

Foias, C. & Temam, R. (1989) Gevrey class regularity for the solutions of the Navier–Stokes equations. *J. Funct. Anal.* **87**, 359–369.

Foias, C., Manley, O.P., & Temam, R. (1983) New representation of Navier–Stokes equations governing self-similar homogeneous turbulence. *Phys. Rev. Lett.* **51**, 617–620.

Foias, C., Manley, O.P., & Sirovich, L. (1989) Empirical and Stokes eigenfunctions and the far-dissipative turbulent spectrum. *Phys. Fluids A* **2**, 464–467.

Foias, C., Manley, O.P., & Temam, R. (1993) Bounds for the mean dissipation of 2-D enstrophy and 3-D energy in turbulent flows. *Phys. Lett. A* **174**, 210–215.

Foias, C., Manley, O.P., Rosa, R., & Temam, R. (2001a) *Navier–Stokes Equations and Turbulence*. Encyclopedia of Mathematics and its Applications, Vol. 83. Cambridge University Press, Cambridge.

Foias, C., Manley, O.P., Rosa, R., & Temam, R. (2001b) Cascade of energy in turbulent flows. *Comptes Rendus Acad. Sci. Paris, Série I* **332**, 509–514.

Foias, C., Manley, O.P., Rosa, R., & Temam, R. (2001c) Estimates for the energy cascade in three-dimensional turbulent flows. *Comptes Rendus Acad. Sci. Paris, Série I* **333**, 499–504.

Foias, C., Jolly, M.S., Manley, O.P., & Rosa, R. (2002) Statistical estimates for the Navier–Stokes equations and the Kraichnan theory of 2-D fully developed turbulence. *J. Stat. Phys.* **108**, 591–645.

Foias, C., Jolly, M.S., Manley, O.P., & Rosa, R. (2003) On the Landau–Lifschitz degrees of freedom in 2-D turbulence. *J. Stat. Phys.* **111**, 1017–1019.

Foias, C., Jolly, M.S., & Manley, O.P. (2005a) Kraichnan turbulence via finite time averages. *Comm. Math. Phys.* **255**, 329–361.

Foias, C., Jolly, M.S., Manley, O.P, Rosa, R., & Temam, R. (2005b) Kolmogorov theory via finite-time averages. *Physica D* **212**, 245–270.

Frisch, U. (1995) *Turbulence: The Legacy of A.N. Kolmogorov.* Cambridge University Press, Cambridge.

Fursikov, A.V. (1999) The closure problem for the Friedman–Keller infinite chain of moment equations, corresponding to the Navier–Stokes system (English summary), in *Fundamental problematic issues in turbulence* (Monte Verita, 1998), 17–24, Trends Math., Birkhäuser, Basel.

Hopf, E. (1951) Über die Anfangswertaufgabe für die hydrodynamischen Grundgleichungen. *Math. Nachr.* **4**, 213–231.

Hopf, E. (1952) Statistical hydromechanics and functional calculus. *J. Rational Mech. Anal.* **1**, 87–123.

Hinze, J.O. (1975) *Turbulence.* McGraw-Hill, New York.

Howard, L.N. (1972) Bounds on flow quantities. *Annu. Rev. Fluid Mech.* **4**, 473–494.

Keller, L. & Friedman, A. (1925) Differentialgleichungen für die turbulente Bewegung einer inkompressiblen Flüssigkeit, in Biezeno, C.B. & Burgers, J.M. (eds.) *Proceedings of the First International Congress for Applied Mechanics, Delft 1924,* Technische Boekhandel en Drukkerij J. Waltman Jr., Delft.

Kerswell, R.R. (1998) Unification of variational methods for turbulent shear flows: the background method of Doering-Constantin and the mean-fluctuation method of Howard-Busse. *Physica D* **121**, 175–192.

Kolmogorov, A.N. (1941) The local structure of turbulence in incompressible viscous fluid for very large Reynolds numbers. *C. R. (Doklady) Acad. Sci. URSS (N.S.)* **30**, 301–305.

Kraichnan, R.H. (1967) Inertial ranges in two-dimensional turbulence. *Phys. Fluids.* **10**, 1417–1423.

Krylov, N. & Bogoliubov, N.N. (1937) La théorie générale de la mesure dans son application à l'étude des systémes dinamiques de la méchanique non linéaire. *Ann. of Math.* **38**, 65–113.

Ladyzhenskaya, O. (1963) *The Mathematical Theory of Viscous Incompressible Flow.* Revised English edition, Translated from the Russian by Richard A. Silverman. Gordon and Breach Science Publishers, New York-London.

Leith, C.E. (1968) Diffusion approximation for two-dimensional turbulence. *Phys. Fluids* **11**, 671–673.

Leray, J. (1933) Étude de diverses équations intégrales non linéaires et de quelques problèmes que pose l'hydrodynamique. *J. Math. Pures Appl.* **12**, 1–82.

Leray, J. (1934a) Essai sur les mouvements plans d'un liquide visqueaux que limitent des parois. *J. Math. Pures Appl.* **13** 331–418.

Leray, J. (1934b) Essai sur les mouvements d'un liquide visqueux emplissant l'espace. *Acta Math.* **63** 193–248.

Lesieur, M. (1997) *Turbulence in Fluids.* 3rd Edition. Fluid Mechanics and its Applications, Vol **40**. Kluwer Academic, Dordrecht.

Lilly, D.K. (1972) Numerical simulation studies of two-dimensional turbulence: II. Stability and predictability studies. *Geophysical & Astrophysical Fluid Dynamics* **4**, 1–28.

Monin, A.S. & Yaglom, A.M. (1975) *Statistical Fluid Mechanics: Mechanics of Turbulence.* MIT Press, Cambridge, MA.

Nicodemus, R., Grossmann, S., & Holthaus, M. (1997) Improved variational principle for bounds on energy dissipation in turbulent shear flow. *Physica D* **101**, 178–190.

Nicodemus, R., Grossmann, S., & Holthaus, M. (1998) The background flow method. Part 1. Constructive approach to bounds on energy dissipation. *J. Fluid Mech.* **363**, 281–300.

Prodi, G. (1960) Teoremi ergodici per le equazioni della idrodinamica, *C.I.M.E.,* Roma.

Prodi, G. (1961) On probability measures related to the Navier–Stokes equations in the 3-dimensional case. *Air Force Res. Div. Contr. A.P.* 61(052)-414, Technical Note no. 2, Trieste.

Ramos, F., Rosa, R., & Temam, R. (2008) Statistical estimates for channel flows driven by a pressure gradient. *Physica D* **237**, 1368–1387.

Reynolds, O. (1895) On the dynamical theory of incompressible viscous fluids and the determination of the criterion. *Phil. Trans. Roy. Soc. London A* **186**, 123–164.

Richardson, L.F. (1922) *Weather Prediction by Numerical Process.* Cambridge University Press. Reprinted in 1965 by Dover Publications, New York.

Rosa, R. (2002) Some results on the Navier–Stokes equations in connection with the statistical theory of stationary turbulence. *Applications of Mathematics* **47**, 485–516.

Rosa, R. (2006) Turbulence Theories, in Franqise, J.P., Naber, G.L., & Tsou, S.T. (eds.) *Encyclopedia of Mathematical Physics.* Elsevier, Oxford, Vol. **5**, 295–302.

Temam, R. (1984) *Navier–Stokes Equations. Theory and numerical analysis.* Studies in Mathematics and its Applications. 3rd edition. North-Holland Publishing Co., Amsterdam-New York. Reprinted in a re-edited edition in 2001 in the AMS Chelsea series, AMS, Providence.

Tennekes, H. & Lumley, J.L. (1972) *A First Course in Turbulence.* MIT Press, Cambridge, Mass.

Vishik, M.I. & Fursikov, A.V. (1977) L'équation de Hopf, les solutions statistiques, les moments correspondant aux systémes des équations paraboliques quasilinéaires. *J. Math. Pures Appl.* **59**, 85–122.

Vishik, M.I. & Fursikov, A.V. (1978) Translationally homogeneous statistical solutions and individual solutions with infinite energy of a system of Navier–Stokes equations. (Russian) *Sibirsk. Mat. Zh.* **19**, 1005–1031.

Vishik, M.I. & Fursikov, A.V. (1988) *Mathematical Problems of Statistical Hydrodynamics.* Kluwer, Dordrecht.

Wang, X. (1997) Time averaged energy dissipation rate for shear driven flows in $\mathbb{R}^n$. *Physica D* **99**, 555–563.

Wang, X. (2000) Effect of tangential derivative in the boundary layer on time averaged energy dissipation rate. *Physica D* **144**, 142–153.

Wang, X. (2008) Stationary statistical properties of Rayleigh-Bénard convection at large Prandtl number. *Comm. Pure Appl. Math.* **61**, 789–815.

Wang, X. (2009) Upper semi-continuity of stationary statistical properties of dissipative systems, *Discrete Contin. Dyn. Sys.* **23**, 521–540.

Printed in the United States
by Baker & Taylor Publisher Services

Printed in the United States
by Baker & Taylor Publisher Services